S0-EAJ-224

TELEMETRY: RESEARCH, TECHNOLOGY AND APPLICATIONS

TELEMETRY: RESEARCH, TECHNOLOGY AND APPLICATIONS

DIANA BARCULO
AND
JULIA DANIELS
EDITORS

Nova Science Publishers, Inc.
New York

For permission to use material from this book please contact us:
Telephone 631-231-7269; Fax 631-231-8175
Web Site: http://www.novapublishers.com

Library of Congress Cataloging-in-Publication Data

Telemetry : research, technology, and applications / editors, Diana Barculo and Julia Daniels.
p. cm.
Includes bibliographical references and index.
ISBN 978-1-60692-509-6 (hardcover)
1. Biotelemetry. I. Barculo, Diana. II. Daniels, Julia.
QH324.9.B5T45 2009
590.72'3--dc22
2008051264

Published by Nova Science Publishers, Inc. ✢ New York

CONTENTS

PREFACE

Telemetry is a technology that allows the remote measurement and reporting of information of interest to the system designer or operator. Telemetry typically refers to wireless communications, but can also refer to data transferred over other media, such as a telephone or computer network or via an optical link. This book provides a brief overview of the telemetric technique and how it has evolved over the years as well as its numerous applications.

By providing the geographical location of animals, and sometimes concurrently allowing the registration of other parameters (e.g. activity, temperature), this technique is often used to study home-range size and shape, habitat selection etc.

Medical telemetry may be defined as "the measurement and recording of physiological parameters and other patient-related information via radiated bi- or unidirectional electromagnetic signals." This book discusses the telemetric devices that can be coupled with biosensors and microsensors that generate electrical signals related to electrochemical processes.

The engagement of fisheries and polar biologists in acoustic telemetry studies on inshore polar fish is explored by sharing their experiences of the issues surrounding environmental constraint, equipment limitations, tracking protocols, choice of species, and safety in the field. Different steps taken for the telemetric system to record the electroencephalogram (EEG) from adult freely moving rats are described. This novel telemetric system was also used to record brain activity in small animals.

The development of a cost-effective telemetric system is addressed, through a combination of a wireless microphone for signal transmission and a computer sound card for recording of signals in the audible range. Finally, the telemetry of body temperature for long-term recordings of breathing is described.

Chapter 1 - Telemetry has now become a vital constituent in the field of medical sciences to remote measurement of biological parameters. Biomedical telemetry provides a means for monitoring and studying human and animal physiologic functions from a remote site with wireless transmission for the goals of minimally disturbing normal activity or free-restraint of target's subject to allow ambulatory freedom. Signals derived from physiologic transducers have been encoded and formatted in many different ways in an effort to improve transmission reliability in air space and water and carrier signals have included radio, sound and light. The long-lived primary and secondary cells have been developed for power source of transmitters. Power can now be transferred at such a radio frequency (RF) across the tissues to implanted biotelemeters using magnetic inductive powering system. Inductive powering of implantable monitoring devices is a widely accepted solution for replacing implanted batteries. Thin film solid state lithium batteries are an attractive to telemetry system of choice, offering high energy density, flexible, lightweight, miniature, rechargeable and longer lifespan so that usable in such applications of implantable medical devices, active radio frequency identification (RFID) tags, flexible displays, and E-paper. The evolution in sophisticated miniaturization provides the improvements of electronic components and assembly capabilities that available to investigators. A new technology of a capsule-camera for endoscopy is now found in sophisticated miniaturized microcontroller implementations. A survey of biotelemetry applications to living subjects in animals and humans is presented and advantages of using biotelemetry systems are described.

Chapter 2 - Understanding the relationships between organisms and their habitat is a central question in ecology. The study of habitat selection often refers to the *static* description of the pattern resulting from the selection process. However the very nature of this habitat selection process is *dynamic,* as it relies on individual movements, which are affected by both internal components (i.e. related to the animal itself, such as its behavior; foraging, resting, etc.) and external components (i.e. related to the composition of the environment). Coupling habitat selection and movement analyses should thus provide new insights into the proximal mechanisms and evolutionary causes of animals' space use.

To achieve this, the introduction of GPS technology in the early 1990s showed great promise, as it facilitates tracking of animals with high fix frequency over long time periods. From a statistical point of view, this led to an increased temporal *autocorrelation* in the positions of successive locations. Whereas classic approaches of habitat selection often relied on the assumption of statistical independence between relocations, the development of newer methods has made possible the use of autocorrelation for more dynamic approaches. As several statistical tools are now available for researchers, autocorrelation can be incorporated successfully into the analysis, instead of being removed or even ignored. The authors emphasize the need to integrate individual behavioral mechanisms in habitat selection studies.

The use of GPS technology in wildlife management issues is, however, often motivated by its technological advantage to produce large amounts of data, rather than biological questions. The authors warn users of GPS devices about the statistical and conceptual changes induced by this technology used for studying habitat selection. The authors encourage a solid biological reflection about the ecological characteristics of studied species and spatial and temporal scales considered, before deciding on which sampling protocol and which telemetry technology to use in accordance with the biological question of interest.

Chapter 3 - Implantable radiotelemetry which allows simultaneous and real-time monitoring of various physiological parameters, such as EEG, ECG and EMG from laboratory animals under unrestrained conditions has provided a major impact primarily in pharmacological, toxicological and basic biomedical research. In this chapter, the authors focus on the use of implantable radiotelemetry in electroencephalography (EEG), especially on electrocorticographic (ECoG) and deep intracerebral EEG recordings, and its crucial role in neurological characterization of transgenic mouse models. This review covers (1) the general historical, financial and ethical aspects of EEG radiotelemetry and its pros and cons; (2) the planning of radiotelemetric experiments, preoperative preparation of mice and anesthesia; (3) different EEG radiotelemetric implantation procedures, examples of physiological and pathophysiological EEGs and strategies of maximizing signal-to-noise ratio; and (4) specific information on postoperative recovery and pain management. Finally, the authors demonstrate current state-of-the-art approaches in EEG data acquisition and analysis, among which the possible pitfalls and artifacts therein are discussed in detail.

Chapter 4 - Telemetry is nowadays one of the most important tools used by ecologists allowing a privileged view of animal life while contributing decisively to its effective conservation. From the smallest rodent to the big blue whale, this is a tool that currently can be used for almost any moving vertebrate, from terrestrial to marine or flying species. Since the early 1960s, when this technique began to be applied, numerous improvements have occurred, mostly at a technological level, but most importantly that allowed increasing the amount of information that might be collected when telemetry studies are implemented. In this chapter the authors will present a brief revision of how this technique has evolved throughout the years and its numerous applications. By providing the geographical location of animals, and sometimes concurrently allowing the registration of other parameters (e.g., activity, temperature, etc.), this technique is often used to study home-range size and shape, habitat selection, activity patterns, den use, migration routes, path tortuosity and many other aspects of an animal's ecology. However, telemetry studies can also be an important source of many other biological data. Because it is an intrusive technique, animals must be trapped live and consequently samples of blood, fur, parasites, as well as morphometric measurements might be acquired. Thanks to today's advances in microbiology and molecular biology, data on kinship, taxonomy, biogeography, presence of virus, bacteria and fungus, and much other biological information might also be derived from samples collected during radio-tracking studies, which can help to improve our knowledge of a population's natural history, ecology and biology. A case study using a Mediterranean population of Eurasian badgers (*Meles meles*) will be presented in order to illustrate the usefulness of this technique.

Chapter 5 - Medical telemetry may be defined as "the measurement and recording of physiological parameters and other patient-related information via radiated bi- or unidirectional electromagnetic signals". Telemetric devices can be coupled with biosensors and microsensors that generate electrical signals related to electrochemical processes. Therefore, latter techniques involve the wireless *in-situ* detection of biologically-active molecules, including nitric oxide (NO), dopamine (DA), glucose, glutamate and lactate, in brain extracellular fluid (ECF), using implanted biosensors and microsensors. NO is a water-soluble free radical that readily diffuses through membranes and its actions in the central nervous system are largely studied. While low concentrations of NO modulate normal synaptic transmission, excess levels of NO may be neurotoxic. The constant-potential oxidation of NO occurs on Nafion poly-ortho-phenylenediamine (p-OPD)-coated carbon fibers at +865 mV vs Ag/AgCl reference electrode (RE) with good selectivity against electrochemically oxidizable anions. Dopamine is a catechol-like neurotransmitter having many functions in the brain including important roles in motor activity, reward and motivation but also in learning and attention. DA is oxidized to the corresponding orthoquinone (DA-OQ) on carbon fibers using different techniques including constant potential amperometry, chronoamperometry and fast scan cyclic voltammetry. However a biosensor is needed when it is not possible to directly oxidize molecules under standard conditions. Glucose, for example, is oxidized by means of the enzyme glucose oxidase (GOx), an oxidoreductase with a covalently-linked flavin adenine dinucleotide (FAD) cofactor. The reconversion of $FADH_2$ to FAD produces H_2O_2 in the presence of O_2. The application of a potential of +700mV to a platinum (Pt) working electrode, relative to RE, causes the generation of a measurable current that is proportional to the H_2O_2 produced. The high specificity and stability of GOx makes this enzyme suitable for biosensor construction when immobilized on the surface of Pt electrodes with p-OPD. The substitution of the GOx with glutamate or lactate oxydases, and the modification of biosensor design, allow detecting, respectively, glutamate and lactate, normally present in the ECF. While glucose and lactate are involved in the neural energetic metabolism, glutamate (GLUT) is considered the widespread excitatory neurotransmitter in the brain. Contrary to expectations, long term potentiation in the glutamatergic transmission determines an increasing of GLUT and NO in ECF with excitotoxic effects. Several biotelemetry systems for electrochemical sensor applications have been developed; in particular, a new parallel-computing embedded biotelemetry system is described in this chapter. The device, capable of driving two independent sensors, consists of a single-supply bipotentiostat-I/V converter, a parallel microcontroller unit (pMCU), a signal transmitter, and a stabilized power supply. The introduction of parallel computing programming techniques, could permit the development of low-cost devices, reduce complex multitasking firmware development, and offer the possibility of expanding the system simply and quickly. The use of CMOS components and careful electronics design will make miniaturization possible and permit implantable devices suitable for *in-vivo* applications.

Chapter 6 - The Shibetsu River located in eastern Hokkaido, Japan was channelized in the 1960s for the purpose of flood control and irrigation. However, a segment of the Shibetsu River was preliminary reconstructed to restore the natural meanders of the rivers and to improve the degraded stream segment to a more natural habitat in 2002. Chum (*Oncorhynchus keta*) and pink salmon (*O. gorbuscha*) are anadromous salmonid species migrating from the ocean to their spawning grounds in the Shibetsu River. To investigate the effects of river re-meandering on upstream migration behavior of chum and pink salmon, their behavior between channelized and reconstructed segments was compared using electromyogram (EMG) transmitters and depth/temperature (DT) loggers from 2001 to 2007. EMG transmitters allowed us to estimate swimming speeds of free-swimming salmon. Swimming behaviors (swimming speed, ground speed and swimming efficiency index) of chum and pink salmon did not differ between the two segments. Further, the authors developed energetic models using a swimming respirometer to estimate energy use swimming for pink salmon from EMG information. These models revealed that energy use during swimming for pink salmon did not differ between the two segments. However, holding time of chum and pink salmon in the reconstructed segment increased in 2005 three years after re-meandering. Measurement of physical conditions in the two segments in 2004 and 2005 revealed that a greater diversity of current speed and water depth was observed in 2005. These results suggest that the physical river conditions developing in the reconstructed segment may be suitable for holding behavior of chum and pink salmon. In addition, chum salmon swimming speed exceeding critical swimming speed (U_{crit}) was observed prior to holding. Swimming depth of chum and pink salmon collected from retrieved DT loggers tended to be deeper during holding behavior than swimming. Since our results suggest that consecutive swimming of chum and pink salmon might be exhaustive during upstream migration, the authors propose that holding behavior in the reconstructed segment may contribute to muscle recovery during upstream migration behavior.

Chapter 7 - Bio-telemetry studies have provided valuable insight into the biology of marine and freshwater fish. The majority of this research has focused on commercially important temperate and tropical species, and the movements and ecology of polar fish remains poorly understood. This lack of bio-telemetry research on polar species in general, and Antarctic fish in particular, is undoubtedly due to logistical constraints imposed by harsh and unpredictable environmental conditions, isolated locations, and a short field season. Moreover, due to the limited capacity of polar science logistic operations, researchers often have only a single opportunity to obtain field data. This essay aims to encourage both fisheries and polar biologists to engage in acoustic telemetry studies on polar fish by sharing our experience of the issues surrounding environmental constraints, equipment limitations, tracking protocols, choice of species, and safety in the field.

Short Communication - The authors examined the patterns of habitat use by 24 adult Japanese fluvial sculpin, Cottus pollux (large-egg type) by direct observation on a single night attaching a set of luminous diode and small lithium battery on the skin just beyond the 1st dorsal fin of fish. Home-range size, which was calculated by the minimum convex polygon method (range: 0.3-79.9 m2; mean: 9.8 m2), was positively correlated with the number of focal points (range: 3-22 points, mean: 10.5 points). Home-range size was not different among three reproductive states (gravid and normal females, non-nesting males). Of four categories of substrate condition (boulder, gravel, sand, and unspecified), boulder was the most common substrate category used by sculpin. Mature females spent most of their time boulder-associated substrate regardless of their reproductive condition. Non-nesting males, on the other hand, exhibited lower dependence of boulder-associated substrate than mature females. Since nesting males spent most of the time in and around their nests (i.e., boulder-associated substrate), lower dependence of boulder-associated substrate of non-nesting males may reflect reproductive state-specific life-history tactic related to foraging during the breeding season.

Short Communication - Wireless technology recording systems are as comfortable as possible to the awake experimental animal and are in many aspects advantageous and more valuable when compared to conventional recording devices. The authors describe here the different steps taken in the realization of a telemetric system to record an electroencephalogram (EEG) from adult freely-moving rats. Our system consists of an implantable transmitter that communicates bidirectionally with a receiver via radio transmission over a distance of up to three meters. The impact of the system on the animal movements was tested and has shown that the device did not restrict limb movements during locomotion. The sleeping posture was also maintained, proving the lack of discomfort produced by our system. The current system is optimized for recording electrical activity from the animal's brain, but can be easily modified to record other physiological parameters. Using our expertise, the authors also discussed here the points that have to be clarified before starting the production of such a system. Building a wireless device requires the consideration of a number of options concerning the experimental design and the recording environment to integrate the right parameters.

Short Communication - This chapter describes the development of a cost-effective telemetric system through a combination of a wireless microphone for signal transmission and a computer sound card for recording of signals in the audible range. Three common communication systems, which are normally used for voice transmission, were compared for data transmission. The final developed telemeter provides a high potential for remote monitoring up to a distance of 30 m with a sampling rate of 10 Hz and 100% accuracy with low noise. The working signal range was from 0 to 2 volts, with resolution of more than a 10 bit A/D. A satisfactorily good precision of 0.1% RSD was achieved. The system works well for wireless monitoring of output from a spectrophotometer and pH meter. This work also demonstrated successful applications of the telemetric system with various chemical analyses in our laboratory.

Short Communication - The barometric methodology is a practical and frequently adopted technique for the measurements of the breathing pattern and pulmonary ventilation in behaving animals. However, one of the problems in its application to long-term studies has been the need of monitoring the animal's body temperature and movements; the former is essential for the computation of tidal volume, the latter for the interpretation of the results. Hence, most commonly the technique is used for intermittent measurements of very short (a few minutes) duration. The availability of commercially available externally powered temperature transmitters, chronically implanted in the abdomen of experimental animals has solved this problem. Their small size is suitable to common laboratory species, like mice and rats. Hence, it is now feasible to obtain uninterrupted measurements of the breathing pattern and pulmonary ventilation lasting hours and days.

In: Telemetry: Research, Technology and Applications
Editors: Diana Barculo and Julia Daniels
ISBN 978-1-60692-509-6

Chapter 1

BIOMEDICAL TELEMETRY: TECHNOLOGY AND APPLICATIONS

Azran Azhim[1*], Yohsuke Kinouchi[2] and Masatake Akutagawa[2]

1. Frontier Research and Development Center, Tokyo Denki University, Ishizaka, Hatoyama, Hiki, Saitama, 350-0394 Japan
2. The Institute of Technology and Science, The University of Tokushima, 1-2 Minamijou Sanjima, Tokushima, 770-8506 Japan

ABSTRACT

Telemetry has now become a vital constituent in the field of medical sciences to remote measurement of biological parameters. Biomedical telemetry provides a means for monitoring and studying human and animal physiologic functions from a remote site with wireless transmission for the goals of minimally disturbing normal activity or free-restraint of target's subject to allow ambulatory freedom. Signals derived from physiologic transducers have been encoded and formatted in many different ways in an effort to improve transmission reliability in air space and water and carrier signals have included radio, sound and light. The long-lived primary and secondary cells have been developed for power source of transmitters. Power can now be transferred at such a radio frequency (RF) across the tissues to implanted biotelemeters using magnetic inductive powering system. Inductive powering of implantable monitoring devices is a widely accepted solution for replacing implanted batteries. Thin film solid state lithium batteries are an attractive to telemetry system of choice, offering high energy density, flexible, lightweight, miniature, rechargeable and longer lifespan so that usable in such applications of implantable medical devices, active radio frequency identification (RFID) tags, flexible displays, and E-paper. The evolution in sophisticated miniaturization provides the improvements of electronic components and assembly capabilities that available to investigators. A new technology of a capsule-camera for endoscopy is now found in sophisticated miniaturized microcontroller implementations. A survey of

* Correspondence: Azran Azhim, Frontier R&D Center, Tokyo Denki University, Ishizaka, Hatoyama, Hiki, Saitama, 350-0394 JAPAN. Tel: +81-49-296-1935, Fax: +81-49-296-2925. Email: azran@frontier.dendai.ac.jp or azran2020@gmail.com.

biotelemetry applications to living subjects in animals and humans is presented and advantages of using biotelemetry systems are described.

1. INTRODUCTION

Biomedical telemetry or biotelemetry is a special field of biomedical instrumentation that often permits transmission of biological information from an inaccessible location to a remote monitoring site. Biotelemetry can be used to obtain a wide spectrum of environmental, physiological, and behavioral data. Biotelemetry, first used in human beings for electrocardiogram monitoring more than 100 years ago, and has evolved as a technology for remote sensing of many applications [1].

Biotelemetry is an important method for monitoring physiological variables by providing a wireless link between the subject and the data collection equipment. Biomedical data has been telemetered through every medium between two sites, including air, space, water and biologic tissue, by using a variety of modulated energy forms like electromagnetic waves, infrared and ultrasound. Physiological measurements are frequently telemetered from a subject. This can be done by a transmitter carried on a belt or in a pocket. However, there are cases in which the transmitter is swallowed or surgically implanted in subject. The transmitting unit can be carried outside the monitored subject as a backpack unit or can be implanted within the subject's body after appropriate miniaturization and sealing against body fluid. An implantable biotelemetry unit is a device usually designed to sense a physiological event and transmit this information, at least over few centimeters of tissue, to an external receiver [2, 3]. Being of varying magnitude, distance is a fundamental but relative concept: in the case of implanted devices, it may be simply the space between the inside and the outside of the body.

Biotelemetry includes the capability for monitoring animals with minimum restraint and for providing a reproduction of the transmitted data. Since the early 1950s biotelemetry has been applied to a wide variety of subjects ranging in size from bees to whales over distances from several feet to thousands of miles. According to this concept, the advantage of biotelemetry is the measurement of physiological variables in conscious, unrestrained animals and humans. The method of biotelemetry is offering wireless, restraint-free, simultaneous and long-term data gathering [4, 5].

Biotelemetry permits the transmission of biological or physiological data from a remote location to a location that has the capability to monitor the vital signs. Wireless technologies usually link the biological source and the information receiver. In principle, any quantity that can be measured is adaptable to biotelemetry, whose main fields of application are described in section 4: Space life sciences research, medical implants, health monitoring, monitoring of animal and fish species in their natural environment (usually known as wildlife biotelemetry), and ambulatory telemetry of bioelectrical and physiological parameters in human.

The use of RF in biomedical telemetry applications has brought numerous advantages such as comfort and mobility, providing the critical information on what animals in the related habitat utilization, improvements in quality of health care, efficiency in hospital administration capabilities and finally reduction at overall medical cost. For these reasons the wireless devices are mostly preferred in such medical industry.

2. TELEMETRY SYSTEM

With the many commercial biotelemetry systems available today, it would be impossible to discuss all the ramifications of each. First of all, a simple system is described to illustrate the basic principle involved in telemetry. The stages in a biotelemetry example are shown in Figure 1. A basic biotelemetry system consists of, besides a transmitter, a receiver an antenna and power source, the basic circuits like oscillators, amplifiers, filter etc, usually present in a communication system. The design of a telemeter (for either backpack or implanted use) is dictated by size, cost, circuit complexity, power requirements (needed operational lifetime), transducers, the nature of the data to be transmitted and performance.

Any quantity that can be measured in the biomedical field is adaptable to biotelemetry. The measurements are divided in two categories: bioelectrical and physiological variables. Bioelectrical variables include measurements of electrocardiogram (ECG), electromyogram (EMG) and electroencephalogram (EEG). Signals are obtained directly in the electric form. Physiological variables such as blood pressure (BP), blood flow (BF), pulse velocity, temperature etc require some excitation or external electrical parameters. Transducers are used for the conversion of physiological parameters into an electrical signal. Parameters are measured as the variation of resistance, capacitance, or inductance.

As shown in Figure 1, physiological and bioelectrical signals are obtained from the subject by means of appropriate, and are fed into the signal conditioner and then to a multiplexer and an encoder, where the encoded is transmitted across transmission media with the help of a transmitter. The modulated carrier or the transmission medium takes the signals to the remote monitoring station where the signals are first detected at a receiver and then passed through a demultiplexer and decoder. The receiver consists of a tuner to select transmitting frequency, a demodulator to separate the signal from carrier wave so as to display or record it.

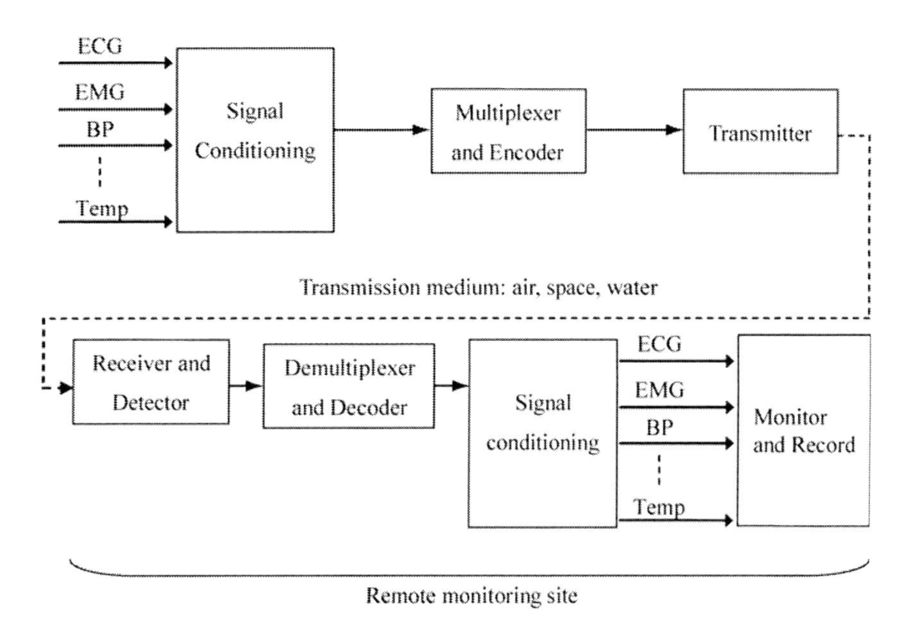

Figure 1. Block diagram of a biotelemetry. ECG: electrocardiogram, EMG: electromyogram, BP: blood pressure and Temp: Temperature.

2.1. Transmitter And Receiver

The stages of a typical biotelemetry system can be divided into functional blocks as illustrated in Figure 2 for the transmitter and in Figure 3 for the receiver [6]. Bioelectric and physiological signals are obtained from the subject by means of appropriate sensors, respectively. Then, the signal is passed through the stages of amplification and processing circuits that include generation of a subcarrier and modulation stage for transmission. The receiver consists of a tuner to select transmitting frequency, a demodulator to separate the signal from the carrier wave as a means of displaying or recording the signal.

The signal transmitted at low power on the FM transmitter is collected by the receiver and tuned to the correct frequency. The subcarrier is removed from RF carrier and then demodulated to reproduce a signal the amplitude and frequency of which can be transformed back to the original data waveform. Later, this signal can be displayed or recorded on chart recorder, oscilloscope and personal computer (PC) [6, 7].

Since most biotelemetry systems involve the use of radio transmission, a brief discussion of some basic concepts of radio (electromagnetic waves) should be helpful to the reader with limited background in this field. Radio-frequency carrier is a high-frequency sinusoidal signal which, when applied to an appropriate transmitting antenna, is propagated in the form of electromagnetic waves. The distance the transmitted signal can be received is called the range of the system. Information to be transmitted is impressed upon the carrier by a process known as modulation. In telecommunications, a high-frequency sinusoid waveform is normally used as carrier signal. The three key parameters of a sine wave are its amplitude (volume), its phase (timing) and its frequency (pitch), all of which can be modified in accordance with a low frequency information signal to obtain the modulated signal. Various methods of modulation are described below. There are two basic systems of modulation; one is amplitude modulation (AM) and the other is frequency modulation (FM) [6, 7].

In AM system, the amplitude of the carrier is caused to vary with the information being transmitted. Standard radio broadcast stations utilize this method of modulation. AM systems are susceptible to natural and man-made electrical inference.

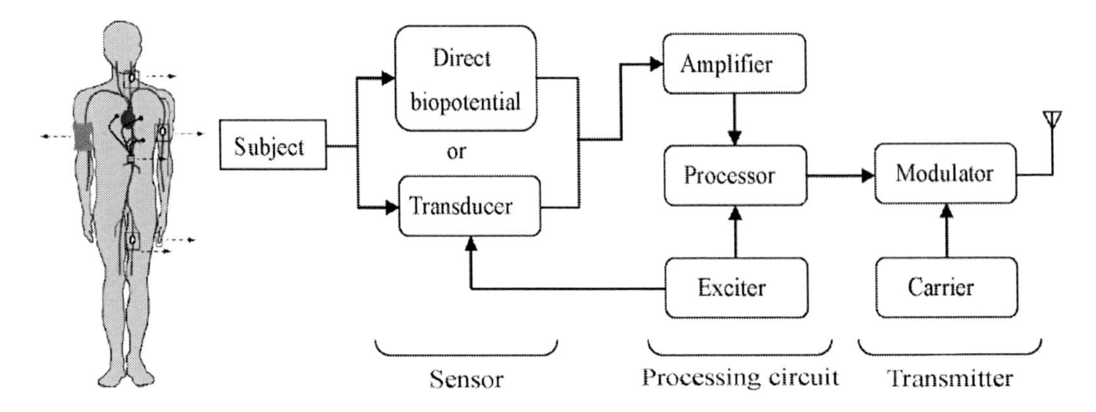

Figure 2. Block diagram of a biotelemetry transmitter.

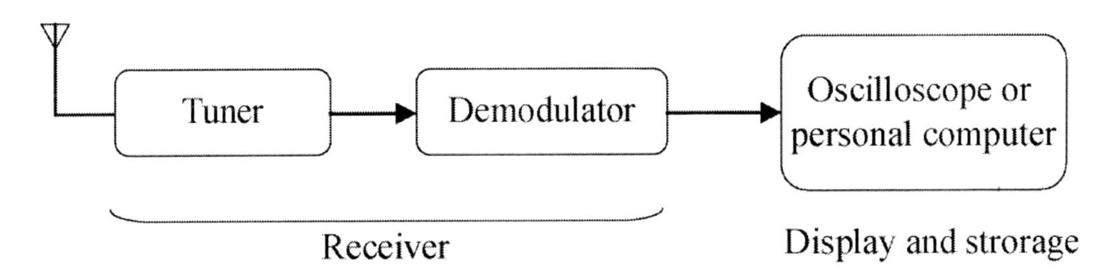

Figure 3. Block diagram for receiver-storage-display units.

In FM system, the frequency of the carrier is caused to vary with the modulated signal. FM system is much less susceptible to interference, because in variations in the amplitude of the received signal caused by interference can be removed at the receiver before demodulation takes place. Because of insusceptible interference, FM transmission is often used for telemetry.

In biotelemetry systems, the physiological signal is sometimes used to modulate a low frequency carrier, sub carrier, often in the audio frequency range. The subcarrier then modulates the RF carrier of the transmitters. If several physiological signals are to be transmitted simultaneously, each signal is placed on a subcarrier of a different frequency and all of the subcarriers are combined to simultaneously modulate the RF carrier. This process of transmitting many channels of data on a single RF carrier is called frequency multiplexing. At the receiver, a multiplexed RF carrier is first demodulated to recover each of the separate sub carriers; it must then the demodulated to retrieve the original physiological signals. Either frequency or amplitude modulation can be used for interesting data on the sub carriers. A system in which the sub carriers are frequency modulated and the RF carriers are amplitude modulated is designated as FM/AM. An FM/ AM designation means that both the sub carriers and the RF carrier are frequency modulated. Both FM/AM and FM/ FM, systems have been used in biotelemetry. FM/FM biotelemeters have been popular in a variety of restraint-free monitoring studies [6, 7].

Most of other approaches use a technique known as pulse modulation, in which the transmission carrier is generated in a series of short pulses. If the amplitude of pulses is used to represent the transmitted information, the method is called pulse modulation (PAM). If the width (duration) of each pulse is varied according to the information, the method is a pulse width modulation (PWM) or pulse duration modulation (PDM) system. In pulse position modulation (PPM), timing of a very narrow pulse is varied with respect to a reference pulse [7].

Other designations are pulse code modulation (PCM) and pulse interval modulation (PIM). In PCM, information is represented by a sequence of coded pulses, which is accomplished by representing the signal in discrete form in both time and amplitude. In PIM, it uses the spacing between constant width (length) pulses to transmit the data [8]. In all these systems, the designations can be defined as PIM/FM, PWM/FM, and so forth. As in amplitude and frequency-modulation systems, multiplexing of several channels of physiological data can be accomplished in a pulse modulation system [8].

However, instead of frequency, time multiplexing is used. In time multiplexing, each of the physiological signals is sampled and used to control either amplitude, width, or position of one pulse, depending on the type of pulse modulation. If sampling rate is several times the highest frequency component of each data signal, no loss of information results from the sampling process.

2.2 Basic Antenna and Frequency Design Considerations

Nearly all antennas have been developed from two basic types, the Hertz and the Marconi. The basic Hertz antenna is 1/2 wavelength long at the operating frequency and is insulated from ground. It is often called a dipole or a doublet. The basic Marconi antenna is 1/4 wavelength long and is either grounded at one end or connected to a network of wires called a counterpoise. The ground or counterpoise provides the equivalent of an additional 1/4 wavelength, which is required for the antenna to resonate. Most mobile transmitting and receiving antennas are quarter-wave antennas. The grounded end of the quarter-wave antenna has a low input impedance and has low voltage and high current at the input end. The directional characteristics of a grounded quarter-wave antenna are the same as those of a half-wave antenna in free space.

The consideration of antennas and electromagnetic wave propagation is essential to the complete designs of radio communication and electronic systems. The antenna converts the energy into radio waves that radiate into space from the antenna at the speed of light. A radio wave travels 300,000,000 meters a second (speed of light, v); therefore, a radio wave of 1 hertz would have traveled a distance (or wavelength) of 300,000,000 meters. Wavelength-to-frequency conversions of radio waves are really quite simple because wavelength, λ and frequency, f are reciprocals: $\lambda = v/f$. The radio waves travel through the transmission medium until they are either reflected by an object or absorbed. If another antenna is placed in the path of the radio waves, it absorbs part of the waves and converts them to energy. This energy travels through another transmission line and is fed to a receiver. From this example, you can see that the requirements for a simple communications system are as follows: Transmitting equipment, transmission line, transmitting antenna, medium, receiving antenna, and receiving equipment.

The distance the transmitted signal can be received is called as the range of the system. The range of the system depends on power and frequency of the transmitter, relative locations of transmitting and receiving antennas, and sensitivity of the receiver. There are several important factors that telemetry users should be considered when using antennas. Some of these factors are as follows: Keep clear of the antenna when taking a bearing, do not stand within 1/2 wavelength of the antenna elements, protect the antenna elements to prevent them from getting bent out of shape, and keep all metal objects from interfering with the antenna. In this way, confusion will be reduced and success rate in tracking will be increased [9]. Transmitting devices built for patients must have light weight and be compact to ensure adequate user comfort. This physical constraint on package volume means that any built-in antenna must be electrically small, with correspondingly low efficiency. Further problems appear as the telemeter is usually worn next to the skin at chest or abdominal level, so the transmitting antenna is at extremely close proximity to body tissue. In practice, the most important operational parameters for a body-surface-worn antenna used for biotelemetry are its radiation efficiency and the radiation pattern in the azimuthal plane. The analysis of electrically small antennas under near-body proximity conditions has received little consideration. The use of numerical electromagnetic modelling methods such as finite difference time domain (FDTD) technique can provide a flexible, more efficient alternative. Application of FDTD to the analysis of body-surface–mounted radiators for use in 418 MHz radio biotelemetry systems has been explained. Numerical simulation of whole body problems using suitable models enables a rapid determination of all critical parameters affecting close coupled antenna performance [2, 10].

In the National Aeronautics and Space Administration (NASA) program for RF biotelemetry, they presented the development of a printed multi-turn loop antenna for contact-less powering and RF telemetry to acquire data from implantable bio-microelectromechanical systems (bio-MEMS) based capacitive pressure sensors. They use the unique aspects of approach are as follows: first, we make use of a multi-turn loop antenna printed on a dielectric substrate with a central annular region. The central annular region facilitates housing of signal processing circuits and thus lowers the height profile of the packaged hand-held unit. Second, the diameter of the loop antenna is significantly smaller which makes the hand-held unit very compact [11].

The range of frequencies used for electromagnetic wireless communications are from 121 MHz, used in the Breitling wrist watch emergency beacon microtransmitter, to 2.5 GHz, commonly used in wireless short-range LAN. Common frequencies for ham radio, also called amateur radio are 141 MHz (2 m wavelength) and 400 MHz (70 cm wavelength). Cellular phone frequencies are in the 900 MHz (30 cm wavelength), 1800 MHz (17 cm wavelength) and 1900 MHz (16 cm wavelength) bands [12].

An argument of operating telemetry units at high frequencies is that the antenna becomes more efficient. In order to telemeter the data in biotelemetry devices, several previous studies have used a wide variety of frequency ranging anywhere from 30 MHz to 3.2 GHz. Primarily, transmitting frequencies were widely used within the 88-108 MHz frequency band, because it was set aside by the U.S. Federal Communication Commission (FCC) for "unlicensed custom-built telemetry radiators for experimentation in educational institutes". However in recent studies, the advantage to selecting a higher frequency, such as in the gigahertz, is that the specific absorption rate (SAR) actually decreases in this range because of the frequency/temperature dependence of the electrical conductivity of tissue. SAR is the method that is used to measure the effects of electromagnetic radiation on the body and is used for all standards laid out by the FCC. Heating occurs at higher temperatures, which decreases the electrical conductivity and lowers the SAR, meaning that choosing a higher frequency (gigahertz over megahertz) may be better for the subject, yet it seems to result in a lower transmission power that is in the microvolt range [13, 14].

In medical telemetry, FCC prescribed frequency bands have been available for the remote monitoring of patients' health through radio technology (nonimplantable communication) and medical implant communication services (MICS) which is used three blocks of 608-614, 1395-1400 and 1427-1432 MHz, and 402-405 MHz, respectively [15]. In Japan, frequency band of medical telemeter radio equipment is prescribed 420-450 MHz [16]. The specifications and standards of MICS will be described further in section 2.4.

It should be noted that, with the transmission of radio-frequency energy, legal problems might be encountered. Many systems use very low power and the signals can be picked up only a few feet away. Such systems are not likely to present problems. However, systems that transmit over longer distances are subject to licensing procedures and the use of the certain allocated frequency bands. Regulations vary from country to country, and in some European countries they are more strict than in the United states.

2.3 Power Sources

In many applications of biotelemetry, long operational life of the remote unit is a prime requirement. Many different types of biotelemetry systems featuring low battery drain have been developed over the years, but adequate battery life can still be a problem [17].

In animal wildlife telemetry, implantable and non-intrusive approach is required, so a primary battery is preferred. Due to the extremely limited space of the device, the shape of the battery should be coin type or low in profile and the volumetric energy density should be as large as possible. The battery should also have a long shelf life of at least ten years [18]. For the transmitter's power sources, lithium and silver oxide batteries are most commonly used. The battery life is proportional to pulse period and inversely proportional to pulse width and signal strength.

Table 1 lists the most commercially available batteries that may meet the needs for this application. The end of life behavior is also of interest, since the device will stay in the animal. The Zinc/Air battery has the highest volumetric energy density (1150Wh/l) among the listed types of batteries. However, the positive electrode of this type of battery is air. An inexhaustible supply of air is necessary to be presented within the battery, which is not suitable for this implantable application. The problem with these widely used batteries is that they are not capable of delivering a high pulse current [19, 20].

Table 1. Comparisons of widely used primarily batteries

Type	Zinc Air	Coin type Manganese Dioxide Lithium	Thionyl Chloride Lithium
Nominal Voltage	1.4V	3V	3.6V
Positive Electrode	Air	Manganese dioxide	Thionyl chloride
Electrolyte Solution	Potassium hydroxide	Organic electrolyte	Non aqueous inorganic
Negative Electrode	Zinc	Lithium	Lithium
Discharge Characteristics	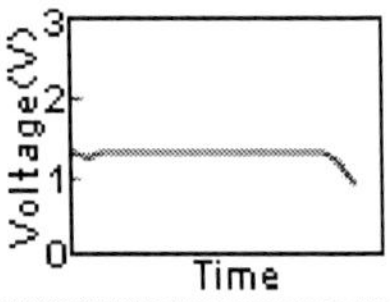	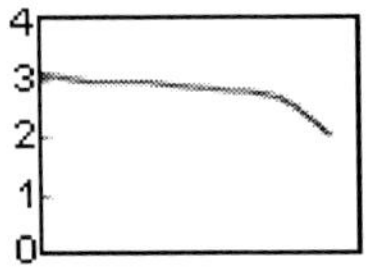	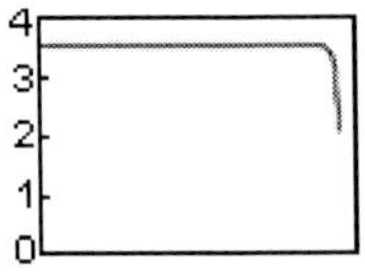
Energy Density	1150Wh/l 390Wh/kg	750Wh/l 260Wh/kg	900Wh/l 500Wh/kg
Temp. Range	-10 to 60	-20 to 85	-55 to 85
Applications	Hearing aids, pagers	Memory backup, wrist watches, PDAs, electric games, cameras, thermometers	Automatic meter reading, memory backup, security/alarm system, RF-ID tag, digital set top boxes, measuring instruments

Although some advanced lithium-based batteries with high power capability and superior long-term shelf life are available, such as Thionyl Chloride Lithium batteries, Sulphur Dioxide Lithium batteries, and Silver Vanadium Oxide Lithium batteries, they are either bio-chemically hazardous or commercially expensive. Batteries are sealed and may be encased with the transmitter. Smaller transmitters and batteries are encased in a waterproof epoxy resin, and larger models are enclosed in an aluminum or nylon case filled with waterproof resin [21].

The long-lived primary batteries that power implantable cardioverter defibrillators (ICDs) differ from batteries used to power implantable pacemakers and most other implantable devices in that they must deliver much higher power. Where pacemaker batteries typically operate for several years at power levels of less than 50 microwatts, batteries for implantable defibrillators must in addition be able to deliver a pulse of about 70 joules to the defibrillation circuit within 7-10 seconds of detection of ventricular fibrillation. Often batteries for defibrillators are designed with two cells in series. In this case, each cell must be capable of power in excess of 3.5-5 watts. In practice, each cell of a two-cell defibrillator battery is typically designed to deliver about 35 joules of energy initially to an impedance-matching load at an average power of 7-l0 watts. In order to achieve this level of power in a small package, the battery must be designed with a chemistry and mechanical configuration conducive to high power. Medtronic has developed a line of two-cell batteries for ICDs that employs a lithium/silver vanadium oxide chemistry [22, 23].

Miniaturization and long-term use of such implant electronic systems for medical applications have resulted in a growing need for an external powering system. Various environmental power sources have been investigated such as light, atomic radiation, radio waves, and biomechanical and biochemical energy converters. Among these, external radio frequency (RF) powering has been shown to be a feasible method to transport electrical energy to the implant unit [24-29]. Although large batteries can be used for long-term implants such as pacemakers, with the limitation of small size and weight, RF powering is, at the present time, the most practical way to power a chronic implant in the human body.

The method of RF powering is to supply power at a frequency f1 and to telemeter the data at another frequency f2. In this case, the difference (f2-f1) should be very large as compared to the lower frequency (fl) to avoid interference and crosstalk. Another approach is used the same RF frequency for powering and transmission, but on a time-sharing basis [30]. When properly designed, this technique eliminates the crosstalk between the powering circuit and the signal circuit and only requires a single coil set for the implant [30]. Two factors in RF-powered telemetry need to be considered: (1) the RF interference of the signal (either on the transducer or on the biological system) and, (2) the biological, particularly the long term, effect of RF radiation on the subject [31].

Another method might be considered passive biotelemetry systems. A passive telemetry system has been developed, operating according to the principle of the impedance transformation of two inductively coupled coils. Signal-dependent modulation of the implanted coil's load, which can be achieved with practically zero power consumption, reduces the data transmission to the impedance measurement of the external receiver coil [32]. Long operating life of implantable electronic circuits can be obtained by using low power transmission techniques. For example, pulse code modulation combined with remote switching systems to turn the circuit on only when monitoring is necessary [33]. A single-transistor low power under damped RF pulse position modulator, with remote switching, for implantable biotelemetry units has been presented [34].

Inductive powering of implantable monitoring devices is a widely accepted solution for replacing implanted batteries. Inductive powering is based on the magnetic coupling between an internal coil and an external coil that is driven by an alternating current [35, 36]. Both coils form a loosely coupled, coreless transformer. Parallel resonance of the internal coil with a capacitor is most often used for higher link efficiency. Such an approach would allow the use of wireless, battery-less sensor modules, with zero maintenance requirements, making them ideally suitable for use in inaccessible areas.

Inductive powering is used extensively in the area of radio frequency identification (RFID) to remotely power tags containing a stored ID and then transmit the data. However in such applications the power levels required to power and transmit the ID are generally very low, being of the order of microwatts. Inductive powering has also been extensively used in the biomedical area, for the powering of systems implanted in the body. Applications include, recharging of an implanted battery as in the case of the artificial heart or an alternative to implanted batteries in the case of long term monitoring, in for example hip prosthesis. There are also, some commercially available, inductively powered sensor modules designed for industrial applications. The two main frequency ranges used for inductively coupled applications are 125Hz and 13.56MHz [27].

Advancement in nanomaterial technology contributes an ever-increasing demand for rechargeable thin film batteries (TFBs) to drive portable microelectronic devices. Lithium batteries are the systems of choice, offering high energy density, flexible, lightweight design and longer lifespan than comparable battery technologies. The performance of thin-film solid state lithium and lithium-ion batteries makes them attractive for application in many consumer and medical products. Manufacturing scale-up is underway at several companies, and at presently estimated production costs, the products targeted first for commercial application include implantable medical devices, complementary metal oxide semiconductor (CMOS)-based integrated circuits (IC), and active RFID tags, flexible displays, E-paper and military. The batteries will be incorporated as separate components, or they will be integrated into devices by direct fabrication on IC packages or chips [37, 38].

The range of possible applications for these batteries derives from their important advantages as compared to conventional battery technologies. They can be made in virtually any shape and size to meet the requirements of each application. The batteries are rechargeable, which means their size need be no larger than is required to satisfy the energy requirements on a single cycle, thus reducing cost and weight, which in itself may give birth to new applications. Normally, TFB thickness is less than 5 μm on a substrate that can be used in microelectromechanical systems (MEMS), microrobots and microsensors, and can also be printed on ICs like any other electronic components.

The thickness of the TFB is less than 10 μm and the final product, including the air-tight package, is less than 0.1 mm thick. The cells have a surface of about 2 cm^2 and may deliver a maximum current density of about 200 $\mu A\ cm^{-2}$ for 10 min under an average potential of 2.0 V. They present a capacity of 5-300 $\mu Ahcm^{-2}$. More than 10 000 charge-discharge cycles have been performed [38]. TFB is now commercially available at Blue Spark Technologies and Infinite Power Solutions, Inc.

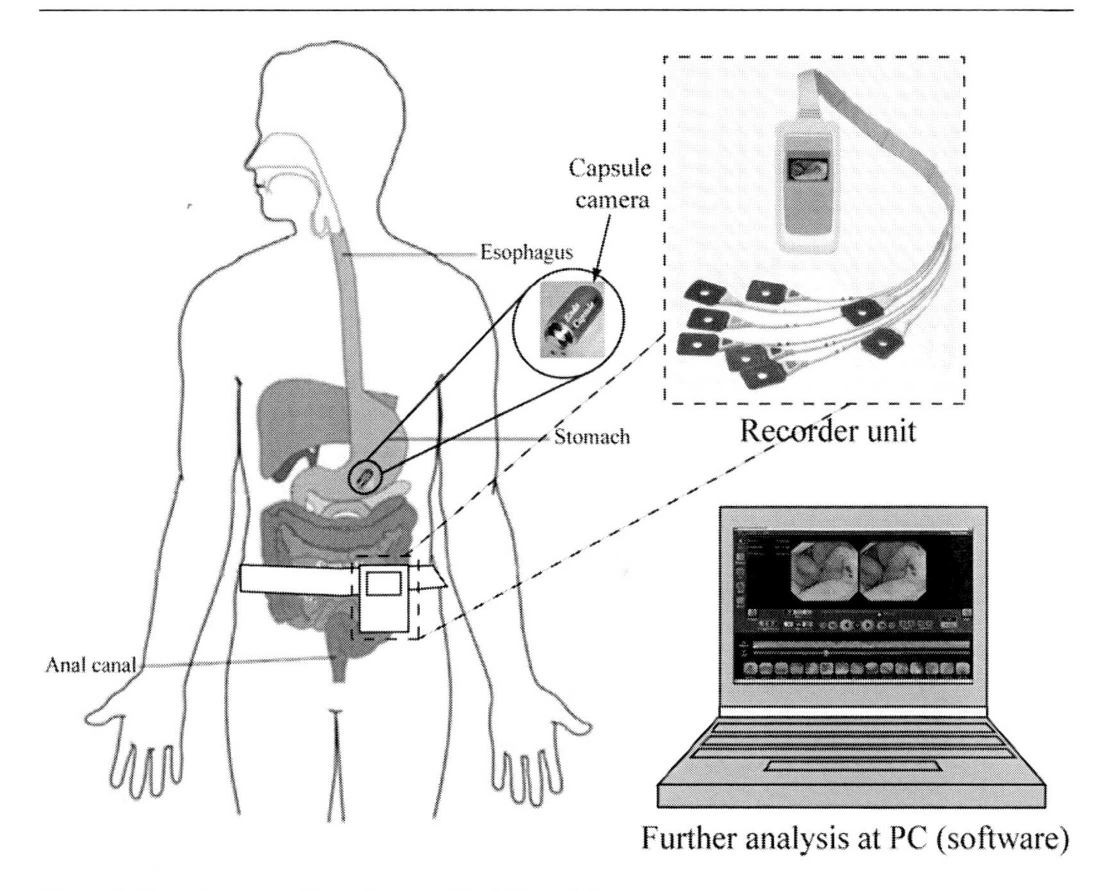

Figure 5. Capsule-camera for endoscopy (EndoCapsule).

2.4. Implanted Telemetry Devices

In some occasions, it is desirable to implant the telemetry transmitter or receiver subcutaneously. The transmitter is swallowed or surgically implanted in the subject. These systems allow monitoring and collecting data from conscious, freely moving subjects. Conscious subjects provide data free from the effects of anesthesia. It has been clearly shown in the literature that anesthetic agents can change blood pressure, heart rate, peripheral vascular resistance, thermoregulation, gastrointestinal function, and other body functions [39-43]. Comparing with the tethered animal model animals instrumented with implanted telemetry is free of exit-site infections. Also, animals instrumented with implanted telemetry are free of infections that result from exteriorized catheters and lead wires that are often required when using jacketed telemetry. In this situation, animals are free of stress and discomfort of carrying instrumentation in tight-fitting clothes. Once the telemetry device is implanted, data can be monitored 24 hours per day without human intervention or contact while the animal remains in its home cage [44, 45].

There are some requirements for the usage of an implantable medical electronic device. The implanted device needs to be highly reliable and have ultra-low power consumption. Recent progress in CMOS IC technology allows high reliable, microminiature, cost effective, ultra low power wireless transceiver designs. Although, implantable medical devices must have relatively miniature and lightweight, the implantable parts of the system must be also encapsulated in a biocompatible material. The outer case and any wiring must be impervious to body fluids and moisture. For many implantable electronic instruments, packaging is important. Requirements of packaging materials, used in current techniques are as follows: (1) epoxy, silicone rubber, and other polymeric material; (2) metal flat-packs with resistive, electron beam, or laser beam welding, or solder sealing; (3) pyrex and ceramic outer cases with epoxy or metal solder sealing [46]. For example, *in situ* measurements in the gastrointestinal (GI) tract, the whole unit of "endoradiosonde" or "radio pill" is encapsulated by a low viscosity epoxy resin (Araldite 2020, Vantigo AG, Switzerland), which cured to a hard chemically resistant transparent polymer. The sensors were protected by a custom made 2 mm high and 3 mm wide polypropylene ''cup'' attached by 3140 silicone rubber (Dow corning, US) [47].

Capsule camera for endoscopy is now available at such Given Imaging, Ltd and Olympus Medical Systems Corp. Figure 5 illustrated the new endoscopy capsule camera (EndoCapsule) system was developed by Olympus Medical Systems Corp. [105]. The system is for capsule endoscopes without batteries (wireless power supply) to be used in all parts of the gastrointestinal tract, including the esophagus, the stomach and the colon, provided high-resolution imaging and real-time viewing including drug delivery system and ultrasound scanning from inside the body.

The medical electronic devices can be classified into two categories in terms of the protocol and standards they use. The first standard is Wireless Medical Telemetry Services (WMTS). WMTS, which is for non-implantable devices, is for the remote monitoring of a patient's health where the RF communication occurs between a patient worn transmitter and a central nursing station. WMTS devices are pre-programmed to transmit at periodic intervals with short transmission duration and a long waiting period. This leads to low power consumption and less probability of interference occurring to receivers in other services. FCC has allocated interference-protected spectrum for the use by licensed physicians, healthcare facilities and certain trained and supervised technicians in 2000. The frequencies currently allocated for WMTS are divided into three blocks: the 608-614 MHz frequency band (which corresponds to UHF TV channel 37 but is not used by any TV station because it is used for radio astronomy) and the 1395-1400 MHz and 1427-1432 MHz frequency bands. The frequencies in the 1427-1432 MHz band are shared by WMTS with non-medical telemetry operations, such as utility telemetry operations. Generally, WMTS operations are accorded primary status over non-medical telemetry operations in the 1427-1429.5 MHz band, but are treated as secondary to non-medical telemetry operations in the 1429.5-1432 MHz band. [48]

The second standard is Medical Implant Communications Service (MICS). MICS devices are implantable medical devices. They communicate with outside control unit periodically or when there is any change in specified parameters that implanted device is monitoring. FCC allocated a frequency band for MICS devices in 1999. The 402-405 MHz frequency band is available for MICS operations on a shared, secondary basis. The FCC determined that, compared to other available frequencies, the 402-405 MHz frequency band best meets the technical requirements of the MICS for a number of reasons. The 402-405 MHz frequencies have propagation characteristics conducive to the transmission of radio signals within the human body [48]. It allows bi-directional radio communication with a pacemaker or other electronic implants. The maximum transmit power is very low, at 25 microwatt of equivalent isotropically radiated power (EIRP), in order to reduce the risk of interfering with other users of the same band. The maximum used bandwidth at any one time is 300 kHz, which makes it a low bitrate system compared with WiFi or Bluetooth. The main advantage is the additional flexibility compared with previously used inductive technologies, which required the external transceiver to touch the skin of the patient. MICS gives a range of a couple of meters. The standard is defined by FCC, European Telecommunications Standards Institute (ETSI), and Ministry of Internal Affairs and Communications, Japan (MIC) [49-51].

In addition, equipment designed to operate in the 402-405 MHz band can fully satisfy the requirements of the MICS with respect to size, power, antenna performance, and receiver design. Further, the use of the 402-405 MHz band for the MICS is compatible with international frequency allocations. Finally, the use of the 402-405 MHz frequency band for the MICS does not pose a significant risk of interference to other radio operations in that band [48].

3. Types of Telemetry

Three methods of wireless transmission to be considered in transmitting biological information are radio-frequency electromagnetic signals, infrared optical signals and acoustic signals. The main focus of most wireless transmission is RF, but optical data transmission can be efficient and interference proof in some applications as presented in early work by Kimmich [52]. Acoustic transmission of data over phone lines from ECG data devices has shown some application [53], and through-water communications of physiological signals have been demonstrated [54]; these modes are discussed further in the subsection as below.

3.1. Radio Telemetry

During the latter part of the 19th century, both distance and time limitations were largely overcome. The invention of the telegraph made possible instantaneous communication over long wires. Then a short time later, man discovered how to transmit messages in the form of radio waves.

In general, biotelemetry systems involve the use of radio transmission. A radio frequency carrier is a high efficiency sinusoidal signal propagated in the form of electromagnetic waves when applied to an appropriate transmitting antenna. Radio frequency energy is the most commonly used to link between biotelemeter and receiver. It is now common use to send data between two sites via satellite.

Radio-based signalling devices will play an important role in future generations of remote patient monitoring equipment, both at home and in hospital [10]. Ultimately, it will be possible to sample vital signs from patients, whatever their location and without them necessarily being aware that a measurement is being taken.

Radio telemetry is an excellent tool for gathering data on the biology of animals and their interactions with the environment they inhabit [55]. The choice of operating frequency has always been the subject of considerable controversy amongst researchers. Many select the radio frequency at which they conduct their studies basing solely upon the availability of equipment at hand or simply on tradition. Researchers are often not fully aware of the proximal impact of frequency choice on overall system performance and its ultimate impact upon the quality of the data that the study generates [56].

3.2. Ultrasound Telemetry

Electromagnetic energy, particularly at radio frequencies, is rapidly absorbed as it passes through even a few centimeters of seawater. Acoustic energy is transmitted with low energy loss through seawater [57]. Acoustic energy is the best available technique for marine fishes to transmit a signal over a distance. Ultrasonic ranges 30-100 kHz are above most animal auditory ranges and are transmitted with low energy loss through seawater [57]. There is a definite need to reveal the daily activities of animals that range out of sight of an underwater observer and to obtain information on behavior beyond release and recapture sites. Ultrasonic telemetry appears to be well suited for such studies as animals can be followed from boats or by other arrangements for up to several kilometers distance [58]. Sound travels at a predictable speed through water. If a pulsed signal is detected at a series of hydrophones at known positions, the position of the sound source can be calculated from the differences in time taken for the sound to reach each of the hydrophones successively more distant from the source. A number of tracking systems based on this principle have been used to track sound-producing animals as well as acoustically tagged individuals. Several transmitters may also be tracked simultaneously using such techniques. Consideration of position-fixing errors associated with fixed-array acoustic tracking techniques has generally centered upon the consequences of inaccuracies in the timing of the arrival of the ultrasonic pulse at each hydrophone [59-63].

The simplest type of ultrasonic transmitter the pinger, which emits pulses of a given ultrasonic frequency and repetition rate to transmitters with sensors for monitoring swimming speed, tail-beat frequency, water depth, and water temperature. [64]. Most acoustically telemetered field data are interval-encoded. Data can also be encoded in the duration of the ping, but power constraints usually mean that "off-time" is a better way to carry signal than "on-time," which is minimized for signaling. Most standard, available acoustic receiving equipment is designed to decode data in this off-interval format. Receivers are usually set to ignore later pulses in a blanking interval, typically 200-300 ms [65].

The most vexing problems with ultrasound result "multipath": the arrival of the signal at the receiver over several different paths of differing acoustic length. This is caused by reflection (off the bottom, off structures, and off the air-water interface with phase inversion), and by refrection (caused by sound velocity varying with temperature, salinity and depth). In shallow water, temperature is the most important contributor to refraction. A thermochile (warm water on top of colder) gives higher velocities near the surface and causes sound emitted near the horizontal to be "bent" down toward the bottom. At the same time, sound emitted more directly upward is reflected off the surface toward the bottom. The net result is a "shadow zone" into which no energy propagates. A transmitter with a 1 km range in unstratified water may be inaudible at 50 m during strong stratification. Even under less stratified conditions, multipath may still be severe if direct, refracted and reflected waves arrive at times different enough to cause destructive interference [66].

A robust telemetry system must tolerate multipath with errors. Single-channel systems are relatively immune because of their simple modulation schemes, but multichannel and digital systems are quite vulnerable. Amplitude modulation is once again a poor choice because constructive and destructive interference between different signal paths also modulate received amplitude. Under severe multipath conditions, frequency-shift keying (FSK) also fares poorly. Copies of the data packet arrive out of register, bursts of frequencies representing "0" and "1" overlap, and PLL decoders get confused. Frequency-hopping PIM has worked quite well, but is limited to modest data rates (approximately 200 baud). In deeper water (oceanography), PSK modulation commonly is used in acoustic modems with quite high data rates [66].

Ultrasonic tracking is a well-established tool in fresh water ecosystems, and is usually the only practical transmission mode in marine and estuarine waters which, being conductive, rapidly absorb radio transmissions. Use of electromagnetic radiation in seawater is limited to very low frequencies and short distances, e.g., Phillips et al.'s system for studying the movements of rock lobsters [68]. Ultrasonic transmission has been used primarily to track a wide variety of fish, as well as cetaceans, pinnipeds, turtles, sea snakes, and alligators [67]. Several species of invertebrates also have been tracked with ultrasonic telemetry [67].

All of these studies provided data on movements of free-ranging animals, from which behavioral information (e.g., habitat utilization) could be derived. Ultrasonic telemetry can be used to provide directly more detailed behavioral information, in addition to location. Data on water temperature, depth, or swimming speed or direction have been telemetered from various vertebrates, providing such information as dive profiles and thermal preferences [67]. In some studies, ultrasonic biotelemetry has been used to record physiological functions of free-ranging animals, particularly heart rate in large vertebrates [67].

Although telemetry of an animal's location provides important data on movement rates, track, and habitat selection, biotelemetry of physiological function along with location can provide critical information on what animals are actually doing and how they are utilizing the habitat.

3.3. Infrared Telemetry

Recent advances in technology of optoelectronics such as photodiodes, and light-emitting diodes (LEDs) lead opportunity for patient monitoring inside single room at hospital using infrared (IR) telemetry system.

The use of modulated infrared (IR) light as an alternative to RF carriers in biotelemetry systems has many advantages: because, it does not require any transmitting and receiving antennas and is not vulnerable to electromagnetic interference (EMI) and man-made noise.

The transmitter worn by the patient is more compact and does not require an inductive coil so that it can be realized in surface-mount technology in a very small size [69] When IR biotelemetry systems are used for monitoring and locating hospital patients, there are no bandwidth restrictions. Some reports suggest that RF transmission causes electromagnetic interference in medical devices such as cardiac pacemakers or infusion pumps [70-71]. The possible effects of electromagnetic interference during wireless connectivity are searched and no evidence was found of electromagnetic interference of IR modems with any of the medical devices. Furthermore, IR modems showed similar performance to a wired system even in an electrically noisy environment. As a result, IR wireless connectivity can be safely and effectively used in operating rooms [72-74].

Infrared telemetry also has a very wide field of application. The IR radiation enables transmission of different physiological parameters from moving subjects like patients in intensive care units, wards, newborn babies in incubators and animals in biological and hospital laboratories. In a typical IR biotelemetry system, the patient carries a battery-powered transmitter and one or more small arrays of infrared light emitting diodes (IRLEDs) that send encoded data to remotely located photo detector– based receivers [75].

IR radiation in information transmission is used in two ways: Narrow beam or direct radiation and diffuse IR radiation. Diffuse IR radiation fills up almost homogeneously the room where the IR transmitter and receiver are located. The mobility of the transmitter worn by the patient is complete within the room, without any restraint. The coverage of the room with IR radiation is based on reflections from the walls, ceiling, and floor. This is the reason for greater application of diffuse IR radiation in biotelemetry than narrow beam or direct IR radiation in biological parameter measurements.

For analyzing an IR telemetry system and its feasibility in a room, it is important to know the voltage amplitude and photo amplifier output related to the noise level voltage in any location at the room. At least one reflection must occur on the pathway between transmitter and receiver to realize a diffuse IR radiation. There are two opposite types of reflection. One is called specular (mirror-like) reflection and the other is reflection from matt surface. In some large rooms there are some locations where IR irradiance is not sufficient. This can happen in some corners with dark background, close to the windows, or in very long rooms. A repeater can solve such problems. The repeater consists of a sensitive enough IR receiver and transmitter; the receiver is situated in a place where there is enough irradiation so that pulses can be received. The efficiency coefficient is the most important parameter that enables the estimation of how large the room can be. Pulse frequency or pulse time interval modulation is used for biopotential or some other biological signal transmission [69, 75].

The disadvantages of IR biotelemetry are as the following: range of IR biotelemetry is short and restricted to a single room. The short range is not particularly troublesome in medical telemetry, since patients are usually confined to small rooms or wards. Diffuse IR biotelemetry system has difficulties with multichannel operation in the same room. Realization of a multichannel operation is not as simple as in radio frequency telemetry. There are two possibilities: frequency multiplexing and time division multiplexing. Transmitters are difficult to separate by wavelength division multiplex (frequency multiplex). It cannot be realized easily because of the broad frequency response characteristics of photodiodes. In IR biotelemetry system power consumption of transmitter is relatively high. Transmitter power consumption may be kept to a minimum by operating the infrared light emitting diodes in a pulsed mode with a favorable duty cycle. The pulse duration is usually set to just a few microseconds, and pulse position modulation (PPM) is used to convey the data. It is only necessary to transmit one IR pulse for each physiological signal or data channel [75].

4. Applications of Biomedical Telemetry

4.1. Space Life Sciences

In the early days of human space flight, The National Aeronautics and Space Administration (NASA) used biotelemetry to provide biomedical data from orbiting astronauts to medical personnel. Biomedical data transmitted to earth from space included astronaut's heart rate, body temperature, ECG, and oxygen and carbon dioxide concentration. Telemetry was employed to establish an understanding and to monitor health and well-being of the astronauts while they were in orbit. Therefore, NASA has been involved in the development and application of biotelemetry since the Agency's beginnings [6, 76]. Because of the great distance from the earth, systems and procedures were developed to support medical operations in flight. All astronauts wore a biosensor harness, which provided for transmitting critical physiological data back to the earth from the space craft and lunar surface. This real-time telemetry was also available to monitor astronauts in the event of illness in flight [76]. In 1962, NASA and Texas Medical Center began working together, when their scientists and physicians collaborated in the use of electroencephalography (EEG) to help monitor brain function during space flight. The first in-flight test of EEG monitoring was performed in December 1965 on Gemini Astronaut Frank Borman [77].

NASA had role in the development of space flight animal habitats and monitoring hardware in 1970s. In response to this development advanced biosensor and bioinstrumentation technologies were required. Miniaturized, specific-application biosensors, biotelemetry, and noninvasive monitoring has become vital to the telemedicine industry. Sensors 2000! (S2K!) program at NASA has implemented a variety of advanced biosensor and bioinstrumentation technologies for space research and ground medical and surgical applications. For these applications, miniaturized implantable biosensor to measure blood pH, ionic sensors for CaC, KC, and NaC, biophysical sensors for flow, pressure, and dimension, and miniature high resolution CO2 sensors, as well as advanced biotelemetry and instrumentation and data systems, are developed [78-80].

Biomedical research is expected to be one of the major scientific activities aboard NASA's redesigned space station, providing scientists with a better understanding of the impact of extended-duration missions on humans and validating countermeasures to lessen the undesirable effects of microgravity. This understanding could also yield fundamental information about processes within the body that can be applied to the diagnosis, prevention and treatment of disease states on Earth. Biotelemetry is now used routinely for monitoring patients in intensive or coronary care units of hospitals, newborns who are being transported from one unit to the next, or emergency patients who are being transported in ambulances. Another dramatic area of technology transfer has been in the development of diagnostic aids such as Computer Assisted Tomography (CAT) scans, Magnetic Resonance Imaging (MRI) and Positron Emission Tomography (PET) that have been instrumental in the detection and early diagnosis of various diseases. These specialized scanning instruments were based on the satellite image-enhancement technology that NASA used in Landsat, a satellite that produces digitized electronic pictures of the Earth's resources [77].

4.2. Health Monitoring

Under current medical systems, early signs of disease are detected by hospital examinations. However, it is difficult to find them if no subjective symptoms are recognized. To perform screening tests frequently in a hospital is impractical. Especially in advanced countries (e.g., Japan), maintaining the health of the increasing proportion of elderly people is becoming an important issue. Growing medical expenditure in those countries is a serious problem as the population ages. Earlier diagnosis will be a solution to decrease such costs. To provide earlier diagnosis, early signs of diseases must be detected before subjective symptoms appear, and health monitoring at home can help achieve this.

Most of the research for health monitoring in out-of-hospital conditions has concentrated on health monitoring at home. Most studies do not provide data about the usefulness of the approach but report initiatives taken to build systems for home monitoring. For example, Ogawa and coworkers developed a fully automated system for monitoring physiological data in the home [106, 107], and Celler et al. developed a remote monitoring system of health status of elderly at home [108]. In these approaches the measurements are mostly based on embedded sensors in the home furnishings and structures. They attempted to develop a home health monitoring system that did not provide any restrictions during sleep, bathing or elimination. The system consisted of monitoring devices and a computer terminal for collecting data. Data were automatically monitored as described below:

1. Monitoring during sleeping [109, 110]; a) thermistors are placed under the bed sheet for monitoring body movements during sleep, and b) ECG are recorded using conductive textile electrodes placed on the bed sheet and pillow,
2. Monitoring in the bathtub [111]; a) ECG are recorded using electrodes immersed in the wall of the bathtub (the ECG waveform can be measured through tap water), and b) Photoplethysmography (PPG) is recorded during bathing using a probe attached to the bottom surface of a bathtub, as PPG can be measured from the buttocks,
3. Monitoring in toilet [112]; a) body weight measurements are recorded using a precision load cell system placed in the toilet seat, b) Measurements of urine and feces weights are recorded similarly, and c) Measurement of the ballistocardiogram are recorded to evaluate information relating to cardiac function,
4. For daily behavior monitoring [113, 114]; a) infrared sensors (human detectors) are used to monitor human presence and movement in each room, b) touch detectors are installed in appliances, c) magnetic switches in doors record opening and closing events, d) thermistors in the kitchen sink are used to detect kitchen use through temperature changes from using hot water, and e) CO2 sensors are used to detect the presence of the subject's exhaled breath.

Together with the advances in electronics and communication technology, the other approaches based on wearable sensors have been introduced. Wearable health sensors may be attached to clothes [115], jewelry (e.g., rings [116]), wristwatches [117], etc., or they may be separately wearable (e.g., Polar Electro's heart-rate monitor requires wearing a chest belt). Environmental sensors may be embedded in the house [106, 107, 118, 119], furniture [107, 110, 120], car, etc., but they still measure the human being. Commonly used separate measurement devices such as personal scales, blood pressure meters, etc., fall in between these categories as these devices are often portable and hence not integrated into the environment, but they are not wearable either.

Although wearable sensors have recently been attracting more interest than the environmental sensors, both approaches have advantages and disadvantages. For wearable sensors, the main advantages are 1) possibility to measure many physiological parameters not otherwise accessible, 2) inherent support for mobility, and 3) support for continuous measurement. The disadvantages include 1) limitations in form factor, power consumption, processing power, and communications; 2) potential obtrusiveness; and 3) high demands for durability of materials etc. For environmental sensors the main advantages are related to less severe limitations with respect to power consumption, form factor, processing power, and communications means, while the disadvantages include 1) need for user identification, 2) limited possibilities for continuous monitoring, and 3) limited access to physiological parameters.

Despite the pros and cons, the list above suggests increasing employment of wearable sensors in health and wellness monitoring. The optimal solutions for multiparametric health monitoring lie in combination of embedded and wearable sensors. In a practical solution some parameters are monitored continuously with wearable sensors (e.g., user activity, heart rate) while environmental sensors are used to record other parameters (body weight, body temperature, even ECG [106, 107]).

Some collaborative projects on health monitoring are organized at a national level, for example in Italy by the MURST (Ministry of Scientific Research and Technology) and in the UK by the Department of Trade and Industry (DTI) and the EPSRC (Engineering and Physical Sciences Research Council) [121]. But a key aspect of research in the European Union is that, next to national or university research programs, research in specific fields is organized at a European level through transnational research projects. There are basically two schemes for cooperative research which apply: the European Commission (EC) Framework Program projects and the EUREKA projects [122]. Recording of physiological and psychological variables in real-life conditions could be especially useful in management of chronic disorders or health problems; e.g., for high blood pressure, diabetes, anorexia nervosa, chronic pain, or severe obesity. Furthermore, real-life long-term monitoring of health could be useful for measurement of treatment effects at home, in situations where the subjects live their daily life.

4.3. Medical Implants

Prior to the establishment of the MICS in 1999, medical implant devices had to be magnetically coupled to external programmers or readers. This magnetic coupling required that the device implanted in the patient be in very close proximity to the external monitoring control equipment, often necessitating body contact for proper operation. Furthermore, since implanted device was magnetically coupled to an external nursing unit, they could easily be affected by electromagnetic interference (EMI). According to the reports of the Center for Devices and Radiological Health under the US Food and Drug Administration (FDA), over 500 incidents were suspected to be attributable to EMI affecting cardiac devices. More than 80 of these reports involve cardiac and other medical device interactions with electronic security systems. EMI represents a major risk to patient safety and medical effectiveness due to the increasing usage of electromagnetic energy radiating devices such as cell phones and security systems [81].

In addition, medical implant devices operate with low data rates, sometimes requiring up to fifteen minutes for the required data transfer. The FCC established the MICS to overcome these limitations of medical implant devices. The FCC concluded, with the agreement of representatives of the medical community and equipment manufacturers, that establishing a MICS would greatly improve the utility of medical implant devices by allowing physicians to establish high-speed, easy-to-use, reliable, short-range (six feet) wireless links to connect such devices with monitoring and control equipment. Whereas magnetic-inductive coupling supports only one-way communication at data rates of around 50 kbit/s over a range of only a few inches, MICS solutions using an RF link can achieve up to 250 kbit/s at a range of around six feet [82].

A minimal weight is required for any implantable package. The pressure (amplitude, duration, etc), produced by the implant on the surrounding tissue may alter the blood circulation at the implant site, possibly affecting tissue reaction. One reason that titanium is used commonly as an implantable metal is because it possesses a low specific gravity and an excellent strength-to-weight ratio compared to other metals such as tantalum, tungsten, and stainless steel. Blunt corners and sharp edges should be eliminated because they irritate tissues locally. A streamlined contour is desirable. Implant location and implant technique also influences the local reaction at the site.

Sensors implantable within the human body for the measurement and wireless transmission of physiological parameters have been discussed for almost fifty years. In 1957, the "endoradiosonde" was reported by Mackay and Jacobson [83]. Several years later, a device for measurement of gastrointestinal (GI) pressure was described in [84]. In this device, a predecessor of today's "smart pills", a capsule containing a pressure sensor was swallowed and used to monitor and wirelessly transmit pressure as it moved through the GI tract. Such pills are available to measure temperature, pH, and enzyme activity also. A landmark effort in miniaturization of these devices, although still achieved using traditional 'macro' fabrication approaches, was discussed by Collins in 1967 [85]. The use of wireless sensing to monitor physiological parameters within the body is gaining widespread interest, due in part not only to advances in medicine (which help identify promising application opportunities), but also to advances in microelectronics and micromanufacturing technology (which have made such devices practical). Passive resonant pressure sensors made from MEMS-based ceramic fabrication approaches, with alumina diaphragms and fired metallic inks, have been utilized in high-temperature environments [86, 87].

Fonseca have reported the design, modeling, fabrication and characterization for flexible wireless implantable pressure sensors intended for biomedical applications [87, 88]. The devices were fabricated from flexible polymer substrates and optionally-incorporated ceramics, through lamination techniques, in order to implement a passive resonant circuit capable of wireless interrogation [88]. In the measurement of physical parameters within living organisms, passive resonant circuit techniques have been used to create sensors that are wireless, do not require power supplies and can be implanted, both acutely and chronically [88]. Passive resonant circuits used to measure physiological parameters date back to 1957 reported by Mackay [83]. In 1967 Collins reported a miniature "Transensor" ranging in size from 2-6 mm in diameter designed to be implanted in the eye [85]. These devices used two oppositely-wound planar spirals, connected at the periphery, to form an LC circuit with distributed inductance and capacitance. More recently, passive sensors have been developed for continuous measurement of intraocular pressure to monitor glaucoma using two types of sensors: a magnetic sensor and a resonant sensor, reported by [89]. The development and clinical demonstration in humans of non-flexible passive sensors for monitoring the pressure of abdominal aortic aneurysms was presented in [90].

Recent advances in neuroscience, microelectronics, and information technology have allowed construction of miniature, but highly intelligent, devices to be implanted within the brain to perform in-vitro diagnostic and therapeutic functions [91]. There has been active research on brain-computer interface (BCI) with implantable devices [92]. Some researches have mainly targeted the access/delivery of meaningful signals from/to the human cortex so that information in a bioelectric form can be converted to/from information in a digital form [92-94]. However, there exists a significant problem in establishing an effective wireless data communication link between brain implants and external computer. In the current experimental settings, wires are utilized. Clearly, this type of connection involves a high risk of infection. Wireless RF connection provides an alternative. However, its feasibility is in serious question due to the following limitations: 1) RF antenna and certain circuit elements (e.g., induction coils) increase the size and the mass of the brain implant; 2) the conversion between signal and RF waves requires a considerable amount of energy which is drained from an internal battery within the implant, and this battery is difficult to recharge or replace; and 3) the ionic fluid of biological tissues, such as the cerebrospinal fluid, is highly conductive [92]. As a result, transmitting an RF signal within the head is similar to transmitting a radio wave through an electrically shielded room. Such transmission is possible only when the RF signal is strong and its frequency is relatively low, which requires more energy consumption. However, the conductive electrical current is capable of passing information by 'volume conduction of biological tissue' [92].

In addition, BCIs can be noninvasive or invasive. Noninvasive BCIs, which derive the user's intent from scalp-recorded EEG activity, are already in use for basic communication and control [95-97]. Invasive BCIs, which derive the user's intent from neuronal action potentials or local field potentials recorded within the brain, are being studied mainly in nonhuman primates [98-101]. These invasive BCIs face substantial technical difficulties and entail significant clinical risks: they require that recording electrodes be implanted in the cortex and function well for long periods, and they risk infection and other damage to the brain. The efforts to develop them, despite these disadvantages, are based on the widespread belief [102-104] that only invasive BCIs will be able to provide users with real-time multidimensional control of a robotic arm or a neuroprosthesis. Wolpaw and McFarland presented that a noninvasive BCI in humans, using sensorimotor rhythms recorded from the scalp, can provide multidimensional control that is within the range reported for invasive BCI studies in monkeys [93].

4.4. Wildlife Biotelemetry

Wildlife radio-telemetry may be defined as the transmission of information from a transmitter on a free-ranging wild animal to a receiver. Wildlife-related telemetry is also known as radio tagging, radio-tracking or simply `tagging' or `tracking'. Advances in the field of wildlife telemetry have made it possible to acquire detailed data on many aspects of wildlife biology, including habitat use, home range size, mortality and survivorship, and migration timing and routes. Since many wildlife species are secretive and difficult to observe, radio-telemetry has provided a valuable tool to learn more about their respective life-histories.

Telemetry allows the remote sensing of the positions, movements, aspect of physiological or behavioral variables of an animal or of environmental conditions around it by means of radio (30-150 MHz) or acoustic signals (20-300 kHz). An individual can be equipped with a transmitter sending a signal which can be carrying information about heart rate or some other measurement of interest. Different transmitters are individualized by different frequencies or coded pulses. The signal is detected by hydrophones for acoustic and antennas for radio signals respectively. The distance at which the signals will be detected will mainly depend on the power radiated by the transmitter, the sensitivity of the receiving station and propagation losses. The propagation of radio signals in air and water has been extensively investigated [123].

For many species, determination of habitat selection is based on habitat-use data obtained through radio telemetry. The effects of habitat patch size, level of telemetry signal inhibition, and selection pattern are observed. Monte Carlo simulations are used to assess the effect of habitat-dependent bias in radio telemetry studies on the assessment of habitat selection. The characteristics of habitat mosaics selected by animals can be studied in this way [124, 125]. Various species of animals are used in biomedical research, and some of the radio telemetry systems consist of implantable transmitter and receiver. Implantable devices used to monitor various physiological parameters in mice, hamsters, rats, rabbits, ferrets, dogs, cows, sheep, bears, and other species including birds and fishes [45, 126-129].

For monitoring aquatic species in their natural environment, telemetry has become an increasingly important tool for studying the behavior of fishes. A combined acoustic and radio transmitting tag employing a dynamic conductivity switch suitable for investigating the migratory behavior of diadromous fish has presented [130]. The unique feature of the transmitter is its ability to sense the electrical conductivity of the ambient water and therefore operate in the appropriate signal mode. Under freshwater conditions the transmitter operates in radio mode, in seawater it operates in acoustic mode [130].

Modern telemetry system is also used to study about sharks because of their size and the need to understand their interactions with humans. The examples show the parallel progression of shark biology and acoustic biotelemetry illustrating that telemetry systems are tools for gathering data [57]. Movement rates of sharks are often estimated from the distance traveled by an animal over a certain time period, where the shark's position is obtained from a telemetry device. This resultant speed is referred to as rate of movement or point-to-point swimming speed. Instantaneous swimming speed requires more complicated and expensive transmitters to be externally attached [131, 132].

By EMG telemetry, implantation of transmitters in fish relay muscular activity to aerial or submerged antennas and receiver systems. Muscular activity rates in free-swimming fish are used to describe upstream migrations, spawning behavior, swimming performance and oxygen consumption, activity associated with stressors such as temperature changes and metabolic rates and to test bioenergetic models [133, 134]. Telemetered EMG signals indicate that muscle activity varied significantly for electrodes implanted at different longitudinal positions along the fish. As a result, electrode placement is an important influence affecting the signals obtained from radio transmitters [135]. The monitoring of neural signals of aquatic animals in the freely behaving condition is essential to understand the neural mechanism of their behavior. Underwater radio telemetry system is developed to receive EEG signals from the fish freely swimming in freshwater areas. The system uses simple and generally available instrumentation and is composed of a transmitter and a receiver. By using the system, EEG signals are successfully received from the fish freely swimming in an outdoor pond [136]. Heart rate telemetry has been utilized as a tool for the assessment of metabolic rate in wild fish by a number of investigators. It is obvious that remote monitoring of heart rate is a good indicator of physiological activity [137]. Animal movement and behavior is remotely assessed within a wide range of environments characterized by water conductivity and depth. The optimal mode of transmission is dependent upon ambient conductivity and water depth and determined by the transmitter's microprocessor and sensing devices.

4.5. Ambulatory Monitoring in Human

Various telemetric techniques have been developed for ambulatory and noninvasive determination of bioelectrical signals (e.g., ECG, EMG and EEG) and physiological signals (e.g., blood pressure, blood flow and velocity, pulse oximetry, and respiratory signals) in human subjects. In many of these techniques, electromagnetic waves of various frequencies are used as carriers of the data transmission. Many researchers have been used ambulatory methods for measuring human bioelectrical and physiological data to evaluate health care, patient care, athlete physical activity and exercise performance so that there is only the representative study of those will be introduced in the section.

The earlier investigation of recording ECG is used string galvanometer as reported by Einthoven in 1903 [1]. He used an immobile equipment required transtelephonic transmission of the ECG from the physiology laboratory to the clinic at the Academic Hospital about a mile away as documented in the 1906 paper on the "télécardiogramme" [138]. For certain cardiac abnormalities, such as ischemic coronary artery disease, diagnostic procedures require measurement of ECG while patient is exercising, usually on a treadmill or a set of steps. For this application, telemetry is often used in conjunction with exercise ECG measurements. To measure cardiac function from ECG's on land and during surface or underwater swimming, an amphibious ECG telemetry has developed by Utsuyama et al [139]. They telemetered the ECG signal using combinations of electromagnetic and conductive transmissions for use in aerial and aquatic environments, respectively.

The most frequently used the present ECG telemetry method by athletes and individuals in exercise condition programs is the chest-strap monitor, which transmits the heart rate signals collected across the chest by an electrode that picks up the potential differences [e.g., Polar (http://www.polarusa.com/)]. The signal is detected within 2 m by a wrist-mounted device, which also serves as a timepiece. Electronic receivers in exercise equipment (e.g., treadmills, cycle ergometer) also detect these signals. Reviews of most of these methods of heart rate measurement are found in the book "Medical Instrumentation and Measurements" [Cromwell 1980] and the "Biomedical Engineering Handbook" [140]. There are five noninvasive methods of measuring blood pressure: a) auscultation, b) palpation, c) flush, d) oscillation, and e) transcutaneous Doppler. The characteristics and limitations of each are discussed in "handbook of blood pressure measurement". Of the five methods, the oscillometric and transcuteneous Doppler can allow remote monitoring by incorporation of sensors for pressure oscillations or Doppler shift in the pressure cuff around the wrist or finger of humans and around other sites in animals (e.g., tail or leg).

It is sometimes difficult to meet these requirements at the doctor's office; also, some patients become nervous when their arterial pressure is taken at the office, causing readings to increase (this phenomenon is called white coat hypertension). Taking blood pressure levels at home or work with a home blood pressure monitoring device may help determine a person's true range of arterial pressure readings and avoid false readings from the white coat hypertension effect. Long term assessments may be made with an ambulatory blood pressure (ABP) device that takes regular arterial pressure readings every half an hour throughout the course of a single day and night. ABP monitoring therapy can help differentiate hypertension from white coat hypertension and the data generated can be used to adjust patient medication levels.

Recently, automatic self-contained blood pressure monitors are available at reasonable prices, some of which are capable of Korotkoff's measurement in addition to oscillometric methods, enabling irregular heartbeat patients to accurately measure their blood pressure at home, which was not possible using the traditional devices.

The 2003 US Joint National Committee recommends the use of self monitoring of arterial pressure, before considering the more expensive ambulatory monitoring of arterial pressure, to improve hypertension management [141]. Both the Joint National Committee and the 2003 guidelines from the European Society of Hypertension and the European Society of Cardiology suggest that self monitoring might also be used as an alternative to ambulatory monitoring for the diagnosis of white coat hypertension [142]. The study showed home arterial pressure monitoring is as accurate as a 24 hours ambulatory monitoring in determining arterial pressure levels [143]. The Japanese Society of Hypertension (JSH) Guidelines for the Management of Hypertension have all emphasized the importance of home BP measurements in clinical practice, clinical research, and clinical epidemiology, since home BP measurements reflect more accurately and reliably target organ damage and the prognosis of cardiovascular diseases [144].

The ability to measure and interpret variations of pressure and flow in humans depends on an understanding of physiologic principles and is based on a heritage well over 100 years old. Studies of pressure preceded those of flow, since reliable tools were available for pressure measurement almost 100 years ago but for flow only 50 years ago [145]. There are two kinds of noninvasive technique to measure blood flow for portable telemetry applications, one is a Doppler ultrasound and the other is an optical method.

The Doppler ultrasound is widely used to measure hemodynamic in blood vessel as arteries that exist in the deep place from the human tissue. Blood flow velocity has firstly measured by ultrasonic Doppler shift effects from the surface of the skin about 50 years ago [146], and has dramatically evolved in clinical modality with its widespread and familiar applications in noninvasive two-dimensional and Doppler imaging for about 30 years ago [147, 148]. Doppler ultrasound device which is usable for measuring arterial blood flow is limited to human physical performances under natural conditions, but still not available for portable telemetric application. The understanding in arterial hemodynamics and circulation regulations during physical stress is still insufficient.

As with many rapidly expanding technologies there have been a considerable number of types of instrument developed and used in their institutions of origin, whereas only a few are in widespread portable device for exercise use [149-153]. A real-time three channels telemetry system was designed for monitoring carotid and brachial blood flow velocities with synchronized of ECG in human during exercise in both aerial and aquatic environments to telemeter both blood flow velocity and ECG during exercise [149]. Multichannel biotelemetry system of blood velocities in carotid, brachial and femoral arteries with synchronization of ECG and blood pressure measurable during postural changes has developed by Azhim et al [150]. Data are transmitted using 315 MHz FM/FSK transmitter which has 28.8 kbps and ~0.5 mV/m (feeble wave) for transmission speed and output, respectively. The biotelemetry system has enough performance to get accurate data for estimation of blood circulation during physical exercise stress in both aerial and aquatic environments [149-153].

Ambulatory telemetry systems allow patients to move around the hospital while certain physiologic parameters are monitored. Traditionally, telemetry systems have used compact transmitters worn by the patient to take readings and transmit them to a central station. More recently, to expand telemetry's capabilities without adding excessive size and weight to the patient-worn transmitters, some suppliers have provided certain parameters using small wireless portable bedside monitors that are wheeled around on a roll stand by the patient. The application of biotelemetry techniques to telemedicine using wireless communications is also widely used. Telemedicine is a useful technique and has been used in such applications as medical care in remote rural areas and patient care (e.g., aftercare for pacemaker-implanted patients). Telemedicine includes transfer of basic patient information over computer networks (medical informatics), diagnosis, treatment, monitoring, and education of patients using systems that allow access to expert advice and patient information. Many studies presented their prototype telemedicine system transferring human biosignal data (e.g., ECG, blood pressure, etc) via mobile transmission, satellite transmission, internet-based applications [154-157].

CONCLUSION

Biomedical telemetry systems have been used for about 100 years and has evolved as an important tool for monitoring and studying human and animal physiologic functions from a remote site with wireless transmission for the goals of minimally disturbing normal activity or free-restraint of target's subject to allow ambulatory freedom. Biomedical telemetry applications has brought numerous advantages such as comfort and mobility, providing the critical information on what animals in the related habitat utilization, improvements in quality of health care, efficiency in hospital administration capabilities and finally reduction at overall medical cost. Biomedical telemetry is a reliable tool for data gathering since the invention of integrated circuit (IC) technology in 1958. IC technology has enormous impacts on the contributions of microelectronics to biomedicine and health care applications. Microminiature and micropower are essential contributions to modern biotelemetry design and construction. Modern biotelemetry began as a single transistor Endoradiosonde but now is found in sophisticated miniaturized microcontroller implementations with wireless RF power supply. The evolutions of semiconductor microcircuit and wireless communication technologies have closely paralleled the improvements in the areas. Power supply can now be transferred at RF across the tissues to power implantable biotelemetry units and to recharge their rechargeable batteries to provide for long operational lifetimes. The rechargeable thin film batteries are available to investigators and attractive for application in many consumer and medical products including implantable medical devices. The field of biomedical telemetry is truly exciting challenging and diverse in applications to living subjects. It seems likely the development will be in further miniaturization and integration of biotelemeters and transducers, by improved power supply and packaging.

REFERENCES

[1] Einthoven W. (1903). Die galvanometrische Registerung des menschlichen Elektrokardiogram: Zugleich eine Beurtheilung der Anwendung des Capillar-Elektrometers in der Physiologie. Pfl¨ugers *Arch ges. Physiol*, 99, 472-480.

[2] Welkowitz, W.; Deutsch, S. and Akay, M. (1976). Biomedical Instruments Theory and Design (2nd edition). San Diego, CA: Academic Press.

[3] Ziaie, B.; Nardin, M. D.; Coghlan, A. R. and Najafi, K. (1997). A single–channel implantable microstimulator for functional neuromuscular stimulation. IEEE *Trans. Biomed. Eng.,* 44(10), 909-920.

[4] Meindl, J. D. and Ford, A. J. (1984). Implantable telemetry in biomedical research. IEEE *Trans. Biomed. Eng*., BME-31(12), 817-823.

[5] Güler, I. and Kara, S. (1996). A low-cost biotelemetry system for long time monitoring of physiological data. *J. Med. Syst.,* 20(3), 151-156.

[6] Cromwell, L.; Weibell, F. J. and Pfeiffer, E. A. Biotelemetry. In: Huebner, V. editor. Biomedical Instrumentation and Measurements. Englewood Cliffs: Prentice-Hall, 1980, 316–343.

[7] Oppenheim, A. V.; Willsky, A. S. and Nawab, S. H. (1983). Communication systems. In: Oppenheim A. V. editor. Signals and Systems. New Jersey: Prentice-Hall; 1983; 582-625.

[8] Haykin, S. Pulse modulation. In: Elliot, S. editor. Communication Systems. New York: Wiley; 1978; 351-406.

[9] Available from http://biotelem.org/appnotes.htm: Grant Systems Engineering Inc. Biotelemetry: Antenna Tips.

[10] Scanlon, W. G.; Evans, N. E. and Burns, J. B. (1999). FDTD Analysis of close-coupled 418 MHz radiating devices for human biotelemetry. *Phys. Med. Biol.,* 44, 335-345.

[11] Simons, R.N.; Hall, D.G. and Miranda, F.A. (2004). Printed multi-turn loop antenna for RF biotelemetry. Antennas and Propagation Society International Symposium, IEEE, 2, 1339-1342.

[12] Budinger, T. F. (2003). Biomonitoring with wireless communications. *Annual Review of Biomedical Engineering*, 5, 383-412.

[13] Solazzo, S. A; Liu, Z.; Lobo, S. M.; Ahmed, M.; Hines-Peralta, A. U.; Lenkinski, R. E. and Goldberg, S. N. (2005). Radio frequency Ablation: Importance of Background Tissue Electrical Conductivity - An Agar Phantom and Computer Model Study. *Radiology,* 236, 495-502.

[14] Available from http://www.ece.duke.edu/undergrads/IndStudy01/AraneeGwD/aranee/aranee.html: Techawiboonwong, A. (2001) Temperature Dependence of Microwave Electrical Properties.
[15] FCC Rules and Regulations, "MICS Band Plan", Part 95, Jan. 2003.
[16] Available from http://uljapan.co.jp/emc/to_kitei_e.html: UL Japan certification service under Japanese Radio Law.
[17] Ko, W. H; Liang, S. P and Fung, C.D.F. (1977). Design of radio-frequency powered coils for implant instruments. *Med. and Biol. Eng. and Comput.*, 15, 634-640.
[18] Sun, C. C. and Dunford, W. G. (2005). Power Management in a Biotelemetry Application. IEEE of Power Electronics Specialists Conference, 1182-1190.
[19] Linden, D. and Reddy, T. B. (2002). Handbook of Batteries. McGraw-Hill Professional.
[20] Available from http://www.maxell.co.jp/e/products/industrial/battery/knowmore/comparison.html. Hitachi Maxell, Ltd. (2000). Comparison of Primary and Rechargeable Batteries.
[21] Takeuchi, E. S. and Thiebolt III, W. C. (1988). The Reduction of Silver Vanadium Oxide in Lithium/Silver Vanadium Oxide Cells. *J. Electrochem. Soc.,* 135(11), 2691-2694.
[22] DiMarco, J. P. (2003). Implantable Cardioverter–Defibrillators. *N Engl. J. Med*, 349, 1836-1847.
[23] Skarstad, P.M. (1997). Lithium/silver vanadium oxide batteries for implantable cardioverter-defibrillators. Battery Conference on Applications and Advances, 151-155.
[24] Ko, W. H. and Neuman, M. R. (1967). Implant Biotelemetry and Microelectronics, Science, 156(3773), 351-360.
[25] Ko, W. H.; Plonsey, R. and Kang, S. R. (1972). The radiation from an electrically small circular wire loop implanted in a dissipative homogeneous spherical medium. *Ann. Biomed. Eng.* 1, 135-165.
[26] Fryer, T. B.; Lund, G. F. and Williams, B. A. (1978). An inductively powered telemetry system for temperature, EKG and activity monitoring. *Biotelemetry and Patient Monitoring*, 5(2), 53-76.
[27] O'Donnell, T.; Chevalerais, O.; Grant, G.; O'Mathuna, S.C.; Power, D. and O'Donnovan, N. (2005). Inductive Powering of Sensor Modules. Applied Power Electronics Conference and Exposition, 3, 2024-2029.
[28] Vanschuylenbergh, K. and Puers, R. (1996). Self-tuning inductive powering for implantable telemetric monitoring systems. Proc. of Int. Conf. on Solid-State Sensors and Actuators Eurosensors IX, 52(1-3), 1-7.
[29] Donaldson N. de N. (1986). Passive signaling via inductive coupling. *Med. and Biol. Eng. and Comput.*, 24, 223-224.
[30] Ko, W. H.; Hynecek, J. and Homa, J. (1979). Single frequency RF powered ECG telemetry system. IEEE *Trans. Biomed. Eng., BME*-26, 460-467.
[31] Ko, W. H. and Liang, S. P. (1980). RF-powered cage systems for implant biotelemetry. IEEE *Trans. Biomed. Eng., BME*-27(8), 460–467.
[32] Towe, B. C. (1986). Passive biotelemetry by frequency keying. IEEE *Trans. Biomed. Eng., BME*-33(10), 905-909.
[33] Leung, A. M.; Ko, W. H.; Spear, T. M. and Bettice, J. A. (1986). Intracranial pressure telemetry system using semicustom integrated circuits. IEEE *Trans. on Biomed Eng, BME*-33, 386-395.

[34] Abatti, P. J. and Pichorim, S. F. (1994). Single transistor underdamped RF pulse position modulator with remote switching for implantable biotelemetry units. *Electron. Lett.,* 30(19), 1564–1565.

[35] Donaldson, N. de N. and Perkins T. A. (1983). Analysis of resonant coupled coils in the radio frequency transcutaneous links. *Med. and Biol. Eng. and Comput.*, 21, 612-627.

[36] Ko, W. H. and Liang, S. P. (1980). RF-powered cage systems for implant biotelemetry. IEEE *Trans. Biomed. Eng., BME-*27(8), 460–467.

[37] Bates, J. B.; Dudney, N. J.; Neudecker, B.; Ueda, A. and Evans, C. D. (2000). Thin-film lithium and lithium-ion batteries. Solid State Ionics, 135, 33-45.

[38] Souquet, J. L. and Duclot, M. (2002). Thin film lithium batteries. Solid State Ionics, 148, 375-379.

[39] Halliwill, J. R. and Billman, G. E. (1992). Effect of general anesthesia on cardiac vagal tone, *Am. J. Physiol. Heart Circ. Physiol*, 262, 1719-1724.

[40] Mapleson, W. W. (1996). Effect of age on MAC in humans: a meta-analysis. *Br. J. Anaesth*., 76, 179-185

[41] Eger, E. I. II. (2001). Age, minimum alveolar anesthetic concentration, and minimum alveolar anesthetic concentration-awake. *Anesth. Analg*., 93, 947-953.

[42] Lerman J.; Gregory G. A.; Eger E. I. II. (1984). Hematocrit and the solubility of volatile anesthetics in blood. *Anesth. Analg*., 63, 911-914.

[43] Weiskopf R. B.; Eger E. I. II, Noorani, M. and Daniel, M. (1994). Fentanyl, esmolol, and clonidine blunt the transient cardiovascular stimulation induced by desflurane in humans. *Anesthesiology*, 81, 1350-1355.

[44] Graichen, F.; Bergmann, G. and Rohlmann, A. (1996). Patient monitoring system for load measurement with spinal fixation devices. *Med. Eng. Phys*., 18(2), 167-174.

[45] Cooke, S. J.; Hinch, S. G.; Wikelski, M.; Andrews, R. D.; Kuchel, L. J.; Wolcott, T. G. and Butler, P. J. (2004). Biotelemetry: a mechanistic approach to ecology, Trends in Ecology and Evolution, 19(6), 334-343.

[46] Ko, W. H. and Smith, S. R. (1980). Packaging of implantable elecronic instruments. IEEE *Trans. Biomed. Eng., BME-*27(9), 533.

[47] Johannessen, E. A.; Wang, L.; Reid, S. W. J.; Cumming, D. R. S. and Cooper, J. M. (2006). Implementation of radiotelemetry in a lab-in-a-pill format. The Royal Society of Chemistry, 6, 39-45.

[48] FCC Rules and Regulations, "WMTS Band Plan", Part 95, Mar. 2003.

[49] Available from http://www.soumu.go.jp/joho_tsusin/eng/Releases/NewsLetter/Vol15/Vol15_23/Vol15_ 23.html: (2005) Technical Requirements of Medical Implant Communications System. News letter of the Ministry of Internal Affairs and Communications, Japan. 15(23).

[50] Available from http://www.soumu.go.jp/joho_tsusin/eng/Releases/Telecommunications/news070411_2.html: Ministry of Internal Affairs and Communications, Japan. (2007). Report from the Radio Regulatory Council Concerning the Draft MIC Ordinance to Amend Part of the Rules for Regulating Radio Equipment, Etc., and Results of Public Comment Procedures Thereon. Press Release-Telecom.

[51] Savci, H. S.; Sula, A.; Wang, Z.; Dogan, N. S. and Arvas, E.. (2005). MICS transceivers: regulatory standards and applications (medical implant communications service). Proc. of IEEE SoutheastCon, 179-182.

[52] Kimmich, H. P. (1982). Biotelemetry, based on optical transmission. Biotelem Patient Monit., 9(3), 129-143.

[53] Orlov, O. I.; Drozdov, D. V.; Doarn, C. R. and Merrell, R.C. (2001). Wireless ECG monitoring by telephone. *Telemed. J. e-Health*, 7, 33-38.

[54] Woodward, B. and Bateman, S. C. (1994). Diver monitoring by ultrasonic digital data telemetry. *Med. Eng. Phys.*, 16, 278-286.

[55] Salvatori, V.; Skidmore, A. K.; Corsi, F. and Van der Meer, F. (1999). Estimating temporal independence of radio-telemetry data on animal activity. *J. Theor. Biol.*, 198 (4), 567-574.

[56] Sisak, M. M. and Lotimer, J. S. (1998). Frequency choice for radio telemetry - the HF vs. VHF conundrum. *Hydrobiologia*, 372, 53-59.

[57] Voegeli, F. A.; Smale, M. J.; Webber, D. M.; Andrade, Y. and O'Dor, R. K. (2001). Ultrasonic telemetry tracking and automated monitoring technology for sharks. *Environ. Biol. Fishes,* 60, 267-281.

[58] Sundström, L. F.; Gruber, S. M.; Clermont, S. M.; Correia, J. P. S.; Marignac, J. R. C.; Morrissey, J. F.; Lowrance, C. R.; Thomassen, L. and Oliveira, M. T. (2001). Review of elasmobranch behavioral studies using ultrasonic telemetry with special reference to the lemon shark, Negaprion brevirostris, around Bimini Islands, Bahamas. *Environ. Biol. Fishes*, 60, 225-250.

[59] Smith, G. W.; Urquhart, G. G.; MacLennan, D. N. and Sarno, B. (1998). A comparison of theoretical estimates of the errors associated with ultrasonic tracking using a fixed hydrophone array and field measurements. *Hydrobiologia,* 371/372, 9-17.

[60] Freitag, L. E. and Tyack, P. L. (1993). Passive acoustic localization of the Atlantic bottlenose dolphin using whistles and echolocation clicks. *J. Acoust. Soc. Am.*, 93, 2197-2205.

[61] Lagardere, J.P.; Ducamp, J. J.; Favre, L.; Dupin, J. M. and Sperandio, M. (1990). A method for the quantitative evaluation of fish movements in salt ponds by acoustic telemetry. *J. Exp. Mar. Biol. Ecol.*, 141, 221–236.

[62] Juell, J. E. and Westerberg, H. (1993). An ultrasonic telemetric system for automatic positioning of individual fish used to track Atlantic Salmon (Salmo Salar L.) in a sea cage. *Aquacult. Eng.*, 12, 1-18.

[63] Sarno, B.; Glass, C. W.; Smith, G. W.; Mojsiewicz, W. R. A. and Johnstone, A . D. F. (1994). A comparison of the movements of two gadoid species in the vicinity of an underwater reef. *J. Fish Biol.,* 45, 811-817.

[64] Voegeli, F. A.; Lacroix, G. L. and Anderson, J. M. (1998). Development of miniature pingers for tracking Atlantic salmon smolts at sea. *Hydrobiologia*, 371/372, 35-46.

[65] Voegeli, F. A. and Pincock, D. G. (1981). Determination of fish swimming speed by ultrasonic telemetry. *Biotelem. Patient Monit.*, 7, 215-220.

[66] Wolcott, T. G. (1995). New options in physiological and behavioural ecology through multichannel telemetry. *J. of Experimental Marine Biology and Ecology*, 193(1-2), 257-275.

[67] Wolcott, T. G. and Hines, A. H. (1989). Ultrasonic Biotelemetry of Muscle Activity from Free-Ranging Marine Animals: A New Method for Studying Foraging by Blue Crabs (Callinectes sapidus). *Biol. Bull*, 176, 50-56.

[68] Phillips, B. F., Joll, L. M. and Ramm, D. C. (1984). An electromagnetic tracking system for studying the movements of rock (spiny) lobsters. *J. Exp. Mar. Biol. Ecol.*, 79, 9-18.

[69] Santic, A. (1991). Theory and application of diffuse infrared biotelemetry, *Crit.l Rev. Biomed. Eng.*, 18(4), 289-309.
[70] Barbaro, V.; Bartolini, P.; Donato, A.; Militello, C.; Altamura, G.; Ammirati, F. and Santini, M. (1995) Do European GSM mobile cellular phones pose a potential risk to pacemaker patients?. *Pacing Clin. Electrophysiol.*, 18(6), 1218–1224.
[71] Hayes, D. L.; Wang, P. J.; Reynolds, D. W.; Estes, M. I.; Griffith, J. L.; Steffens, R. A.; Carlo, G. L.; Findlay, G. K. and Johnson, C. M. (1997). Interference with cardiac pacemakers by cellular telephones. *N. Engl. J. Med.*, 336(21), 1473-1479.
[72] Hagihira, S.; Takashina, M.; Mori, T.; Taeneka,N.; Mashimo, T. and Yoshiya, I. (2000). Infrared transmission of electronic information via LAN in the operating room. *J. Clin. Monit. Comput.*, 16, 171-175.
[73] Gfeller, F. R. and Bapst, U. (1979). Wireless in-house data communication via diffuse infrared radiation. *Proc. Conf. of IEEE*, 67, 1474-1486.
[74] Hof, A. L.; Bonga, G. J.; Swarte, F. G. and de Pater, L. (1994). Modular PPM telemetry system with radio, infrared and inductive loop transmission. *Med. Biol. Eng. Comput.*, 32(1), 107-112.
[75] Weller, C. (1985). Modulation scheme suitable for infrared biotelemetry. *Electron. Lett.,* 21(14), 601-602.
[76] Available from http://www.quasar.org/21698/nasa/history.html: Charles Doarn, NASA-History Biotelemetry.
[77] Available from http://www.nasa.gov/centers/johnson/news/releases/1993_1995/93-055.html: Watson, C. (1993). NASA, Texas Medical Center Formalize Cooperation.
[78] Hines, J. W. (1996). Medical and surgical applications of space biosensor technology. *Acta Astronautica*, 38, 261-267.
[79] Hines, J.W.; Somps, C.J. and Madou, M. (1997) Space biosensor systems: implications for technology transfer. *Proc. Conf. of the IEEE Eng. in Med. and Biol. Soc*, 2, 740-743.
[80] Beach, R. D.; Von Kuster, F. and Moussy, F. (1999). Subminiature implantable potentiostat and modified commercial telemetry device for remote glucose monitoring. IEEE *Trans. Instrum. Measur.*, 48, 1239-1245.
[81] Witters, D.; Portnoy, S.; Casamento, J.; Ruggera, P. and Bassen, H. (2001). Medical device EMI: FDA analysis of incident reports, and recent concerns for security systems and wireless medical telemetry. IEEE *Int. Symp. on Electromagnetic Compatibility*, 2, 1289-1291.
[82] Falcon, C. (2004). Low-power transceivers get patients mobile [wireless medical applications]. *Communications Engineer*, 2(3), 40-43.
[83] Mackay, R. S. and Jacobson, B. (1957). Endoradiosonde. *Nature*, 175, 1235-1240.
[84] Ferrar, J. T. and Zworykin, V. (1959). Telemetering of Gastrointestinal Pressure in Man by Means of an Intraluminal Capsule Energized from an External Power Source. *The Physiologist,* 2, 37.
[85] Collins, C.C. (1967). Miniature Passive Pressure Transensor for Implanting in the Eye. IEEE *Transactions on Biomedical Engineering*, 14(2), 74-83.
[86] English, J. M. and Allen, M. G. (1999). Wireless Micromachined Ceramic Pressure Sensors. *Proc. Conf. of IEEE Microelectromechanical System*, 511-514.

[87] Fonseca, M. A.; English, J. M.; von Arx, M. and Allen, M. G. (2002). Wireless micromachined ceramic pressure sensor for high-temperature applications. *Journal of Microelectromechanical Systems*, 11(4), 337-343.

[88] Fonseca, M. A.; Allen, M. G.; Kroh, J. and White, J. (2006). Flexible Wireless Passive Pressure Sensors for Biomedical Applicatons. Solid-State Sensors, Actuators, and Microsystems Workshop, 37-42.

[89] Lizón-Maritinez, S.; Giannetti, R.; Rodrígez-Marrero, J. L. and Tellini, B. (2005). Design of a System for Continuous Intraocular Pressure Monitoring. *IEEE Transaction of Instrumentation and Measurement*, 54(4), 1534-1540.

[90] Allen, M. G. (2005). Micromachined Endovascularly-Implantable Wireless Aneurysm Pressure Sensor: From Concept to *Clinic. Proc. Transducers*, 1, 275-278.

[91] Sun, M.; Mickle, M.; Liang, W.; Liu, Q. and Sclabassi, R. J. (2003). Data Communication between Brain Implants and Computer. IEEE Transac on *Neural Systems and Rehabilitation Engineering*, 11(2), 189-192.

[92] Berger, T. W.; Baudry, M.; Brinton, R. D.; Liaw, J.-S.; Marmarelis, V. Z.; Park, Y. A.; Sheu, B. J. and Tanguay Jr., A. R. (2001). Brain-implantable biomimetic electronics as the next era in neural prosthetics. Proc. of IEEE, 89, 993-1012.

[93] Wolpaw, J. R. and McFarland, D. J. (2004). Control of a two-dimensional movement signal by a noninvasive brain-computer interface in humans. *Proc. of National Academy of Sciences of the USA*, 101(51)17849-17854.

[94] Schwartz, A. B.; Cui, X. T.; Weber, D. J. and Moran, D. W. (2006). Brain-Controlled Interfaces: Movement Restoration with Neural Prosthetics. *Neuron*, 52, 205-220.

[95] Birbaumer, N.; Ghanayim, N.; Hinterberger, T.; Iversen, I.; Kotchoubey, B.; Ku¨bler, A.; Perelmouter, J.; Taub, E.. and Flor, H. (1999). A brain-controlled spelling device for the completely paralyzed. *Nature*, 398, 297-298.

[96] Pfurtscheller, G.; Neuper, C.; Guger, C.; Harkam, W.; Ramoser, H.; Schlogl, A.; Obermaier, B. and Pregenzer, M. (2000). Current trends in Graz brain-computer interface (BCI) research. IEEE *Trans. on Neural Systems and Rehabilitation*, 8 (2), 216-219.

[97] Wolpawa, J. R.; Birbaumer, N.; McFarland, D. J.; Pfurtscheller, G. and Vaughan, T. M. (2002). Brain–computer interfaces for communication and control. *Clinical Neurophysiology*, 113, 767-791.

[98] Chapin, J. K.; Moxon, K. A.; Markowitz, R. S. and Nicolelis, M. A. L. (1999). Real-time control of a robot arm using simultaneously recorded neurons in the motor cortex. *Nat. Neurosci.*, 2, 664-670.

[99] Wessberg, J.; Stambaugh, C. R.; Kralik, J.; Beck, P. D.; Laubach, M.; Chapin, J. K.; Kim, J.; Biggs, J.; Srinivasan, M. A. and Nicolelis, M. A. (2000). Real-time prediction of hand trajectory by ensembles of cortical neurons in primates. *Nature*, 408, 361-365.

[100] Serruya, M. D.; Hatsopoulos, N. G.; Paminski, L.; Fellows, M. R. and Donoghue, J. P. (2002). Brain-machine interface: instant neural control of a movement signal. *Nature*, 416, 141-142

[101] Taylor, D. A.; Helms Tillery, S. I. and Schwartz, A. B. (2002). Direct cortical control of 3D neuroprosthetic devices. *Science*, 296, 1829-1832.

[102] Fetz, E. E. (1999). Real-time control of a robotic arm by neuronal ensembles, *Nat. Neurosci.*, 2, 583-584.

[103] Chapin, J. K. (2000). Neural prosthetic devices for quadriplegia. *Curr. Opin. Neurobiol*, 13, 671-675.
[104] Donoghue, J. P. (2002). Connecting cortex to machines: recent advances in brain interfaces. *Nat. Neurosci.*, 5, 1085-1088.
[105] Available from http://www.olympus-global.com/en/news/2004b/nr041130capsle.cfm: Olympus Medical Systems Corp. (2004). Development of Capsule Endoscopes and Peripheral Technologies for further Expansion and Progress in Endoscope Applications. New release of Olympus.
[106] Ogawa, M.; Tamura, T. and Togawa, T. (1998). Automated acquisition system for routine, noninvasive monitoring of physiological data. *Telemed. J.*, 4(2), 177-195.
[107] Tamura, T.; Togawa, T.; Ogawa, M. and Yoda, M. (1998). Fully automated health monitoring system in the home. *Med. Eng. Phys.*, 20(8), 573-579.
[108] Celler, B.G.; Earnshaw, W.; Ilsar, E. D.; Betbeder-Matibet, L.; Harris, M. F.; Clark, R.; Hesketh, T. and Lovell, N. H. (1995). Remote monitoring of health status of the elderly at home. A multidisciplinary project on aging at the University of New South Wales. *Int. J. Bio-Med. Comp.*, 40, 147-155.
[109] Tamura, T.; Zhou, J.; Mizukami, H. and Togawa, T. (1993). A system for monitoring temperature distribution in bed and application to the assessment of body movement. *Physiol. Meas*, 14, 33-41.
[110] Ishijima M. and Togawa, T. (1989). Observation of electrocardiogram through tap water. *Clin. Phys. Physiol. Meas.,* 10, 171-175.
[111] Ogawa, M. and Togawa, T. (2003). The concept of the home health monitoring, Enterprise Networking and Computing in Healthcare Industry, 71-73.
[112] Yamakoshi, K.; Kuroda, M.; Tanaka, S.; Yamaguchi, I. and Kawarada, A. (1996). Nonconscious and automatic acquisition of body and excreta weight together with ballistocardiogram in a lavatory. Proc. of Conf. IEEE Eng. in *Med. Biol. Soc.*, 1, 67-68.
[113] Suzuki, R.; Ogawa, M.; Tobimatsu, Y. and Iwaya, T. (2001). Time-course action analysis of daily life investigations in the Welfare Techno House in Mizusawa. *Telemedicine Journal and E-Health*, 7(3), 249-259.
[114] Barger, T. S.; Brown, D. E.. and Alwan, M. (2005). Health-status monitoring through analysis of behavioral patterns. IEEE Transactions on Systems, Man and Cybernetics, 35, 22-27.
[115] Rantanen, J.; Impio¨, J.; Karinsalo, T.; Malmivaara, M.; Reho, A.; Tasanen, M. and Vanhala, J. (2002). Smart Clothing Prototype for the Arctic Environment. *Journal of Personal and Ubiquitous Computing*, 6(1), 3-16.
[116] Rhee, S.; Yang, B-H. and Asada, H. (1998). The Ring Sensor: A new ambulatory wearable sensor for twenty-four hour patient monitoring. Proc. Conf. of the IEEE *Eng. in Med. and Biol. Soc.*, 1906-1909.
[117] Available from http://www.suunto.com: Suunto Oy.
[118] Available from: http://www.anchortrust.org.uk/publications/telecare.html: Porteus, J. and Brownsell, S. (2000). Using Telecare: Exploring Technologies for Independent Living for Older People. *Anchor Trust* [ISBN 0 906178568].
[119] Sixsmith, A. J. (2000). An evaluation of an intelligent home monitoring system. *J. Telemed. Telecare,* 6, 63-72.

[120] Alihanka, J. and Vaahtoranta, K. (1979). A static charge sensitive bed. A new method for recording body movements during sleep. Electroencephalogr. *Clin. Neurophysiol.*, 46(6), 731-734.

[121] Foote, P.D. (1999). Structural health monitoring: Tales from Europe. Structural Health Monitoring 2000, Stanford University, California, pp. 24-35.

[122] Auweraer, H. V. der and Peeters, B. (2003). International research projects on structural health monitoring: an overview, Structural Health Monitoring, 2 (4), 341-358.

[123] Velle, J. I.; Lindsay, J. E.; Weeks, R. W. and Long, F. M. An investigation of the loss mechanisms encountered in propagation from a submerged fish transmitter. In: Long F. M. editor. Proc. of the Second Int. Conf. on Wildlife Biotelemetry. University of Laramie Press: Laramie, Wyoming; 1979; 228-237.

[124] Rettie, W. J. and McLoughlin, P. D. (1999). Overcoming radiotelemetry bias in habitat-selection studies. *Can. J. Zool.*, 77(8), 1175-1184.

[125] Rettie, W. J. and Messier, F. (2000). Hierarchical habitat selection by woodland caribou: Its relationship to limiting factors. *Ecography*, 23(4), 466-478.

[126] Schwarz, C. J. and Seber, G. A. F. (1999). Estimating Animal Abundance: Review III. *Statistical Science*, 14(4), 427-456.

[127] Mackay, R. S. (1964). Deep Body Temperature of Untethered Dolphin Recorded by Ingested Radio Transmitter. *Science*, 144(3620), 864-866.

[128] Kulik, I. L.; Karaseva, E. V.; Litvin, V. Y. (1967). New techniques in studying home ranges of small mammals. *Zoologicheskii Zhurnal*, 46, 264-71. [Translation No. L. Trans 66, Elton Library, Oxford.]

[129] Stoddart, D. M. (1970). Individual range, dispersion and dispersal in a population of water voles (Arvicola terrestris (L.)). *Journal of Animal Ecology*, 39(2), 403-425.

[130] Niezgoda, G. H.; Mckinley, R. S.; White, D.; Aderson, G. and Cote, D. (1998). A dynamic combined acoustic and radio transmitting tag for diadromous fish. *Hydrobiologia*, 371-372, 47-52.

[131] Sundstro¨ m, L. F.; Gruber, S. M.; Clermont, S. M.; Correia, J. P. S.; Marignac, J. R. C.; Morrissey, J. F.; Lowrance, C. R.; Thomassen, L. and Oliveira, M. T. (2001). Review of elasmobranch behavioral studies using ultrasonic telemetry with special reference to the lemon shark, Negaprion brevirostris, around Bimini Islands, Bahamas. *Environ. Biol. Fishes,* 60, 225-250.

[132] Dagorn, L.; Josse, E. and Bach, P. (2001). Association of yellowfin tuna (Thunnus albacares) with tracking vessels during ultrasonic telemetry experiments. *Fishery Bull.*, 99 (1), 40-48.

[133] Bunt, C. M. (1999). A tool to facilitate implantation of electrodes for electromyographic telemetry experiments. *J. Fish Biol.,* 55, 1123-1128.

[134] Briggs, C. T. and Post, J. R. (1997). Field metabolic rates of rainbow-trout estimated using electromyogram telemetry. *J. Fish Biol.*, 51(4), 807-823.

[135] Beddow, T. A. and Mckinley, R. S. (1999). Importance of electrode positioning in biotelemetry studies estimating muscle activity in fish. *J. Fish Biol.*, 54(4), 819-831.

[136] Kudo, Y., Satou, M., Kitamura, S., Iwata, M., and Takeuchi, Y., Underwater radio-telemetry of electroencephalographic activity from the Hime salmon, landlocked sockeye-salmon Oncorhynchus-Nerka. *Fisher. Sci.* 63(5):687–691, 1997.

[137] Anderson, W. G.; Booth, R.; Beddow, T. A.; Mckinley, R. S.; Finstad, B.; Okland, F. and Scruton, D. (1998). Remote monitoring of heart rate as a measure of recovery in angled Atlantic salmon, Salmo salar. *Hydrobiologia*, 371/372, 233-240.
[138] Einthoven W. (1906). Le télécardiogramme. Arch Int Physiol, 4, 132-164.
[139] Utsuyama, N.; Yamaguchi, H.; Obara, S.; Tanaka, H.; Fukuta, S.; Nakahira, J.; Tanabe, S.; Bando, E and Miyamoto, H. (1988). Telemetry of human electrocardiograms in aerial and aquatic environments. IEEE *Trans on Biomed. Eng*, 35, 881-884.
[140] Bronzino J.D., (2000). Biomedical Engineering Handbook (2nd edition). Boca Raton, FL: CRC.
[141] Chobanian, A. V.; Bakris, G. L.; Black, H. R.; Cushman, W. C.; Green, L. A.; Izzo, J. L.; Jones, D. W.; Materson, B. J.; Oparil, S.; Wright, J. T.; Roccella, E. J. and the National High Blood Pressure Education Program Coordinating Committee. (2003). Seventh report of the Joint National Committee on Prevention, Detection, Evaluation, and Treatment of High Blood Pressure. *Hypertension*, 42(6), 1206-1252.
[142] 2003 European Society of Hypertension-European Society of Cardiology guidelines for the management of arterial hypertension. *J. Hypertens*. 21(6), 1011-1053.
[143] Niiranen, T. J.; Kantola, I. M.; Vesalainen, R.; Johansson, J. and Ruuska, M. J. (2006). A comparison of home measurement and ambulatory monitoring of blood pressure in the adjustment of antihypertensive treatment. *Am. J. Hypertens*., 19(5), 468-74.
[144] Imai, Y.; Otsuka, K.; Kawano, Y., Shimada, K.; Hayashi, H.; Tochikubo, O.; Miyakawa, M; Fukiyama, K. and Japanese Society of Hypertension. (2003) Japanese society of hypertension (JSH) guidelines for self-monitoring of blood pressure at home. *Hypertens Res*., 26(10), 771-782.
[145] Nichols, W. W. and O'Rourke, M. F. (2005). McDonald's Blood Flow in Arteries: Theoretic, Experimental and Clinical Principles (5th edition). London: Hodder Arnold.
[146] S. Satomura (1959). Study of the flow pattern in peripheral arteries by ultrasonics. *J. Acoust Soc. Jpn,* 15, 151-158.
[147] Goldberg, B. B.; Merton, D. A. and Deane, C. R. (1997). An Atlas of Ultrasound Color Flow Imaging. London: Martin Dunitz.
[148] Evans, D. H.; McDicken, W. N.; Skidmore, R. and Woodcock, J. P. (1989). Doppler Ultrasound: Physics, Instrumentation, and Clinical Applications. Chichester: Wiley.
[149] He, J.; Pan, W. A.; Ozaki, T.; Kinouchi, Y. and Yamaguchi, H. (1996). Three Channels Telemetry System: ECG, Blood Velocities of the Carotid and the Brachial Arteries. *Biomed. Eng. Appl. Bas. Comm*., 8(4), 364-369.
[150] Azhim, A.; Katai, M.; Akutagawa, M.; Hirao, Y.; Yoshizaki, K.; Obara, S.; Nomura, M.; Tanaka, H.; Yamaguchi, H. and Kinouchi, Y. (2008). Measurement of Blood Flow Velocity Waveforms in the Carotid, Brachial and Femoral Arteries during Head-up Tilt. *J. Biomed. and Pharm. Eng*., 2(1), 1-6.
[151] Azhim, A.; Katai, M.; Akutagawa, M.; Hirao, Y.; Yoshizaki, K.; Obara, S.; Nomura, M.; Tanaka, H.; Yamaguchi, H. and Kinouchi, Y., (2007). Exercise improved age-associated changes in the carotid blood velocity waveforms. *J. Biomed. and Pharm. Eng*., 1(1), 17-26.
[152] Jiang, Z-L.; He, J.; Yamaguchi, H.; Tanaka, H. and Miyamoto, H. (1994). Blood flow velocity in common carotid artery in humans during breath-holding and face immersion. *Aviat. Space Environ. Med*., 65, 936-43.

[153] Jiang, Z-L.; Yamaguchi, H.; Takahashi, A.; Tanabe, S.; Utsuyama, N.; Ikehara, T.; Hosokawa, K.; Tanaka, H.; Kinouchi, Y. and Miyamoto, H. (1995). Blood flow velocity in the common carotid artery in humans during graded exercise on a treadmill. *Eur. J. Appl. Physiol,* 70(3), 234-239.

[154] Chouinard, J. (1983). Satellite contributions to telemedicine: Canadian CME experiences. *Can. Med. Assoc. J.*, 128, 850-855.

[155] Murakami, H.; Shimizu, K.; Yamamoto, K.; Mikami, T.; Hoshimiya, N. and Kondo, K. (1994). Telemedicine using mobile satellite communication. IEEE *Trans. on Biomed. Eng.*, 41(5), 488-497.

[156] Conner, D. A.; Grimes, G. J. and Goldman, J. (2000). Issues and techniques in networked-based distributed healthcare: Overview. *J. Syst. Integration*, 10, 81-94.

[157] Sargsyan, A. E.; Doarn, C. R. and Simmons, S. C. (1999). Internet and World-Wide-Web technologies for medical data management and remote access to clinical expertise. *Aviat. Space Environ. Med.*, 70(2), 185-190.

In: Telemetry: Research, Technology and Applications
Editors: Diana Barculo and Julia Daniels
ISBN 978-1-60692-509-6

Chapter 2

ON THE USE OF TELEMETRY IN HABITAT SELECTION STUDIES

Jodie Martin[1,2,3,1], Vincent Tolon[1,3,4], Bram Van Moorter[1], Mathieu Basille[1,5] and Clément Calenge[3]

[1] Université de Lyon, F-69000, Lyon ; Université Lyon 1 ; CNRS, UMR5558, Laboratoire de Biométrie et Biologie Evolutive, F-69622, Villeurbanne, France.
[2] Department of Ecology and Natural Resource Management, Norwegian University of Life Science, P.O. Box 5003, NO-1432 Ås, Norway
[3] Office National de la Chasse et de la Faune Sauvage, 85 bis avenue de Wagram, BP 236, F-75822 Paris Cedex 17, France
[4] Laboratoire d'Ecologie Alpine CNRS UMR5553, Université de Savoie, Bâtiment Belledonne, F-73376 Le Bourget-du-Lac, France.
[5] Centre for Conservation Biology, Department of Biology; Norwegian University of Science and Technology; NO-7491 Trondheim, Norway

ABSTRACT

Understanding the relationships between organisms and their habitat is a central question in ecology. The study of habitat selection often refers to the *static* description of the pattern resulting from the selection process. However the very nature of this habitat selection process is *dynamic,* as it relies on individual movements, which are affected by both internal components (i.e. related to the animal itself, such as its behavior; foraging, resting, etc.) and external components (i.e. related to the composition of the environment). Coupling habitat selection and movement analyses should thus provide new insights into the proximal mechanisms and evolutionary causes of animals' space use.

To achieve this, the introduction of GPS technology in the early 1990s showed great promise, as it facilitates tracking of animals with high fix frequency over long time periods. From a statistical point of view, this led to an increased temporal *autocorrelation* in the positions of successive locations. Whereas classic approaches of habitat selection often relied on the assumption of statistical independence between relocations, the

---

[1] Corresponding author e-mail address : martin@biomserv.univ-lyon1.fr, jodiem@free.fr

development of newer methods has made possible the use of autocorrelation for more dynamic approaches. As several statistical tools are now available for researchers, autocorrelation can be incorporated successfully into the analysis, instead of being removed or even ignored. We emphasize the need to integrate individual behavioral mechanisms in habitat selection studies.

The use of GPS technology in wildlife management issues is, however, often motivated by its technological advantage to produce large amounts of data, rather than biological questions. We warn users of GPS devices about the statistical and conceptual changes induced by this technology used for studying habitat selection. We encourage a solid biological reflection about the ecological characteristics of studied species and spatial and temporal scales considered, before deciding on which sampling protocol and which telemetry technology to use in accordance with the biological question of interest.

INTRODUCTION

Understanding the relationships between organisms and their habitat is a central question in ecology. The habitat defines the available range of resources and living conditions for a species (Hall et al., 1997). Thus, the habitat potentially has an important impact on vital rates, such as survival and reproduction, which are directly related to population dynamics and evolution (Caswell, 2000). It is therefore crucial to study the mechanisms of habitat selection, i.e. the process by which animals actually choose specific habitat components within their environment. This process is recognized to be highly scale dependent; habitat selection at a given scale can be seen as the result of individual behavioral processes at finer scales (like movements), while at the same time it will be constrained by larger scale processes (like the geographical range of a species). In order to clarify the study of habitat selection, Johnson (1980) distinguished four levels or orders of selection, from the species' geographical range to the selection of food items, through individual home range establishment and patch selection within the home range. The higher levels, like the species' range, are governed mainly by population processes, whereas at lower levels individual choices are more important. To understand animal space use, ecologists have to investigate these multiple scales, often simultaneously (Johnson, 2002).

It has been recognized that both spatial and temporal scales are positively correlated (Holling, 1992). For example, distributional ranges of populations are necessarily defined at large spatial scales and the study of factors affecting large scale distribution of a given population requires knowledge of long-term population dynamics at large temporal scales. On the contrary, individual movement processes at fine spatial scales (e.g. within a home range) are considerably faster than population processes, and therefore need to be studied at much finer temporal resolutions.

Although "habitat selection" has been defined as a process, the study of habitat selection often refers to the *static* description of the pattern resulting from this process, i.e. "space occupancy". For example, many individual-level studies dealt with the characteristics of the environment within home ranges (e.g. Mc Loughlin 2002 and 2004; Mitchell and Powell 2007). However, the very nature of the habitat selection process is *dynamic,* as it relies on individual movements, which are affected by internal and external components (such as individual state or vegetation types). Hence, habitat selection and movement processes are intimately related, because movement partly is driven by habitat selection, whereas habitat selection is a consequence of movements. Coupling habitat selection and movement analyses should thus provide new insights into the proximal mechanisms and evolutionary causes of animals' space use.

To study the mechanistic aspects of individual habitat selection, it is therefore advisable to obtain and use detailed information on movements of individuals, based on a high frequency of relocations. The transition from a static to a more dynamic approach to habitat selection therefore can be achieved by increasing serial (or temporal) *autocorrelation* in the positions of successive locations (Swihart and Slade, 1985). The level of serial autocorrelation in datasets describes the importance of temporal structure in individual movements. In a dynamic approach of habitat selection, the spatio-temporal structure of relocations arising from more or less complex movements should be considered and studied (Frair et al. 2005), whereas in the static approach all relocations are considered to be temporally independent (Otis and White 1999). The choice to analyze habitat selection from a static or a dynamic point of view should be motivated by the biological questions of interest. Commonly, when studying habitat selection in a static way, autocorrelation is considered a nuisance that should be overcome (Morrison et al. 1998). The use of autocorrelation in habitat selection studies reflects a shift from the more classic static approach to a dynamic approach of habitat selection.

To study mechanisms of individual habitat selection, one possibility is to record movements by directly observing animals in their environment (e.g. Shine et al. 2004; Klaassen et al. 2006). However, this method of tracking animals requires many hours spent in the field, and cannot be done for elusive species, which are difficult to observe in their natural environment. Since its appearance in the early 1960s (Figure 1), VHF technology facilitated tracking of wildlife by providing researchers with a new tool to remotely detect individuals and then directly measure their location (consisting of a set of coordinates in the form of latitude x, longitude y, and time t) by homing or triangulation. This technology has been used successfully on elusive species. However, using this technology to estimate individual small-scale movements still requires a great deal of field effort (in terms of field work) to allow the collection of frequent relocations, and therefore often results in short-duration tracking (e.g. 1 location every 10 minutes for a couple of days, Nicholls and Racey 2006).

The introduction of GPS technology in the early 1990s, and its generalization at the turn of the century (Figure 1), showed great promise for studying processes of habitat selection. Thousands of relocations can now be stored on-board or even directly transmitted to the user, without relying on much human intervention after an animal is equipped with the tracking device. GPS technology, therefore, facilitates fine-scale tracking (relocations can be delivered up to every second, Fritz et al. 2003) of animals over longer time periods (e.g. 1 location every 3 hours for a whole year, Johnson et al. 2002). Although the accuracy of individual relocations obtained by GPS does not equal direct observations and measurements in the field, the accuracy obtained with recent GPS remains acceptable for most research purposes (around 20 meters). Hence, GPS technology facilitates the study of animal movement both at small and large spatial scales during long periods.

The purpose of this chapter is twofold. The first part is dedicated to the role of autocorrelation in habitat selection studies from the past till now. We demonstrate that, whereas temporal autocorrelation was first considered as a problem when studying habitat selection in a static way, the progressive combination of habitat selection and movement concepts have led to an increased use and interest in autocorrelation.

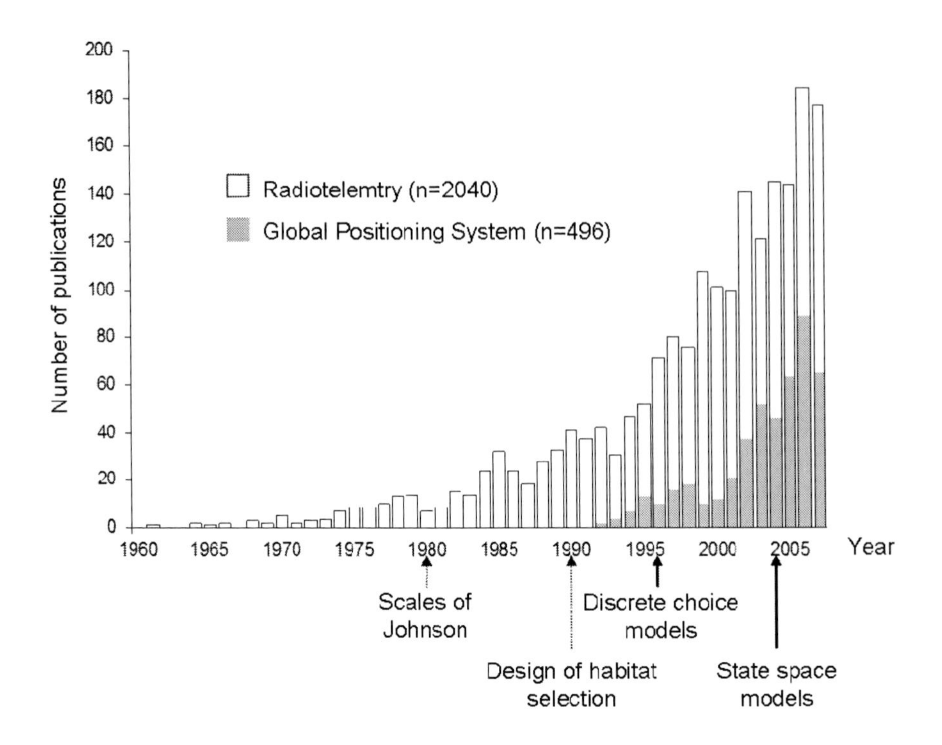

Figure 1. Trend in the number of publication including « radio telemetry » and « global positioning system ». The review was performed using the ISI Web of Knowledge with all data bases and was restricted to zoology, biodiversity and conservation, evolutionary biology and behavioral sciences. Dashed arrows represent the appearance of important concepts in habitat selection studies and black arrows the appearance of analytical methods.

As several statistical tools are now available for researchers, autocorrelation can be incorporated successfully in the analysis, instead of being removed or even ignored. We emphasize the need to integrate individual behavioral mechanisms into habitat selection studies. In the second part, we stress the importance of thorough reflections about biological questions before heading out into the field to deploy GPS collars. We provide some guidelines for the choice of monitoring technology in the context of habitat selection, regarding considerations of biological questions, spatio-temporal scales, and research costs.

1. Towards the Use of Autocorrelation in Individual Habitat Selection Studies

Johnson (1980) defined the selection of a habitat component as *"the process in which an animal actually chooses that component"*. Thus, tests of nonrandom habitat selection by individual animals usually compare used habitat components (the relocations of individuals, i.e. their actual habitat choice) with a null model describing the habitat components that could have been used alternatively by the individuals under the hypothesis of absence of habitat selection (i.e. random habitat use; Pendleton 1998, Millspaugh and Marzluff 2001, Boyce et al. 2002). The latter components are called "available" components (Manly 2002). Rejection of this null model for the observed pattern allows the conclusion that a nonrandom process generated the observed data. At the scale of individual habitat selection, it is therefore crucial to define precisely what is available to animals, because the null model will depend directly on it.

Availability can be defined differently depending on whether the animals are identified or not (see Thomas and Taylor 1990, 2006, for an overview of the different study designs). In the case of identified animals (typically the case with telemetry), availability can be related to the selection order of concern: for second-order selection availability is defined at a population level, whereas for third-order selection availability is defined at an individual level (Aebischer et al. 1993). Several other factors may further affect availability. For example, movement capacity of animals is a factor that can restrict availability (Martin et al. 2008). The behavioral state of an animal is also a potential factor that can limit access to a particular area (e.g. resting animals). These factors therefore should be taken into account when determining which habitat components are available for use.

Classic statistical methods used to test habitat selection (e.g. logistic regression, log-linear regression, χ^2) assume independence between locations of a given individual (Johnson 1980, Thomas and Taylor 1990, Swihart and Slade 1997, Aebisher et al 1993, Alldredge and Ratti 1992, Pendleton et al 1998). In other words, individual relocations must not be spatially or temporally correlated. Dependency between relocations produces more similar values than expected by chance; as such, positive autocorrelation should result in underestimating the true variance. This induces an increased probability of type I error by inflating the number of degrees of freedom (Legendre 1993, Lennon 1999, Diniz-Filho et al. 2003, Martin et al. 2008), i.e. the null hypothesis (random habitat use) is rejected too frequently. The independence between relocations is often ensured by adopting a sufficiently large time lag between successive relocations, which circumvents the problem of autocorrelation. However, the growing use of GPS technology has led to a decrease in time intervals, and resulted in an increased serial autocorrelation.

1.1. From the Past. The Null Model as Random Locations: Autocorrelation as a Problem

Few statistical tools have been developed for individual data analyses; most of them being adapted from methods developed for second-order selection by unidentified individuals (Calenge 2005). For example, Manly et al. (2002) recommended estimating a Resource Selection Function (RSF) for each animal and then combining the results to infer conclusions at the population level. Means and variances for each individual are therefore estimated without considering autocorrelation between relocations and then averaged across animals to estimate population level selection parameters (Thomas and Taylor 2006). As a matter of fact, several authors considered autocorrelation not to be a concern if the statistical unit is the animal and not the animal's relocations (Alldredge and Ratti 1992, Aebisher 1993, Otis and White 1999). Indeed, there are several ways to deal with pseudo-replication when pooling a collection of relocations from several animals in analyses. However, temporal autocorrelation between individual relocations still remains a problem. When using relocations as sampling units, autocorrelation in the data makes variances and hypothesis tests no longer valid.

Therefore, tools that have been developed to test habitat selection at the population level still assume independence between relocations, i.e. a comparison of independent use points and independent available points (that is, randomly sampled in the study area; Figure 4a).

This problem is critically important in third-order selection, where the null model commonly is built using points randomly sampled within what is considered to be available. This will result, when important autocorrelation exists in the animal's relocations, in the comparison of used locations containing this autocorrelation structure with random available locations lacking such structure. Therefore, comparing autocorrelated data with uncorrelated data is not valid (Martin et al. 2008). An empirical demonstration of the effects of not considering autocorrelation in habitat selection studies has been conducted by Martin et al. (2008) on brown bears (*Ursus arctos*). They compared two extreme approaches to test third-order selection by individuals with autocorrelated data. Both approaches relied on the comparison of two datasets; an observed dataset of habitat used by individuals (through individual trajectories) and a simulated dataset (with the same number of relocations as the observed dataset) under the hypothesis of random habitat use. For each dataset, they estimated marginality (an index of the eccentricity of the used environmental conditions relative to the average available environmental conditions) and tolerance (an index of the range of used environmental condition compared to the range of available conditions), two measures of the strength of habitat selection (Doledec et al. 2000). In the first approach, simulated datasets did not take into account movement constraints affecting individuals, i.e. random relocations where points are randomly and independently sampled in the home range (corresponding to the classic approach). In the second approach, simulated datasets accounted for the observed shape of individual trajectories, i.e. simulations were carried out by randomly rotating the observed trajectories within the home range. The results obtained from the two approaches differed dramatically (Figure 2). The second approach (with unchanged trajectory shape) led to the conclusion that bears showed no selection of the variables considered in the analysis (i.e. the marginality of the observed dataset was not different from marginalities estimated from simulated datasets). On the contrary, the classic approach concluded that there was strong habitat selection (i.e. marginality of the observed dataset was significantly different from marginalities estimated using random locations from the home range) (see Figure 2). However, as noted by the authors, both approaches have their limitations, because neither of them took into account the nature of the autocorrelation between relocations. Indeed, the first one did not take any movement constraints into account and the second one kept the shape of the trajectory unchanged; in this later case both movement constraints and some actual habitat selection were included in the null model, which results in a too conservative testing of habitat selection. Therefore, this study highlights the risk of not accounting for autocorrelation in animal relocations by comparing them with an inappropriate null model.

As autocorrelation affects our ability to perform standard statistical procedures (Legendre 1993), it has often been advised to avoid autocorrelation in individual relocations (Morrison et al. 1998). Swihart and Slade (1985) developed a framework for analyzing independence between successive relocations in order to determine the time interval necessary to achieve statistical independence. They proposed using Schoener's (1981) ratio statistic (t^2/r^2; where t is the average distance between successive observations and r the average distance to the center of activity) to estimate the Time To Independence (TTI), i.e. the smallest time lag necessary to consider successive relocations as independent.

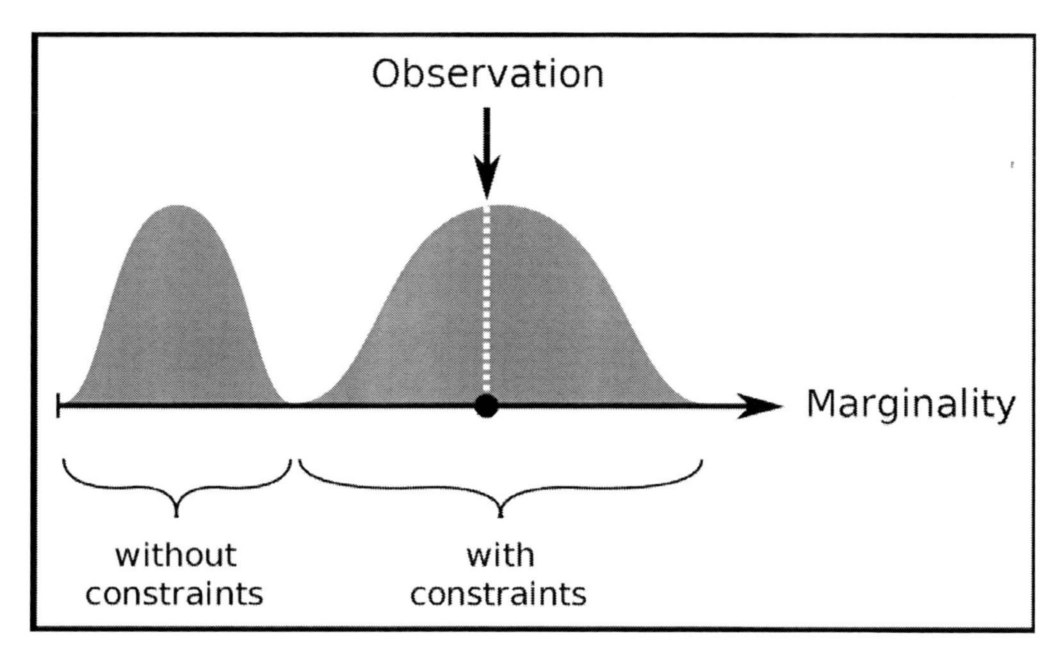

Figure 2. Martin et al. (2008) tested the marginality (deviation from the average conditions in the area) in habitat selection of female brown bears, following two tests. Both relied on the comparison of the observed dataset with datasets simulated under the hypothesis of random habitat use. The first analysis did not take movement constraints into account (simulations were carried out by randomly distributing a set of points in the home range), whereas the second analysis accounted for these constraints (simulations were carried out by building random trajectories within the home range). In the first case, the observation is out of the range of the simulations and would be considered significant, whereas it is not the case while taking into account constraints (after Martin et al. 2008).

Often, the TTI between consecutive fixes is considered as the time lag required by an animal to cross its entire home range (Swihart and Slade 1985, White and Garrot 1990). The problem of autocorrelation can then be effectively circumvented by sub-sampling data (Boyce et al. 2002) or adopting a sampling regime that uses the TTI as a criterion for independence between relocations. Unfortunately, subsampling data (which is not a problem in itself, because it only removes redundancy) inherently brings with it the loss of data which are, as every field biologist knows, expensive to collect. Moreover, several authors have shown that attempting to obtain independent data is not always possible and can lead to a loss of biological meaning (Rooney et al. 1998, De Solla et al. 1999).

Serial autocorrelation is linked intimately with the definition of availability for individuals. Hence, independence between individual relocations assumes that animals are free to move between two relocations across the area that the researcher considered available, meaning that this whole area (often the home range) is available at each step. But even if the time lag between two relocations is long enough for the animal to cross this area, behavioral constraints (e.g. need for rest, need for foraging, movement constraints) result in the fact that animals are not free to move everywhere every time in this area. An extreme example illustrates this; an animal never wakes up at the other side of its home range! Therefore, even if we statistically reach the TTI, we seldom reach the biological independence between relocations with telemetry data. Moreover, when there is no stable home range, there is no TTI, or the TTI might become very large, in which case subsampling will not provide a solution to the problem. There is thus a need to create a statistical framework that allows taking into account spatio-temporal structures of individual relocations, i.e. explicitly incorporating autocorrelation into models (Legendre 1993). Today, with the increased use of intensive sampling protocols, we can no longer consider fixes as independent relocations, but instead should consider them as trajectories. Analyses of animal movements are therefore needed in order to proceed with habitat selection.

1.2. The Present. The Null Model is a Random Walk: Accounting for Serial Autocorrelation

1.2.1. Analyses of Animal Movement

An animal's movement is a continuous path in space and time, but a discrete representation of the path facilitates its analysis (Figure 3a; Turchin 1998). Often, relocations are recorded with a fixed time interval and the straight-line moves between consecutive fixes are referred to as steps. The sequence of steps then provides the basic units for further analysis of the path (Turchin 1998). Several descriptors then can be used to describe the step series; most frequently used are step lengths and turning angles (see Figure 3b) and their distributions can be used to characterize animal movement paths. Accordingly, movement types can be identified based on these distributions. For example, intensive search movements can be characterized by short steps and low directionality of the turning angles, whereas exploratory movement steps will be long and have a high directionality in their turning angles.

The statistical framework for animal movement path analyses is based on the comparison of empirical data with a theoretical null model. Several theoretical movement models have been developed that could be used as reference. Each of these models has different statistical properties, and often different assumptions about the independence (i.e. absence of serial autocorrelation) of specific movement descriptors (like step length or relative turning angles). The observed movement characteristics are then compared with the theoretical ones and departure from these models allows inference of biological conclusions about animal behavior (Franke et al. 2004, see below). The mere random walk (RW) is the simplest null model of movement which can be used to model animal movement through a homogeneous environment (Turchin 1998; Figure 4b). It relies on the independence of all descriptors, and thus it assumes the absence of serial correlation between successive steps. Therefore a random walk does not take into account the natural tendency of animals to go forward and is therefore not very accurate to represent most of animal movement, especially at short time intervals (Turchin, 1998; Bovet and Benhamou, 1988). Today, the most widely used theoretical model is the correlated random walk (CRW), which differs from the RW in that absolute angles are generally dependent, whereas the relative or turning angles are assumed to be independent (Turchin 1998; Bovet and Benhamou, 1988). The distribution of turning angles is centered on 0, resulting in a forward persistence in the direction of movements. The direction of the previous step thus influences the direction of the following step.

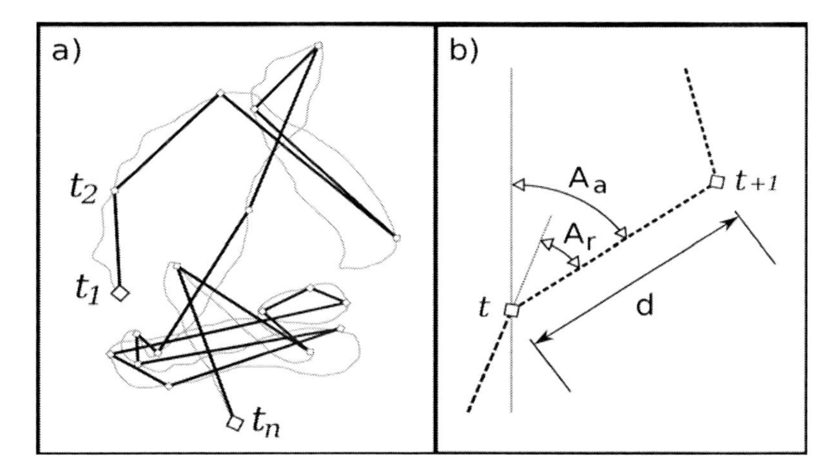

Figure 3: a) Representation of a movement path. In gray: actual path; in black: discrete representation of the path. A movement path is defined by a set of successive relocations, characterized by their position (generally latitude and longitude). Each movement between two successive relocations (i.e. between time t and t+1) describes steps. b) Examples of movement path descriptors: d is the distance between 2 relocations; Ar is the turning angle (or relative angle), i.e. the angle between the direction of the previous step (small dashed line) and the actual one; Aa is the absolute angle, i.e. the angle between a given direction (gray line) and the direction of the actual step.

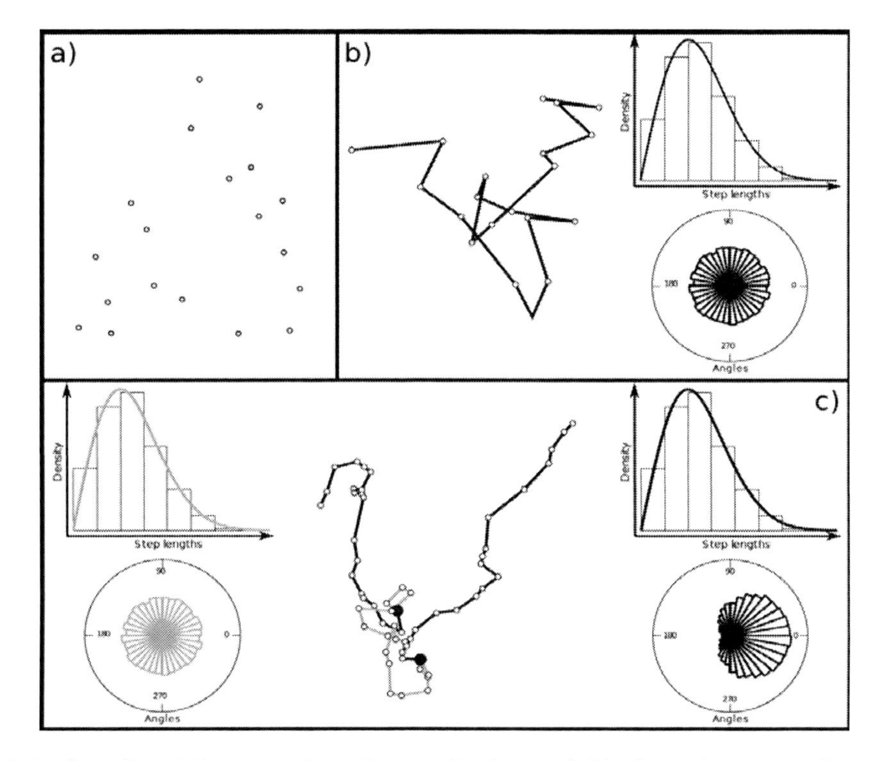

Figure 4: a) The null model as a set of locations randomly sampled in the study area; b) The null model as a random walk, where the whole trajectory can be characterized by step length distribution (top right) and turning angle distribution (lower right); c) The null model as a mixture of two random walks. In gray, a mere random walk, characterized by a uniform distribution of turning angles and normally distributed step lengths; in black a correlated random walk characterized by a distribution of turning angles centered around 0 (forward persistence) and normally distributed step length.

1.2.2. Discrete-Choice Models

Some authors have started to take serial autocorrelation into account by defining habitat availability separately for each relocation in the so-called "discrete-choice models" framework (Arthur et al. 1996, Hjermann 2000, Fortin et al. 2005, Rhodes et al. 2005). The test for nonrandom habitat selection is derived from a comparison of random locations from this fix-specific availability with the actual chosen locations. In this case, the null model consists of random locations with serial autocorrelation, similar to the class of random walk models. Different authors have used different methods to determine availability, corresponding to different types of random walks. The simplest approach involves determining a circle around a fix of available locations (e.g. the availability radius of Arthur 1996); in this case there is no directional persistence. Alternatively, the observed distributions of step lengths and turning angles have been used to define availability. Even more complex relationships are possible with a dependence of availability on the time interval between fixes or habitat characteristics (Hjermann 2000).

It is therefore assumed that for a given time lag between two relocations, the animal has access only to areas close to the current position, and not to its complete home range (Arthur et al. 1996). In the same perspective, Cooper and Millspaugh (1999) adopted a statistical technique derived from the field of economics that allows researchers to define availability separately for each animal observation. At each relocation, a unique set of habitat or resources is available, called the "choice set". Fortin et al. (2005) also developed a simple statistical approach that incorporates movement into a logistic regression framework. This method, called Step Selection Function, considers steps (displacement between two relocations) as sampling units, each of them being contrasted with n random steps, which are defined using the observed distributions of step lengths and turning angles.

This approach partially resolves the statistical and biological issues of serial autocorrelation of relocation data. Indeed, this procedure only considers first-order autocorrelation, i.e. dependence between relocations at time t and t-1 only, and deems the successive steps as independent (Martin et al. 2008). However, the nature of the dependence between all the relocations making up the whole trajectory is seldom analyzed, as noted by Calenge (2005). Martin et al. (2008) stressed that positions of individuals are the results of three effects: (i) intrinsic constraints (e.g. movement capacity, activity patterns, internal state), (ii) extrinsic constraints (e.g. environmental or artificial barrier) and (iii) habitat selection behavior (the animal is in a particular habitat because it is "suitable" for it at this moment). Therefore, testing differences between used and available points suppose knowledge of the processes that generate data without any habitat selection, i.e. null model under the hypothesis of no habitat selection, by taking into account the internal constraints of animals that partly shape the trajectory. However, habitat selection analyses rarely take into account intrinsic and extrinsic constraints (Martin et al. 2008; see Matthiopoulos (2003) for an example accounting for movement and extrinsic constraints). In general, researchers aggregate data from different behavioral states and, therefore, their conclusions on habitat selection result from the joint selection of both activity and habitat components (Cooper and Millspaugh 2001). As habitat selection and animal behavior are closely related, there is need for a statistical approach that includes spatio-temporal aspects of individual behavior (Thomas and Taylor 2006, Martin et al. 2008).

1.3. Into the Future. The Null Model is a Mixture of Random Walks: Autocorrelation as a Paradigm

1.3.1. Nonstationarity, an Interesting Property of Many Animal Trajectories

As movements and activities are closely related, movement processes tend to be different according to the animal's behavioral state. For instance, during foraging activity an animal may have shorter and more sinuous movements than during transitions between patches of resources, where it should have directed and faster movements. As an illustration, consider a bee foraging on a patch of flowers. Movements between flowers will be short and sinuous compared with movement between patches of flowers or between the patch and the hive. Franke et al. (2004) used differences in movement characteristics to differentiate behavioral states of woodland caribou; they distinguished bedding, feeding and displacements. Indeed, for given spatial and temporal scales, we can consider most animal trajectories as a succession of different types of movement corresponding to specific activities, each of them being characterized by its own statistical properties (Figure 4c). However, trajectories can remain unchanged for different activities defined at very fine temporal scales (animals can switch quickly between foraging and vigilance while keeping the same type of movement). But at longer temporal scales, major activities, such as foraging, exploring, or resting, often correspond to specific movement types.

A statistical process is said to be nonstationary if the statistical properties of the process generating the trajectory change over time. Therefore, a trajectory composed of different movement types may be considered to have been generated by a nonstationary process. This implies that the definition of availability should be different according to the state of the individual. Indeed, an animal that is foraging or resting does not have the same available habitat as when it is searching for mates or patrolling its territory. Therefore, this nonstationarity is of major interest, because it provides information on animal behavior and activities. This nonstationarity of the process is often the cause of the autocorrelation in the data.

1.3.2. Building Movement Models as Mixtures of Random Walks

Each movement type trajectory potentially can be represented by a different theoretical model based on its properties. The trajectory can then be modeled as a succession of these movement models (Figure 4c). It is therefore important to partition the whole trajectory into different pieces of stationary paths, with stable mathematical properties. Typically, each of these stationary paths corresponds to a certain type of behavior. To date, several methods have been developed and used for this partitioning of trajectories, for instance First Passage Time (Fauchald and Tveraa 2003), fractal dimension (Nams and Bourgeois 2004), and State Space Modeling (Patterson et al., 2007). First Passage Time, for instance, has been used as a method to detect Area-Restricted Search (ARS) behaviors, which can occur when an animal encounters a food-rich resource patch. More recently, state-space models based on hidden Markov models have become more popular for extraction of behavioral states from movement paths (reviewed by Patterson et al. 2007). For example, Morales et al. (2004) employed state-space models to highlight a biphasic movement for elk (*Cervus elaphus*); the "encamped" movement with small movements and sharp turns and the "exploratory" movements with longer directed movements. They modeled elk movement by fitting a mixture of random walk models with different properties, each model corresponding to a different behavioral state. However, these approaches often assume a constant probability of animal behavior changes ("switching probability"), or at least assume prior knowledge of factors that could potentially affect this switching probability (e.g. constraints on hourly activity patterns or environmental features, Morales et al. 2004). Such methods therefore require prior exploratory analyses of factors potentially influencing the shape structure of the trajectory.

1.3.3. Some Recommendations

In order to test third-order habitat selection with highly autocorrelated data, we stress that an in-depth analysis of the characteristics of individual trajectories is an important step towards a more accurate analysis of habitat selection process. Autocorrelation between relocations should not be removed or avoided, but rather integrated into a statistical framework. Discrete-choice models are a first step toward this integration, but only consider the first degree of autocorrelation to create the null model. Therefore, they do not take into account the behavioral state of individuals, which potentially may affect habitat selection behavior. We emphasize the need to analyze the rules of animal movement using partitioning methods based on the division of the whole trajectory into homogeneous movement bouts. Each of these movement bouts can then be characterized by a probability distribution for each descriptor (e.g. step length, turning angles). Then, for each relocation belonging to a particular behavioral state, availability can be estimated more precisely using the corresponding movement characteristics for this state. These state-movement analyses are required to build more realistic null models of random habitat selection that take into account behavioral constraints.

2. GPS Technology: A Double-Edged Sword

The appearance of new technologies has resulted in important advances in many scientific fields, as it offers new opportunities to answer more questions. Since the early 1990s, GPS technology has facilitated the measure of fine-scale movements of elusive animals in their natural environments over long time periods. Especially, it aids in our understanding of the link between fine-scale behavioral movement mechanisms and the actual distributions of animals. However, scientists, wildlife managers, and conservationists should be careful before adopting this technology. Even though GPS technology is appealing, the choice of the monitoring tool should be the consequence of a well-defined biological question. Such careful planning could aid the avoidance of mismatches between the question of interest and the type of data collected using a particular tool. Compared to older technologies (especially VHF) the use of GPS technology offers many advantages, but also induces different constraints and is not necessarily the adequate tool for every question about animal movements or distributions. Above all, the choice of the appropriate tool to record an animal's locations should be directly dependent on the sampling protocol defined itself by the question of interest. Irrespective of the biological question or sampling protocol, it is an accepted fact that the number of equipped individuals should be as large as possible in order to increase the generality of the findings. Two other parameters can then vary according to the question of interest: the time lag between relocations and the study period. In the following, we will discuss the relationships between biological question, sampling protocols, and choice of the adequate tool to record animal locations.

2.1. Individual Variability

The ecological characteristics of the focus species and especially the ratio of inter/intra-individual variability in habitat selection can help determine the best sampling protocol and therefore the right tool to track animals in their environment. Girard et al. (2006) showed in their study that the number of animals is more important to assess habitat selection than is the number of fixes per animal. They obtained accurate habitat selection by moose with fix frequencies of only 1-7 per week, whereas often more than 15 individuals were needed to generalize the findings over the population. This suggests that, for generalist species with high inter-individual variability, it is important to obtain data from many individuals to assess habitat selection. On the contrary, if there is a strong intra-individual variability, the number of relocations obtained for each individual can become as important as the number of tracked individuals. Thus, knowledge about the biology of the species will influence the choice of sampling protocol (number of samples per individual versus number of sampled individuals) and, by consequence, the monitoring technology.

2.2. A Matter of Scale

The fix frequency of a telemetry protocol and the study period are generally a direct consequence of the scale of the study: questions regarding small-scale movements require high fix frequency (e.g. 1 point every second for a fractal analysis of albatrosses' movements, Fritz et al. 2003). In the beginning of this chapter we discussed the interest to combine fine-scale movement analyses with habitat selection studies. In this context, increased fix frequency may facilitate a more profound investigation of animal behavior. Moreover, the study duration may be long enough to explore the link between fine-scale behaviors and habitat selection at longer time scales.

However, many questions in ecology do not rely directly on fine-scale behaviors of animals. Large-scale distributional questions can be answered with a lower fix frequency (e.g. 1 point every 3 weeks for an analysis of home range composition and habitat use, McLoughlin et al. 2005). Important in the context of conservation and population dynamics in general is the linkage between habitat and animal performance (McLoughlin et al. 2005), for example to identify critical habitats for population viability in conservation biology (e.g. Akçakaya 1995). Mere occurrence has been shown to be misleading in some situations. For instance, an attractive sink is a habitat that is selected despite the lower performance experienced by the animals occupying it (Delibes et al. 2001). It can, therefore, be argued that to assess the existence of such attractive sinks, the measurement of performance is critical.

Performance of animals, e.g. lifetime reproductive success, however is defined over quite long time scales and often is related to large-scale distribution patterns (e.g. habitat use, Conradt et al. 1999, home range composition, McLoughling et al. 2007). These measures of space occupancy do not require fine-scale measures of animals' movements. Instead, a high number of individuals is often needed to highlight relationships between their performance and habitat. In this context it might be more interesting to invest less in fix frequency and more in number of tracked individuals.

2.3. Main Costs and Benefits of GPS Technologies

Tools and sampling protocols are intimately related, they often involve trade-offs between costs and perfect match. Prior to the selection of a telemetry technology, researchers should think about the associated costs and benefits (in terms of correspondence with the defined protocol) of the use of different tools within the biological context of the question. In theory, any study using VHF tracking by triangulation could be done with GPS tracking instead. That is, GPS collars potentially can deliver the exact same data (regarding frequency and time-lag) as VHF collars, whereas the reverse is not necessarily true over a long period. That being said, a study using telemetry will cover three main budget compartments; capture, equipment, and operation costs. We will consider only the case of nonlimiting captures, as it potentially involves limiting equipment and operating costs directly related to the monitoring technology to be chosen.

The cost of equipment to monitor one animal is considerably higher with GPS technology than with VHF technology; GPS-tracking devices are approximately 10 times more expensive than VHF devices. However, with GPS devices the collection of data is automated, whereas VHF devices require human intervention to be effective; relocations are usually collected by triangulation of the signal, which implies a relatively high time and financial investment in the field. This results in limited operation costs of GPS technology (even with the use of a GSM device to download the data) as compared to VHF technology, especially for high fix frequencies. At a fixed cost, GPS technology, therefore, leads to an increase in sampling intensity compared to VHF technology. However, when the budget of the study is limited, this is often at the cost of lower numbers of individuals monitored (Figure 5). It is thus important to take into account these considerations of costs before deciding upon the adequate tool to choose for the sampling protocol.

2.4. Some Recommendations

When a large number of relocations per animal must be collected, GPS technology can provide adequate data, thanks to the automation of the process. This technology is particularly useful when fine-scale movements of animals need to be recorded, especially over long periods of time. This tool is, therefore, most appropriate to study dynamic aspects of habitat selection, as reported in this chapter. However, when the number of individuals is critical (as for generalist species, see above), and no intensive tracking is required, other tools can be more appropriate (e.g. VHF). For example, when ecologists must link habitat selection and animal performance on long time scales, it is better to invest in more individuals, tracking duration, and field personnel to obtain the essential animal performance measures (source of mortality, breeding, litter size, etc.). The GPS technology does not seem to be the best tool for this task and VHF tools combined with direct observations can provide sufficient locations per animals to estimate their habitat selection at a large spatio-temporal scale. In this case, the savings made on equipment can be used to increase the number of tracked individuals. In general, at constant costs, what is gained in sampling intensity with GPS technology, on one hand is lost in generality on the other hand.

It should be noted, however, that the increased fix frequency obtained by using GPS-tracking might prove useful in determining animal performance using their movement patterns. GPS-tracking can be used to assess foraging success in particular cases. For instance, it is now commonly used to determine kill sites by large predators (e.g. cougar (*Felis concolor*), Anderson and Lindzey 2003; wolf (*Canis lupus*), Sand et al. 2005). These applications require previous calibration and validation of the models on the field. These examples are likely only the top of the proverbial iceberg of potential applications; we can expect more applications with our increasing knowledge of how movement patterns change with specific factors, like the presence of offspring etc.

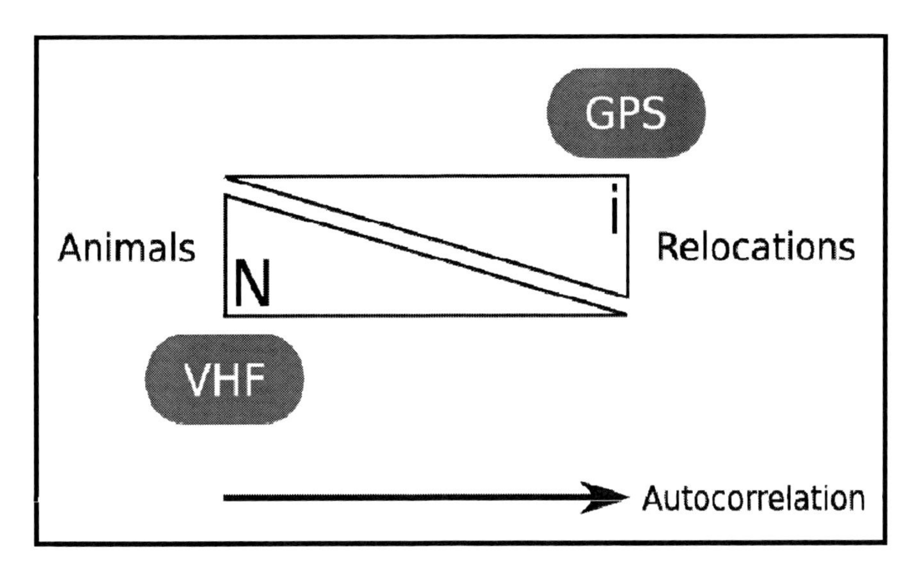

Figure 5. The sampling strategy, at a fixed cost, is the result of a trade-off between the number of individuals marked (N) and the number of relocations per individual (i). Typically, VHF monitoring allows a large number of individuals with fewer relocations per individual (thus no autocorrelation), whereas GPS monitoring allows fewer individuals with more relocations (thus autocorrelation).

CONCLUSION

Despite some budget limitations, GPS technology offers interesting avenues for our understanding of the habitat selection process. The shorter time intervals between relocations allow for the study of more rapidly changing decisions of habitat selection at small spatio-temporal scales. The increased temporal autocorrelation also allows the detailed investigation of different behavioral states with different movement characteristics. Whereas *static* approaches of habitat selection often rely on the assumption of statistical independence of relocations, the development of newer methods, driven by the collection of relocations with a shorter time lag, now allow the use of autocorrelation for more *dynamic* approaches. We expect such dynamic habitat selection studies to become even more common in the near future, as ecologists become more familiar with the use of methods from time-series analysis like state-space models (Patterson et al. 2008). Coupling habitat selection and movement analyses should provide new perspective to understand how individuals react to environmental heterogeneity during their lifetime.

The use of GPS technology in wildlife management or conservation issues, however, often is motivated by technological advantage rather than biological questions. The appealing nature of GPS data (through higher precision and frequency, as well as automation of the data collection) often results in sampling strategies targeting large amounts of data. This often leads afterwards to data-dredging that can cause mismatches between the scale of the monitoring and the scale of the biological processes of interest. We warn every user of GPS devices about the statistical and conceptual changes induced by this technology used for studying habitat selection (Figure 5). In answer to the potential increase in autocorrelation due to higher fix frequency following technological developments in telemetry, we call for a paradigmatic shift in the study of animal habitat selection from static patterns towards dynamic processes. Especially with critical management or conservation issues, we stress that every ecologist should take care to start from the biological question at hand in making his choice of the most appropriate monitoring technology.

ACKNOWLEDGEMENTS

Financial support has been provided by the ANR (project "Mobilité" ANR-05-BDIV-008). We thank Jean-Michel Gaillard (Laboratoire de Biométrie et Biologie Evolutive; Université Lyon1, France) and our referee Eloy Revilla (Estación Biológica de Doñana, Sevilla, Spain) for their useful comments on the manuscript. We also are grateful to the working group "Trajectometry" (Laboratoire de Biométrie et Biologie Evolutive; Université Lyon1, France) for earlier discussions on the topic.

REFERENCES

Aebischer, N.J., Robertson, P.A., and Kenward, R.E. (1993) Compositional analysis of habitat use from animal radio-tracking data. *Ecology*, 74, 1313-1325.

Alldredge, J.R. and Ratti, J.T. (1992) Further comparison of some statistical techniques for analysis of resource selection. *Journal of Wildlife Management*, 56, 1-9.

Akçakaya, H.R., McCarthy, M.A. and Pearce, J.L. (1995) Linking landscape data with population viability analysis: Management options for the helmeted honeyeater *Lichenostomus melanops cassidix*. *Biological Conservation*, 73, 169-176.

Anderson, C.R. and Lindzey, F.G. (2003) Estimating cougar predation rates from GPS location clusters. Journal of Wildlife Management, 67, 307-316.

Arthur, S.M., Manly, B.F.J., McDonald, L.L., and Garner, G.W. (1996) Assessing habitat selection when availability changes. *Ecology*, 77, 215-227.

Bovet, P. and Benhamou S. (1988) Spatial analysis of animals' movements using a correlated random walk model. *Journal of Theoretical Biology*, 131, 419-433.

Boyce, M.S., Vernier, P.R., Nielsen, S.E., and Schmiegelow, F.K.A. (2002) Evaluating resource selection functions. *Ecological Modelling*, 157, 281-300.

Calenge, C. (2005) Des outils statistiques pour l'analyse des semis de points dans l'espace écologique., Université Claude Bernard Lyon 1, Lyon.

Caswell, H. 2000 Matrix population models: construction, analysis, and interpretation, second edition. Sinauer, Sunderland, Massachusetts, USA

Conradt, L., Clutton-Brock, T.H. and Guinness, F.E. (1999) The relationship between habitat choice and lifetime reproductive success in female red deer. *Oecologia*, 120, 218-224.

Cooper, A.B. and Millspaugh, J.J. (1999) The application of discrete choice models to wildlife resource selection studies. *Ecology*, 80, 566-575.

Cooper, A.B. and Millspaugh, J.J. (2001). Accounting for variation in resource availability and animal behavior in resource selection studies. In Radio tracking and animal population (eds J.J. Millspaugh and J.M. Marzluff). Academic Press, San Diego, California, USA.

Dahle, B. and Swenson, J.E. (2003) Seasonal range size in relation to reproductive strategies in brown bears *Ursus arctos*. *Journal of Animal Ecology*, 72, 660-667.

De Solla, S.R., Bonduriansky, R., Brooks, R.J.D.-S.S.R., R, B., and J, B.R. (1999) Eliminating autocorrelation reduces biological relevance of home range estimates. *Journal of Animal Ecology*, 68, 221-234.

Delibes, M., Gaona, P., and Ferreras, P. (2001) Effects of an attractive sink leading into maladaptive habitat selection. *The American Naturalist*, 158, 277-285.

Diniz, J.A.F., Bini, L.M., and Hawkins, B.A. (2003) Spatial autocorrelation and red herrings in geographical ecology. *Global Ecology and Biogeography*, 12, 53-64.

Doledec, S., Chessel, D., and Gimaret-Carpentier, C. (2000) Niche separation in community analysis: a new method. *Ecology,* 81, 2914-2927.

Erickson, W.P., McDonald, T.L., Gerow, K., Kern, J., and Howlin, S. (2001). Statistical issues in resource selection studies with radio-marked animals. In Radio Tracking and Animal Populations. (eds J.J. Millspaugh and J.M. Marzluff), pp. 209-242. Academic Press, San Diego, California, USA.

Fauchald, P. and Tveraa, T. (2003) Using first-passage time in the analysis of area-restricted search and habitat selection. *Ecology*, 84, 282-288.

Ferreras, P., Delibes, M., Palomares, F., Fedriani, J.M.,, Calzada, J. and Revilla, E. (2004) Proximate and ultimate causes of dispersal in the Iberian lynx *Lynx pardinus*. *Behavoral Ecology*, 15 31-40.

Fortin, D., Beyer, H.L., Boyce, M.S., Smith, D.W., Duchesne, T., and Mao, J.S. (2005) Wolves influence elk movements: behavior shapes a trophic cascade in Yellowstone National Park. *Ecology,* 86, 1320-1330.

Frair, J.L., Merill, E.H., Visscher, D.R., Fortin, D., Beyer, H.L. and Morales, J.M. (2005) Scales of movement by elk (*Cervus elaphus*) in response to heterogeneity in forage resources and predation risk. *Landscape Ecology*, 20, 273-287.

Franke, A., Caelli, T., and Hudson, R.J. (2004) Analysis of movements and behavior of caribou (*Rangifer tarandus*) using hidden Markov models. *Ecological Modelling*, 173, 259-270.

Fritz, H., Said, S., and Weimerskirch, H. (2003) Scale-dependent hierarchical adjustments of movement patterns in a long-range foraging seabird. Proceedings of the Royal Society of London Series B-Biological Sciences 270 1143-1148.

Girard, I., Dussault, C., Ouellet, J.-P., Courtois, R., and Caron, A. (2006) Balancing number of locations with number of individuals in telemetry studies. *Journal of Wildlife Management*, 1249-1256.

Hall, L.S., Krausman, P.R., and Morrison, M.L. (1997) The habitat concept and a plea for standard terminology. Wildlife Society Bulletin, 25, 173-182.

Harris, S., Cresswell, W.J., Forde, P.G., Trewhella, W.J., Woolard, T., and Wray, S. (1990) Home-range analysis using radio-tracking data - a review of problems and techniques particularly as applied to the study of mammals. Mammal Review, 20, 97-123.

Hirzel, A.H., Hausser, J., Chessel, D., and Perrin, N. (2002) Ecological-niche factor analysis: How to compute habitat-suitability maps without absence data? *Ecology*, 83, 2027-2036.

Hjermann, D.O. (2000) Analyzing habitat selection in animals without well-defined home ranges. *Ecology*, 81, 1462-1468.

Johnson, D.H. (1980) The comparison of usage and availability measurements for evaluating resource preference. *Ecology*, 61, 65-71.

Johnson, C.J., Parker, K.L., Heard, D.C., and Gillingham, M.P. (2002) Movement parameters of ungulates and scale-specific responses to the environment. *Journal of Animal Ecology*, 71, 225-235.

Klaassen, R.H.G., Nolet, B.A., and Bankert, D. (2006) Movement of foraging tundra swans explained by spatial pattern in cryptic food densities. *Ecology*, 87, 2244-2254

Legendre, P. (1993) Spatial autocorrelation: trouble or new paradigm? *Ecology*, 74, 1659-1673.

Lennon, J.J. (1999) Resource selection functions: taking space seriously? *Trends in Ecology and Evolution*, 14, 399-400.

Manly, B.F.J., McDonald, L.L., Thomas, D.L., MacDonald, T.L., and Erickson, W.P. (2002) Resource selection by animals. Statistical design and analysis for field studies Kluwer Academic Publisher, London.

Martin, J., Calenge, C., Quenette, P.-Y., and Allainé, D. (2008) Importance of movement constraints in habitat selection studies. *Ecological Modelling*, 213, 257-262.

Matthiopoulos, J. (2003) The use of space by animals as a function of accessibility and preference. *Ecological Modelling*, 159, 239-268.

McLoughlin, P.D., Dunford, J.S. and Boutin, S. (2005) Relating predation mortality to broad-scale habitat selection. *Journal of Animal Ecology*, 74, 701-707.

McLoughlin, P.D., Gaillard, J-M. Boyce, M.S., Bonenfant, C., Messier, F., Duncan, P., Delorme, D., Van Moorter, B., Saïd, S., and Klein, F. (2007) Lifetime reproductive success and composition of home range in a large herbivore. *Ecology* 88, 3192–320.

Millspaugh, J.J. and Marzluff, J.M. (2001) Radio tracking and animal populations. Academic Press, San Diego, California, USA.

Mitchell M.S. and Powell R.A. (2007) Optimal use of resources structures home ranges and spatial distribution of black bears. *Animal Behaviour*, 74, 219-230.

Morales, J.M., Haydon, D.T., Frair, J., Holsiner, K.E., Fryxell, J.M., and Holsenger, K.E. (2004) Extracting more out of relocation data: Building movement models as mixtures of random walks. *Ecology*, 85, 2436-2445.

Morrison, M.L., Marcot, B.G., and Mannan, R.W. (1998) Wildlife-habitat relationships: concepts and applications. University of Wisconsin Press.

Nams, V.O. and Bourgeois, M. (2004) Fractal analysis measures habitat use at different spatial scales: an example with American marten. *Canadian Journal of Zoology*, 82, 1738-1747.

Nicholls, B. and Racey, P.A. (2006) Contrasting home-range size and spatial partitioning in cryptic and sympatric pipistrelle bats. *Behavioral Ecology and Sociobiology* 61, 131-142.

Otis, D.L. and White, G.C. (1999) Autocorrelation of location estimates and the analysis of radiotracking data. *Journal of Wildlife Management*, 63, 1039-1044.

Patterson, T.A., Thomas, L., Wilcox, C., Ovaskainen, O., and Matthiopoulos, J. (2008) State-space models of individual animal movement. *Trends in Ecology and Evolution*, 23, 87-94.

Pendleton, G.W., Titus, K., DeGayner, E., Flatten, C.J., and Lowell, R.E. (1998) Compositional analysis and GIS for study of habitat selection by goshawks in southeast Alaska. Journal of Agricultural, Biological, and Environmental Statistics, 3, 280-295.

Rhodes, J.R., McAlpine, C.A., Lunney, D., and Possingham, H.P. (2005) A spatially explicit habitat selection model incorporating home range behavior. *Ecology*, 86, 1199-1205.

Rooney, S.M., Wolfe, A. and Hayden, T.J. (1998) Autocorrelated data in telemetry studies: time to independence and the problem of behavioural effects. Mammal Review, 28, 89-98.

Sand, H., Zimmermann, B., Wabakken, P., Andren, H., and Pedersen, H.C. (2005) Using GPS technology and GIS cluster analyses to estimate kill rates in wolf-ungulate ecosystems. Wildlife Society Bulletin, 33, 914-925.

Schoener, T.W. (1981) An Empirically Based Estimate of Home Range. *Theoretical Population Biology*, 20, 281-325.

Shine, R., Lemaster, M., Wall, M., Langkilde, T. and Mason, R. Why did the snake cross the road? Effects of roads on movement and location of mates by garter snakes (*Thamnophis sirtalis parietalis*). Ecology and Society, 9, 9. http://www.ecologyandsociety.org /vol9/iss1/art9.

Swihart, R.K. and Slade, N.A. (1985) Testing for independence of observations in animal movements. *Ecology*, 66, 1176-1184.

Swihart, R.K. and Slade, N.A. (1997) On testing for independance of animal movements. *Journal of Agricultural, Biological, and Environmental Statistics*, 2, 48-63.

Thomas, D.L. and Taylor, E.J. (1990) Study designs and tests for comparing resource use and availability. *Journal of Wildlife Management*, 54, 322-330.

Thomas, D.L. and Taylor, E.J. (2006) Study designs and tests for comparing resource use and availability II. *Journal of Wildlife Management*, 70, 324-336.

Turchin, P. (1998) Quantitative analysis of movement: measuring and modelling population redistribution in animals and plant. Sinauer Associates, Sunderland, MA.

White, G.C. and Garott, R.A. (1990). Habitat analysis. In Analysis of Wildlife Radio-Tracking Data. (ed U.K. Edition), pp. 183-205. Academic Press, Inc.

In: Telemetry: Research, Technology and Applications
Editors: Diana Barculo and Julia Daniels
ISBN 978-1-60692-509-6

Chapter 3

RADIOTELEMETRIC EEG RECORDINGS IN SMALL RODENTS – A POWERFUL ANALYTICAL TOOL IN BASIC NEUROLOGICAL RESEARCH

Marco Weiergräber [1,3]*, ***Matthew S.P. Ho,*** [2] ***Jürgen Hescheler,*** [1,3] ***and Toni Schneider*** [1,3]

1. Institute of Neurophysiology, University of Cologne, Germany
2. Center for Biochemistry, University of Cologne, Germany
3. Center for Molecular Medicine Cologne (CMMC), Medical Faculty, University of Cologne, Germany

ABSTRACT

Implantable radiotelemetry which allows simultaneous and real-time monitoring of various physiological parameters, such as EEG, ECG and EMG from laboratory animals under unrestrained conditions has provided a major impact primarily in pharmacological, toxicological and basic biomedical research. In this chapter, we focus on the use of implantable radiotelemetry in electroencephalography (EEG), especially on electrocorticographic (ECoG) and deep intracerebral EEG recordings, and its crucial role in neurological characterization of transgenic mouse models. This review covers (1) the general historical, financial and ethical aspects of EEG radiotelemetry and its pros and cons; (2) the planning of radiotelemetric experiments, preoperative preparation of mice and anesthesia; (3) different EEG radiotelemetric implantation procedures, examples of physiological and pathophysiological EEGs and strategies of maximizing signal-to-noise ratio; and (4) specific information on postoperative recovery and pain management. Finally, we demonstrate current state-of-the-art approaches in EEG data acquisition and analysis, among which the possible pitfalls and artifacts therein are discussed in detail.

---

* To whom correspondence should be addressed: Marco Weiergräber, Institute of Neurophysiology and Center for Molecular Medicine (CMMC), Medical Faculty, University of Cologne, Robert Koch-Str. 39, 50931 Cologne, Germany Phone: 0049 (0)221 478 6968, Fax: 0049 (0)221 478 6965. Email: akp74@uni-koeln.de.

Keywords: *biopotential, deep electrodes, electrocorticogram, electroencephalogram, epilepsy, sleep, stereotaxic implantation*

ABBREVIATIONS

BMB	bicucullinemethobromide
ECG	electrocardiogram
ECoG	electrocorticogram
EEG	electroencephalogram
EMG	electromyogram
EOG	electrooculogram
GBL	γ-hydroxybutyrolactone
KA	kainic acid
NSAID	Non-steroidal anti-inflammatory drugs
REM	rapid-eye movement
SWD	spike-wave discharge

1. INTRODUCTION TO RADIOTELEMETRY

1.1. The Importance of Radiotelemetry in Animal Experimentation in Biomedical Sciences

Current biomedical research witnesses the advancement of animal experimentation in investigating human disease status which circumvents the problems of limited availability of native human samples (Kramer and Kinter, 2003). However, ethical issues involving the good practice of animal handling cannot be ignored. Strong effort has been put on establishing new strategies, first among all, to improve postoperative care and pain management, and to optimize the number of animals used and the amount of data acquired from each individual animal, and ultimately to justify the animal models used can resemble the human disease status (Morton et al., 2003). In general, the use of animals as human disease models is warranted if they fulfill three different criteria: homology, isomorphism and predictability. First, this implies that fundamental anatomical, pathobiochemical and pathophysiological parameters, i.e., the etiopathogenesis of the disease is homolog to the human correlate. Second, the animal disease phenotype should be comparable, i.e. isomorphic to the human one and finally, animal models are supposed to predict the efficacy and effects of drug treatment in humans. Thus, various new tools were devised within the last two decades to characterize precisely those parameters in animals models (Poindron et al., 2008). Among the new tools, radiotelemetry emerges as a powerful one and is the state-of-the-art application in precise measurement of basic physiological parameters.

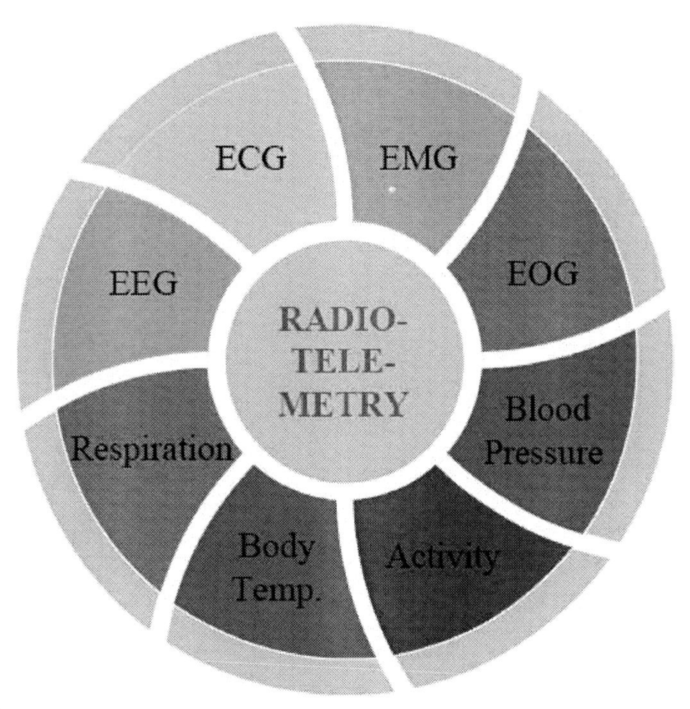

Weiergraber et al., 2008.

Figure 1. Implantable radiotelemetry – fields of application. Current radiotelemetric approaches combine recording and analysis of biopotentials. The major fields of application include the electroencephalogram (EEG), electrocardiogram (ECG), electromyogram (EMG) and electrooculogram (EOG). Various other parameters, e.g. intracardial and intrapleural pressure, locomotion, body core and surface temperature as well as metabolic parameters, such as pH can be monitored in different species ranging from mice to monkeys.

1.2. Radiotelemetric Research – Fields of Application

Current radiotelemetry setups represent a wireless reporting technology, ranging from non-implanted systems, partially-implanted systems, e.g. back-pack systems with transcutaneous connections, to fully-implanted systems in which transponders are powered by an implanted battery or external inductance. The fully implantable radiotransmitters themselves are transducer-radios, hermetically sealed with a biocompatible material, such as silicone, including the power supply. The transmitted signal entities are captured by measuring devices at a remote station, where they are recorded and analyzed.

Various behavioral and physiological parameters, such as the electroencephalogram (EEG) (Cotugno et al., 1996; Tang and Sanford, 2002; Bastlund et al., 2004; Weiergräber et al., 2005a), electrocardiogram (ECG) (Kramer et al., 1993; Gehrmann et al., 2000; Mills et al., 2000; Butz and Davisson, 2001; Weiergräber et al., 2005b), electromyogram (EMG) (Ko and Neuman, 1967; Herzog et al., 1993; Nesathurai et al., 2006), electrooculogram (EOG), arterial and venous blood pressure (Kramer et al., 2000; Mills et al., 2000; Swoap et al., 2004), intraventricular pressure, ocular pressure, intrapleural pressure and respiratory rate (Lewanowitsch et al., 2004), body temperature and locomotion (Tang and Sanford, 2002) in conscious unrestrained animals of various sizes, can be assessed by implantable radiotelemetry (Kramer et al., 2001a) (Figure 1). Thus, implantable radiotelemetry is important in research in pharmacology, toxicology, neurology, cardiovascular diseases, cancer development, regenerative stem-cells, infection models and evaluation of new biomarkers, as well as developing new therapeutic strategies (Kramer and Kinter, 2003). The recent development of small implantable radiotelemetry probes that can be implanted in mice and rats is particularly significant in biomedical research because of an increasing number of rodent models awaiting characterization for proximate physiological parameters of human diseases. Therefore, application of radiotelemetry in rodents provides valuable insight into etiopathogenic mechanisms.

1.3. Restraining and Non-restraining Approaches: Pros and Cons of Implantable Radiotelemetry

The foremost prerequisite for obtaining reliable physiological data in the whole animal is the measurement under unrestrained conditions. In the past, recording of biopotentials from rodents was mainly limited to spatially immobilized animals under conscious conditions, e.g., by using plexiglas tubings, or to immobilized animals under anesthetic conditions (Soltysik et al., 1996). Obviously, this setup raises questions about the validity of data, mainly due to the interwoven complication of stress- and anesthesia-induced physiological changes on basic biopotentials, such as the EEG and ECG, as well as other parameters including heart rate, blood pressure, body temperature, and food intake (Kramer and Kinter, 2003). Typical restraining procedures for acquisition of EEG data from small rodents are, for example physical restraint methods, worn in jacket recorder systems (Bertram and Lothman, 1991) or non-implanted radiotransmitters. The so-called *tethered* approach is still a rather common technique for EEG recording in rats, in which electrodes are connected to a miniature socket anchored to the skull and exposed for attachment of a cable system (Bertram and Cornett, 1994; Bertram et al., 1997; Bertram, 1997a; Bertram, 1997b). In mice, there are definitely limitations for this method due to the small size of the animals (Pitkänen, Schwartzkroin, Moshé, 2005). Although tethered systems allow animals a certain degree of free movement, there is a risk of infection at the electrode implantation site.

Implantable radiotelemetric EEG recording overcomes most of these disadvantages and is considered the state-of-the-art for monitoring physiological functions in awake, freely moving laboratory animals with substantially minimized stress-related artifacts (Matthew, 1997; Weiergräber et al., 2005a). The reduction of distress compared to conventional measurement techniques represents the most humane method for monitoring physiological parameters and elimination of restraints alleviates a potential source of experimental artifacts and inter-animal variability (Morton et al., 2003; Printz, 2004). Implantable radiotelemetry records the complete repertoire of physiological behaviors including resting, locomotor activity and sleep, such as REM and slow-wave sleep, from the animals and it allows time-series measurements over days, weeks and months without any special animal care. These large amount of data obtained from one single animal can reduce the overall number of experimental animals used. In fact, a rough estimation suggested that a reduction of animal use by 60 – 70 % in a single study and by more than 90 % in multiple studies can be achieved (Kramer et al., 2001b; Stephens et al., 2002).

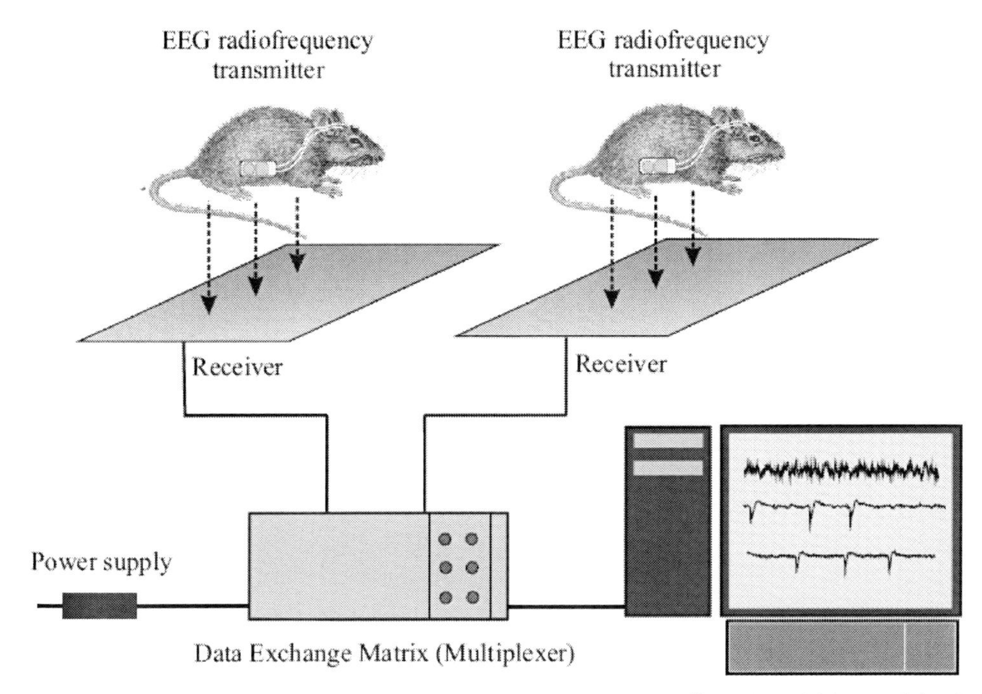

Reprinted from Brain Research Protocols, 14, Weiergräber M, Henry M, Hescheler J, Smyth N, Schneider T, Electrocorticographic and deep intracerebral EEG recording in mice using a telemetry system, p154-164, 2005, with permission from Elsevier.

Figure 2. Radiotelemetry setup for EEG monitoring in small rodents. Principal built-up of a laboratory radiotelemetry setup: The transmitter-implanted animals (mice or rats) housed in cages are either placed on a stationary receiver plate or close to a mobile receiver device. The transmitted and received EEG signal is processed by a multiplexer and forwarded to the data acquisition and analysis system. Various commercial systems are available including the complete hardware and software packages to perform proper and sophisticated EEG analysis.

Arguably, in vitro extracellular field potential recordings in the murine and rat brain slices allow the possibility of sequential drug application and washout studies; such highly standardized experimental procedures are hardly possible in the radiotelemetric approach. However, implantable EEG radiotelemetry allows precise and highly sensitive measurement of acute and long-term effects under various physiological and pathological conditions and different stages of consciousness on a whole organism level (Morton et al., 2003).

Although wireless radiotelemetry technology for monitoring large laboratory animals has existed for some time, only recently it has become more affordable, reliable, and relatively easy-to-use, even for monitoring mice (Weiergräber et al., 2005a). Small transmitters that can be used in mice and rats are now commercially available (e.g., DSI, TSE) and can be implanted in rodents, particularly in mice heavier than 20 g (~ 10 weeks of age), so that animal size is no longer an absolute experimental limitation (Weiergräber et al., 2005a) (Figure 2). So far, several companies offer commercially available telemetry systems, including both hardware, e.g. transmitters, receivers and multiplexers and software packages for data acquisition and analysis, which makes this approach technically accessable to many potential users (Figure 2). However, commercial telemetry systems still have rather high start-up, but also maintaining costs which is also dependent on the species investigated.

Currently, the radiotelemetric approach is predominantly carried out in mice due to the strong need for characterization of transgenic mouse models, more rarely however, in rats. In other species, e.g. cats or primates, radiotelemetry can be more cost intensive (Kramer and Kinter, 2003). As the primary goal is to reduce the number of animals necessary to obtain reliable data, strong efforts are carried out to further miniaturize the instrumentation needed for telemetry or non-hard-wired monitoring, to increase the number of deflections, i.e. biopotential channels and to reduce energy consumption of the transmitter unit.

1.4. Critical Aspects of Radiotelemetric Application in Current Biomedical Sciences

Obtaining and analyzing radiotelemetric data from laboratory animals is indeed a challenging task as it requires a comprehensive, integrative and critical view from the experimenters on the measurement techniques and an overall evaluation of the data in the context of animal physiology, structural and functional neuroscience, pharmacology, toxicology, statistics, biomedical computation and electronics, technical implications and possible restrictions from the influence of the experimenters and the environment. In short, one has to deal with different facets of a living animal. Thus, it is important to develop an internal validation protocol in order to obtain reliable data. Analysis of complex integrated systems, as well as complex diseases, relies on the validity and accuracy of the trait measurement. There are many chances for inadequate, inappropriate or erroneous phenotyping using radiotelemetry, such as problems due to the experimenter and environment as well as the test subjects themselves. Whereas in human studies the intellectual capabilities of the test subject mark the experimental limits, in animals the opposite can turn out to be a major problem as one cannot achieve information about feelings and attitudes in test animals (Printz, 2004).

2. Implantable EEG Radiotelemetry in Mice and Rats

2.1. Functional Application of the EEG

In clinics and biomedical research the electroencephalogram (EEG) is mainly used for diagnosis of epilepsy or epileptic syndromes and sleep related disorders. It can also be a useful tool in determining the depth of anesthesia, CNS-related effects of toxicants, the confirmation of CNS death and as a diagnostic element in sensory (e.g. auditory) evoked potentials. Sophisticated surface and deep EEG recordings can provide precious insight into rhythmicity of brain waves in different brain regions (Boutros, 1992; Rosen, 1997; Ondze et al., 2003; Vanhatalo et al., 2005). Epilepsy research, however, is still the major field of application in implantable EEG radiotelemetry.

In contemporary society, the incidence, impact and sociocultural influences of epilepsy cannot be overestimated. Manifest epilepsy affects around 0.5 – 1 % of the population in

Europe and North America (Forsgren et al., 2005a; Forsgren et al., 2005b; Forsgren et al., 2005c), 5 % of the population perceive a single seizure in their lifetime and about 10 % display an increased seizure susceptibility. Epilepsy is characterized by a huge heterogeneity regarding its etiology, age of onset, seizure type, responsiveness to pharmacological treatment, prognosis and comorbidity. Irrespective of its underlying pathophysiological mechanisms, the fundamental electrophysiological basis for hyperexcitability associated with ictogenesis or epileptogenesis are paroxysmal depolarization shifts (PDSs) within groups of neurons.

In the past, various technical approaches for acute or chronic EEG recordings in epileptic mice have been carried out (Nakano et al., 1994). These non-telemetric techniques represented either monopolar or bipolar deflections, using silver / stainless steel wires or stainless steel / Gold screws as electrode material (Zornetzer and McGaugh, 1970; Valatx, 1970; Valatx, 1975; Noebels and Sidman, 1979; Kaplan et al., 1979; Krauss et al., 1989; Noebels et al., 1990; Gioanni et al., 1991; Bennet and Scheller, 1993; Ishida et al., 1993).

Current radiotelemetric EEG studies using transgenic mouse models have elicited that voltage- and ligand-gated ion channels are the major candidates in the etiopathogenesis of epilepsies (Kullmann and Hanna, 2002; Kullmann, 2002; Turnbull et al., 2005). Thus, this emerging technique has offered a dramatic new pathophysiological insight in the broad field of epilepsy, providing further knowledge in epileptogenesis and the potential antiepileptic treatment in humans as well.

Given the deluge of transgenic mouse models exhibiting epilepsies, the need to document those often subtle epileptic phenotypes, or sometimes very bizarre but prominent behavioral patterns that might or might not be related to epilepsy, clearly requests for better combination of tools (Hunter et al., 2000; Yang and Frankel, 2004).

Simultaneous video-EEG or video-EEG-EMG monitoring, combined with the analysis of other physiological parameters, such as activity or body temperature helps to fix the overall goal of correlating an observed behavioral phenotype with the recorded EEG (Pitsch et al., 2007; Hoffmann et al., 2008). Thus, radiotelemetric EEG recordings in combination with video monitoring provide a useful tool in recording pharmacologically induced seizures and for documentation of spontaneous seizures in small rodents.

2.2. Planning for EEG Radiotelemetry

An essential and often overlooked aspect in performing radiotelemetry is the necessity to clearly outline the study plan that one needs to address specific scientific questions. Though this sound cliché, neglecting this point can generate non-interpretable results, let alone the time, effort and high cost spent. Here are some examples of questions that one needs to ask oneself before starting radiotelemetry implantation: Do the animals need a specific pre-treatment prior to implantation or prior to a pharmacological injection experiment? Which physiological parameters do I want to record and which radiofrequency transmitter is appropriate? How to manage postoperative recovery and postoperative care? Do I have the appropriate software for data acquisition and especially analyses of, for example, EEG data?

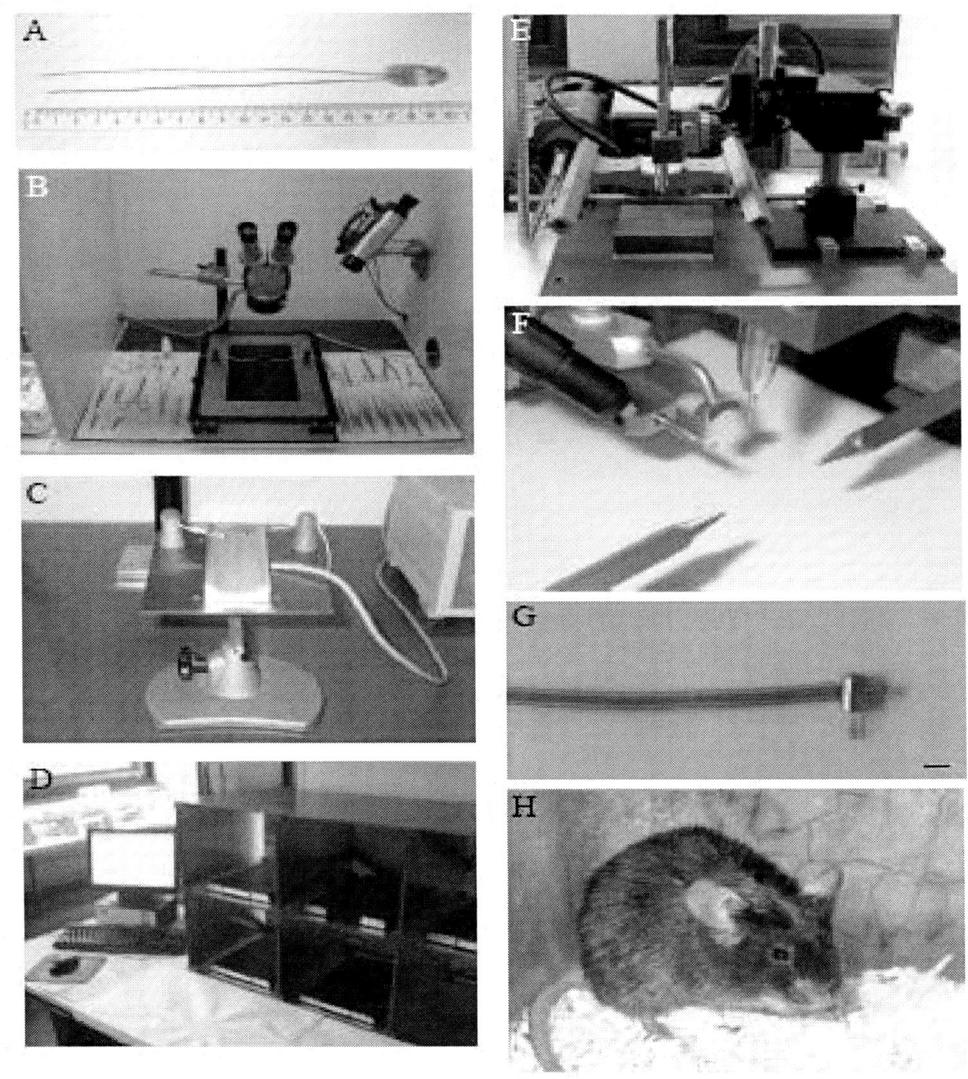

Reprinted from Brain Research Protocols, 14, Weiergräber M, Henry M, Hescheler J, Smyth N, Schneider T, *Electrocorticographic and deep intracerebral EEG recording in mice using a telemetry system*, p154-164, 2005, with permission from Elsevier.

Figure 3: EEG radiotransmitter implantation in mice. (A) Example of a TA10ETA-F20 transmitter (DSI) used in our lab (technical specification: volume 1.9 cc, weight 3.9 g, lead length: 20 cm). (B) Surgical equipment, dissecting microscope, external temperature support with base plate, fixators, elastomers, and wound retractors and a video monitoring system. (C) Alternatively, a small animal operating table can be used. It is mounted on a tripod base by means of a ball and socket joint allowing to swivel the table into any convenient position. The table is equipped with fixators and wound retractors and allows for electronically controlled heating. (D) Multiple receiver plates connected to a data exchange matrix (multiplexer, DSI). Each housing chamber is shielded with stainless steal to avoid crosstalk. (E) Stereotaxic device with base frame, ear bars, nose clamp, and a dentral drill mounted on a calibrated 3-axis micromanipulator. Several drill heads ranging from 0.2 to 0.7 mm in diameter can be used in this setup. (F) Closer view of stereotaxic setup. Ear bars were covered with cotton prior to positioning of the mouse head. (G) Stainless steel pin mechanically attached to the sensing lead used for deep intracerebral EEG recording (bar: 1 mm). (H) C57Bl/6 mouse after subcutaneous pouch implantation of the radiofrequency transmitter four weeks post surgery.

Experimentally, a primary goal is to reduce the number of research animals to the least, insofar without compromising the reliability of results, and to optimize the experimental approach for yield of data acquisition. One should make sure that the animals used are sex- and age-matched, in order to obtain reliable data of physiological evaluation in the laboratory animals. As has been reported in literature repetitively, sex or age can severely influence radiotelemetric and non-telemetric EEG outcomes in rodents, such as pharmacologically induced seizure susceptibility testing (Mejias-Aponte et al., 2002). Furthermore, the route for drug administration is another important issue for consideration. In contrast to electrophysiological experiments in vitro, like patch-clamp or extracellular field potential recordings, which allow consecutive application and washout of several drugs, such approaches in the whole-animal experiment demand the experimenter to consider the pharmacokinetic and pharmacodynamic properties of each drug used, insofar as the drug can pass through the blood-brain-barrier (BBB). Otherwise, one has to consider the use of microdialysis or brain infusion systems via osmotic pumps to target the drug of choice to a specific brain region or the cerebrospinal fluid (CSF). Any compensatory changes related to the experimental regime have to be critically evaluated in data analysis.

Several publications have commented on experimental parameters of mice and rats that are critical to the planning of experiments using telemetry. Biotelemetry transmitter implantation for example can influence growth and circadian rhythms, further affecting body temperature and general motor activity (Leon et al., 2004). In addition, the choice of transmitter placement, i.e. subcutaneously or intraperitoneally, can affect basic cardiovascular parameters (Gehrmann et al., 2000) and is likely to influence brain related electrical activity as well. Detailed investigation further revealed that laboratory rodents require at least 14 days to regain their pre-surgical body weight (Leon et al., 2004) and it is generally concluded that investigators should start recordings 10 - 14 days post-surgery to obtain reliable data (Kramer and Kinter, 2003; Weiergräber et al., 2005a). These examples illustrate that various aspects should be considered by the experimenter before initiating radiotelemetric studies.

2.3. Commercial Versus Self-Constructed Telemetry Systems

Although wireless radiotelemetry technology for monitoring laboratory animals has existed for at least 50 years, affordable, reliable, and easy-to-use commercial products have been available only in the last 10 years [Biomedic Data Systems (BMDS), Seaford, DE 19973; Data Sciences International (DSI), St. Paul, MN, 55126; Konigsberg Instruments, Pasadena, CA 91107; Mini Mitter, Sunriver, OR 97707; Technical and Scientific Equipment (TSE), Bad Homburg, Germany and Star Medical, Arakawa-ku, Tokyo 116, Japan]. Strong efforts have been carried out to further miniaturize the instrumentation needed for telemetry or non-hard-wired monitoring, increase the number of deflections and prolong battery lifetime.

Nowadays, several companies offer commercially available radiotelemetry systems, including both hardware, e.g. transmitters, receivers and multiplexers, and software packages for data acquisition and analysis, which make this approach technically more accessible to potential users. They are available for use in almost all laboratory species, ranging from mice to monkeys, and even in fish (Kramer and Kinter, 2003; Snelderwaard et al., 2006).

Our studies have been carried out using a telemetry implant from Data Sciences International (DSI), the PhysioTel® transmitter TA10ETA-F20 (technical specification: 3.9 g, 1.9 cc) capable of measuring one biopotential (EEG, ECG or EMG), physical activity and temperature in mice and rats (Figure 3A) (Weiergräber et al., 2005a; Weiergräber et al., 2006a; Weiergräber et al., 2006b; Weiergräber et al., 2007). A second transmitter type, the F20-EET (DSI) with two biopotential channels can also be used for combined EEG / EMG recording and sleep analysis in rodents. The signals are transmitted to a receiver (RPC-1, DSI), which picks up the telemetered data from the implant and forwards to a Data Exchange Matrix (DSI) serving as a multiplexer (Figure 3D). The data are then further processed through an acquisition and analysis system (Figure 2). For all commercially available systems the start-up and maintenance costs can be high depending on the species being analyzed. The advantage of commercial systems is that they have *plug and play* character, i.e. after a relatively short training period, the system and experimenter are ready to go. However, one should always be aware of potential shortcomings and inherited problems associated with such procedure (as outlined in section 2.2).

Recent publication demonstrates that home-made radiotelemetry devices can also effectively be applied in biomedical research (Grohrock et al., 1997; Lapray et al., 2008). They are often cheaper than commercial ones. The drawbacks are their relatively large size which is problematic, especially in small rodents, and their short battery life-time. The most demanding issue is that the home-made devices are often incompatible with the commercial acquisition and analysis software and the experimenters inevitably have to build up their own (Sundstrom et al., 1997). However, different data acquisition and analysis programs create difficulty in comparing data from different research groups for evaluation. Thus, for the good of overall scientific community, commercially available systems, though expensive, are likely to be the better and more practical choice.

2.4. General Aspects of Surgical Precautions in Small Rodents

Generally, mice pose an anesthetic challenge because of their small size and varied responses to anesthetic drugs among strains and between genders (Hedrich, 2004). Any kind of disease state in mice can cause sudden and sometimes lethal peri-anesthetic complications. Under anesthesia, mice quickly exhibit hypothermia due to their high surface to body mass ratio. Therefore, during and after surgery, warming plates are necessary in order to maintain their body temperature which should be 37 - 38°C (98.6 - 100.4°F) (Figure 3C). This is an important issue in mice, yet rats should also be temperature-monitored. Apart from body temperature, one should carefully keep track of their heart rate and respiration rate during surgery. As outlined below, occurrence of tachycardia, bradycardia or respiratory alterations, like gasping, require quick intervention to let the animal survive the implantation procedure (Colby and Morenko, 2004). To avoid corneal desiccation, eyes should be covered with dexpanthenole (Bepanthen®, Hoffmann-La Roche AG) during the whole implantation period and early recovery state, as the blinking reflex is impaired under anesthesia. Furthermore, mice should be housed under stable environmental conditions prior to any type of manipulation, either surgical or non-surgical.

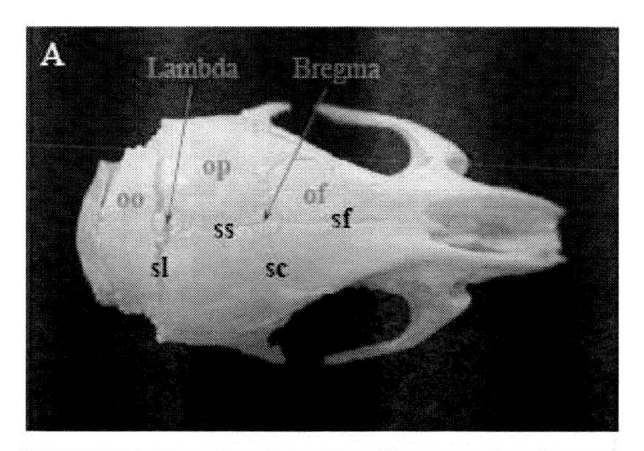

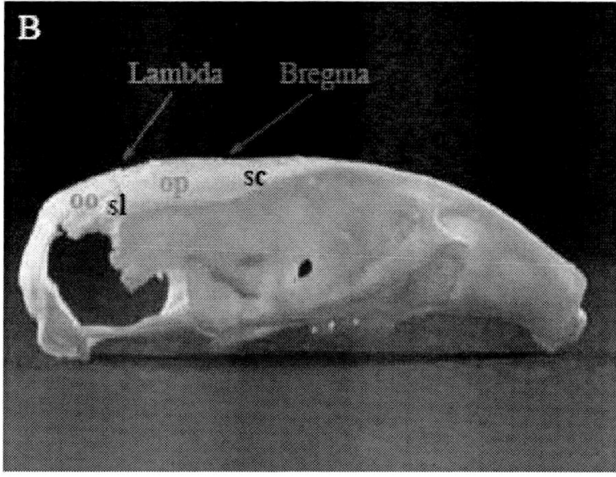

Reprinted from Brain Research Protocols, 14, Weiergräber M, Henry M, Hescheler J, Smyth N, Schneider T, Electrocorticographic and deep intracerebral EEG recording in mice using a telemetry system, p154-164, 2005, with permission from Elsevier.

Figure 4: Anatomic structures and craniometric landmarks of the murine skull. Apical (A) and lateral view (B) of a C57Bl/6 mouse skull which has been prepared in 30 % H_2O_2. Marks on cranial bones: os frontale, of; os parietale, op; os occipitale, oo and on sutures: sutura frontalis, sf; sutura saggitalis, ss; sutura coronaria, sc; and sutura lamdoidea, sl. The sutures determine the major craniometric landmarks bregma and lamda, necessary for stereotaxic lead placement. The murine neurocranium is extremely thin. The thickness of the cranial bones mostly varies from 250 to 450 μm. In contrast to the rigid rat skull, the mouse neurocranium is therefore deformable and flexible.

2.5. Anesthesia

The primary objective of anesthesia when performing radiotransmitter implantation is analgesia, i.e. reduction or absence of pain recognition and responsiveness (Kramer and Kinter, 2003; Hanusch et al., 2007; Wang and Slikker, Jr., 2008; Davis, 2008). In addition, the anesthetic regime should be accompanied with reduction of locomotor activity by inducing muscle relaxation and reduction of basal motor tone of major muscle groups. Generally, administration of anesthetics can be performed via the lateral tail vein (intravenous, iv.), intramuscularly (im.), subcutaneously (sc.), e.g. in the loose skin over the neck and shoulders or intraperitoneally (ip.), and the latter path is probably the most common approach in rodents. Intraperitoneal injections are normally carried out in the lateral caudal part of the abdomen in order to avoid injury or injection into intestinal organs, such as the bowel, the bladder, liver or major blood vessels. Apart from injection, inhalation of narcotics, such as halothane or isoflurane, can also be used as anesthesia. However, the inhalation anesthesia is technically inconvenient during the stereotaxic radiotelemetry implantation process (Figure 3E,F; Figure 4) (Richardson and Flecknell, 2005; Terrell, 2008).

Irrespective to the administration route, the appropriate anesthetic dosage is drug dependent, further influenced by mouse stain, age and sex. Thus, testing of anesthetic drug dosages prior to implantation experiments is still highly recommended, especially for genetic engineered animals in which alterations in anesthetic tolerance might occur. Furthermore, the optimal dosage can vary not only in animals from strain to strain but also from individual to individual. Large amount of adipose tissue in individual mice results in high apparent volume of distribution which in turn causes slow introduction of and slow recovery from anesthesia. In slender animals, however, the induction of and relief from anesthesia can be extremely fast.

A well-established, quick anesthetic regime that proved very useful in EEG radiotransmitter implantation is a combination of ketamine hydrochloride (100 mg/kg) and xylazine hydrochloride (10 mg/kg) (Weiergräber et al., 2005a). This dosage is usually sufficient for EEG radiotransmitter implantation lasting for about 45 min with xylazine providing a good muscle relaxation. Apart from ketamine / xylazine, various other anesthetic drugs or drug combinations can be used in mice and rats, such as pentobarbital, thiopental, propofol, etomidate, ketamine / acepromazine, ketamin / diazepam and urethane. Sometimes more sophisticated anesthetic regimes are applied in mice and rats starting with a premedication of atropine or fentanyl / droperidol followed by induction of anesthesia using, e.g. ketamine / xylazine, ketamine / diazepam, pentobarbital or inhalation narcotics, such as halothane or isoflurane (Flecknell, 1996). To maintain anesthesia, the same drugs used can be further administered; however, multiple intraperitoneal or subcutaneous injections should generally be avoided.

For both rats and mice it is necessary to check for CNS related responses to anesthesia, e.g., the lack of recognition or response to pain stimuli of pinching at skin, tail, feet or toe prior to the implantation procedure. Typical signs for progressive anesthesia include sedation, loss of consciousness, recumbency, and immobilization. If anesthesia is insufficient, animals will react with limb withdrawal, muscle tremor, vocalization, and irregular respiration or heart rate. As anesthesia is accompanied by hypothermia, particularly in small rodents, measuring body temperature by placement of a thermometer or thermistor probe in the rectum gives hints whether the implantation procedure should proceed.

The anesthetic status should also be controlled from time to time during the implantation procedure because anesthetic regimes can sometimes cause life-threatening side-effects in experimental animals, e.g., some rat strains will exhibit hypertensive complications following ketamine, tiletamine, pentobarbital or xylazine administration (Brookes et al., 2004). Therefore, monitoring heart rate and / or blood pressure during implantation via the tail vein is recommended. Monitoring the anesthetic status of mice is difficult because of their small size. However, one should be aware that in mice, ketamine can cause an increase in heart rate, whereas xylazine can dose-dependently decrease heart beat frequency and respiratory rate. Pentobarbital, on the other hand, which is often used as an anesthetic in mice, can dose-dependently result in increased heart beat frequency and respiratory depression.

Artificial respiration, e.g. endotracheal intubation for airway support is a sophisticated yet challenging procedure in rats, especially in mice (Vergari et al., 2004; Zhao et al., 2006; Spoelstra et al., 2007). According to our experience, airway support is not required for successful implantation.

2.6. Intraperitoneal Radiotransmitter Implantation

Almost all commercially available and self-made radiotransmitter implants are composed of a transmitter unit and a battery, both of which are coated with a biocompatible material. They are implanted into the research animals either in their abdominal cavity or in a subcutaneous pouch on the back or lateral flank of the animals.

For intraperitoneal implantation, the body hair at the abdomen has to be shaved by a clipper and the shaved areas are scrubbed with a disinfectant such as Mercuchrom® (Merbromin, Krewel Meuselbach, Eitorf, Germany). No depilation cream should be applied to avoid skin irritation and possible defective wound healing later on. Animals are placed on a sterile drape on a heating pad (37 - 38°C) under an operation microscope (Figure 3B). All surgical instruments necessary for implantation should be heat sterilized. Experimenter should wear masks, surgical gloves and surgical gown to further heighten the sterile conditions during implantation.

An abdominal incision of about 1.5 – 2 cm is first made at the midline of the abdomen. Then the abdominal skin and wall is opened and both incision borders are held aside using a 5 mm wound retractor together with elastomers, fixators and a base frame (Fine Science Tools, FST, Heidelberg, Germany). The transmitter is carefully placed into the peritoneal cavity on top of the gastrointestinal tract. Care must be taken to avoid impingement of the gastrointestinal organs or the diaphragm because compression of the upper abdominal organs can cause vomitus, impinged stomach will cause cessation of food uptake, and compressed diaphragm will cause irregular respiration. Similarly, increased pressure in the lower abdominal cavity can cause cessation of urination and defecation. If the experimenter is careless at this stage, a mechanical-obstructive or paralytic ileus will cause death within 2 – 3 days following implantation.

Anatomical forceps are preferred to surgical forceps, as they create less damage to the tissue when manipulating the bowel. In case of local bleeding, a thermocauter (e.g. Heiland, Germany) should be used.

The next step is to tunnel both sensing leads through the abdominal wall to the cranial part of the incision (made by a 14-Gauge needle). A trocar together with a plastic sleeve can run subcutaneously along the lateral thoracic wall to the intended position at the neck, where a second skin incision has been made beforehand. The trocar can then be withdrawn from the sleeve and the leads are tunneled through it. Alternatively, a thin trocar with a short silicon tubing attached at its tip can be used to direct the leads subcutaneuosly to the neck. In this case, the sensing electrodes are tied to the silicon tubing mechanically or by sticking with glue. The latter method of directing the leads to the neck is much faster to perform and causes less damage to the subcutaneous tissue than using trocar and plastic sleeve, and thus allows faster postoperative recovery.

Leads are secured on the underlying muscle layer using a non-absorbable interrupted stay suture (Ethilon*II, 4-0, M-2, Ethicon, Germany). If not tacked, the intrinsic elasticity of the electrodes will make implantation of the electrodes difficult. The transmitter was fixed in the peritoneal cavity using its suture tab to avoid intra-abdominal movement. Finally, the abdominal wall was closed using non-absorbable suture material (Ethilon® II, 4-0, M-2, Ethicon, Germany) and the ventral skin incision was closed by using wound clips (Michel, 7.5 X 1.75 mm, Heiland, Germany), as mice tend to bite sutures. The wound clips will be dropped off later during wound healing process. The wound area is cleaned with a iodine-based disinfectant, e.g. Mercuchrom®.

Current commercially available radiotransmitters for mice and rats have a volume of up to 1.9 cm^3, e.g. the TA10ETAF20, DSI. Thus, the technical challenges of putting the device into a 20 g mouse (corresponds to 10 – 12 weeks of age) is more demanding than putting it into a 500 g rat.

2.7. Subcutaneous Radiotransmitter Implantation

Alternatively, transmitters can be implanted subcutaneously. This procedure, particularly in EEG radiotelemetry has several advantages. First, the longitudinal incision of the scalp accessing the neurocranium can serve as an entry site for inserting the transmitter under the skin. Second, formation of a subcutaneous pouch by using a trocar as well as fixation of the transmitter at the dorsal skin using its suture tab is easier and faster to perform. Since opening of the abdominal cavity is not necessary, time for transmitter implantation is largely reduced and so is the risk of infection.

Comparing both techniques, the authors favor the subcutaneous approach characterized by quicker implantation and post-surgical recovery, though both ways of implantation have their biological challenges and drawbacks. The major biological challenge of intraperitoneal implantation is the possible physiological complications induced, such as vagal nerve stimulation and alteration of blood flow because of increased intra-abdominal pressure. Such systemic consequences may inevitably influence the electroencephalography. The limitation of subcutaneous pouch implantation becomes obvious when the animals are too small and the underneath pressure caused by the implantation can result in skin necrosis. In rare cases, the elasticity of the sensing leads creates rooms for air within the subcutaneous pouch and when such air space persists, it causes influx of interstitial fluid and fibrin, subsequently resulting in a subcutaneous sterile swelling known as a seroma.

Another concern for transmitter implantation in a subcutaneous pouch on the back is reduced signal strength due to increased distance from the receiver plate on the floor. One way to solve this problem is to implant the transmitter subcutaneously at the flank close to the ventral abdominal region and fix the transmitter at the skin by a single stitch (Figure 3H). At this position, there is no reduction in signal strength when compared to the intraperitoneal implantation procedure. The major drawback for the subcutaneous placement is that the telemetered temperature values are no longer representing the temperature of the core body but that of the body surface.

2.8. Stereotaxic Device

In humans the EEG is normally recorded from the surface using scalp electrodes. These electrodes register changes in the electrical field mainly caused by excitatory postsynaptic potentials (EPSPs) of thousands to ten thousands of pyramidal neurons within the neocortex. Such changes in the extracellular field potentials must be strong enough to reach the recording electrode via the conductor which includes the scalp, the neurocranium, connective and adipose tissue, the meninges, blood vessels and sinuses. Scalp recordings in mice and rats are practically not possible, as the signal is simply too weak to discriminate from background noise. Thus, placing the sensing EEG electrodes directly onto the surface of the brain or inserting them into deep brain structures is necessary for EEG recordings in small rodents. In this way, the experimenter has to perform a craniotomy with the help of a stereotaxic device (Figure 3E,F).

Standard stereotaxic frames that fit the mouse and rat skull anatomy are commercially available. Depending on the mouse and rat strains and their age, the teeth holder, nose clamp, and ear-bars of the frames can be adjusted to achieve the opitimal holding position. A stereotaxic frame can be combined with a high-speed dental drill (e.g. 20000 rpm, KaVo EWL, Germany) mounted to a three-axis micromanipulator (Figure 3E,F). This manipulator enables precise electrode placement according to the stereotaxic coordinates marked the mouse or rat brain atlases (Franklin and Paxinos, 2008; Watson and Paxinos). Manual handling of a dental drill in mice is not recommended due to their small and thin neurocranium, whereas in rats, the stereotaxic device can be used to help mark the position of the holes and drilling can be done manually.

One special feature of the murine cranial bones is their elasticity. Therefore, the holes should be drilled at maximum velocity in order to avoid a tonic applanation of the skull which results in a sudden breakthrough of the drillhead and severe brain damage. The rat skull, being different from the murine one, does not possess such elastic features and applanation of the skull does not occur in both juvenile and adult rats. Therefore, drilling speed can be reduced when working on rats. During the drilling process, one has to take meticulous care of avoiding over-generation of heat which can cause thermal brain damage and of avoiding positions with underlying blood vessels and sinuses.

Another special feature of the murine cranial bones is their non-uniform thickness. The following values are observed from adult C57Bl/6 mice of 3 - 6 months old: *os frontale*: midline section: 320 – 390 μm, lateral section: 300 – 430 μm; *os parietale*: midline section: 210 – 250 μm, lateral section: 200 – 210 μm; *os occipitale*: midline section: 600 – 730 μm, lateral section: 380 – 420 μm (Figure 4). Therefore, one has to be adept in maneuvering the depth of drilling when working on different region of the brain.

The head has to be fixed tightly during the drilling process, but one has to take care not to push the ear-bars too hard as this will damage the inner ears of the animals and cause respiratory alterations. One preventive measure is to cover the ear-bars with cotton balls. This precaution is particularly important for those experiments screening for audiogenic seizures or audiogenic seizure susceptibility of the animals.

2.9. Surface Electrode Implantation for Electrocorticographic Recordings (ECoG)

For putting the leads in the head region, a midline skin incision, 10 mm on the head and 5 – 10 mm down to the neck, is made and the subcutaneous tissue is bluntly separated. The periosteum is cleaned using sterile cotton tips without damaging the temporal and occipital muscles. The superficial thin layer of the skull is pretreated with 3 % H_2O_2 or diluted (65 %) phosphoric acid. This pretreatment removes any tissue remaining on the skull and causes superficial decalcification and roughens the surface, allowing strong adhesive contact of the cement used for fixing the leads. This pretreatment also accentuates the sutures, the bregma, and the lambda landmarks of the skull, which are essential for precise stereotaxic lead placement (Figure 4). Another benefit of doing such pretreatment is that after H_2O_2 or phosphoric acid application to the calvaria, small vessels like meningeal arteries or superficial cerebral arteries or veins can easily be observed under an operation microscope as the calvaria becomes rather transparent, even in adult mice. Thus, damage of blood vessels or sinuses during drilling can be avoided. The rat neurocranium, in contrast, is much thicker, so there is no chance to localize the course of superficial arteries, veins or sinuses. Nevertheless, this pretreatment should be done carefully, as it can produce severe oxidative damage to the surrounding tissue and at higher concentrations, the reagent can penetrate through the murine skull and cause severe oxidative damage on the cortical surface.

Depending on the number of EEG electrodes being positioned, burr holes are drilled using an electric high-speed dental drill at the coordinates of choice. The diameter of the holes should be reduced to minimum for better postoperative recovery and for reducing the risk of damaging meningeal arteries or veins. Hemorrhage from intraosseous bleeding at the cutting edge of the skull can be easily controlled using cotton tampons. Fatal bleeding can be avoided by positioning of the electrodes away from the dural sinuses, such as the superior sagittal sinus. Interestingly, bleeding from sinuses often turned out to cause higher lethality than bleeding from meningeal or superficial brain arteries.

Once the holes are drilled, the experimenter can make use of two different technical approaches for the placement of the leads. The first approach is the direct implantation of the sensing leads of the transmitter: the silicone insulation is first removed from the terminal part of the sensing leads and both electrode ends are shortly bend at the tip and placed directly onto the dura mater (epidural lead placement) in a bipolar deflection, e.g. at the following coordinates: (+)-lead: caudal to bregma 1 mm, lateral of bregma 1mm (right hemisphere); (-)-lead: caudal to bregma 1 mm, lateral of bregma 1 mm (left hemisphere, see recordings from this deflection). The stereotaxic coordinates described above represent an example of a bipolar deflection and correspond to the transition region of the primary to secondary motor cortex. Whenever highly synchronized EEG activity is observed, bipolar deflection can result in possible extinction of EEG discharges. If the experimenter expects such kind of activity, it is recommended to use one electrode as a differential electrode and the second one as a (pseudo) reference electrode, which is placed into an electrically "silent" brain area, such as the cerebellum.

Lead tips are fixed with glass ionomer cement (Kent Dental®, Kent Express Ltd., UK) which is extremely hard when dried and gives strong adhesion to the underlying neurocranium. After the cement has dried (3 - 5 min), the scalp is closed by doing over-and-over sutures using non-absorbable 6-0 suture material (Ethilon® polyamid, Ethicon,

Germany). No severe inflammatory reactions have been observed in the cerebrum, the cerebellum, the meninges or skin so far. However, skin necrosis due to an increased pressure caused by the underlying cement can be occasionally observed during the observation period of 4 - 6 weeks. Thus, minimum amount of cement should be used. Generally, it is possible to leave the scalp incision open as it is often done in *tethered* EEG recording setups. However, there is a risk that the animal scratches at the electrode insertion site and the sensing leads by grooming-like behavior or driven by a pruritus associated with the wound healing process.

The second approach is the use of cortical screws: this approach allows screws made of inert material, such as stainless steel or gold, to be gently implanted together with the sensing leads till they touch the surface of the brain. The sensing leads have to be mechanically fixed or attached to the screws by clipping. Loitering screws and sensing lead tips should be avoided, as this procedure can introduce significant noise to the EEG recordings. In principle, both direct and indirect techniques can be used in mice and rats.

2.10. Deep Electrode Implantation for Intracerebral EEG Recordings

The stereotaxic lead insertion for deep implantation is similar to that described above. Take an example for targeting the region of CA3 in hippocampus, with holes drilled and leads positioned at the following coordinates: (+)-lead: caudal to bregma 2 mm, lateral of bregma 2 mm (right hemisphere), dorsoventral (depth) 2 mm (related to the dorsal surface of the calvaria corresponding to the final targeting region of CA3). The (-)-lead is positioned at: caudal to bregma 6.2 mm, lateral of bregma 0 mm; dorsoventral 0 mm (epidural reference electrode localized on the cerebellum). The electrodes are bent at 90° for insertion. This is easy to achieve as the sensing leads consist of high-grade stainless steel helices which provide flexibility and resilience. Proper lead placement has to be confirmed histologically at postmortem. Electrodes are fixed and the scalp is closed as described above.

More precise intracerebral EEG recordings can be achieved by using stainless straight steel electrodes coated with a metal spraying varnish. The coated electrode is mechanically attached to the sensing leads of the transmitter by clipping and mounted on a stereotaxic device (Figure 3G). Once the cranial holes are drilled, the electrode is lowered deep into the brain at the desired depth, targeting to, for example, a specific brain nucleus. The EEG is finally recorded from the non-isolated tip of the sensing lead.

Both radiotelemetric surface and deep EEG recordings can also be performed in adult rats and to a limited extent also in rat pups. Given the rapid growth of the young rat heads, the implanted electrodes are likely to change their positions over time, limiting the long-term use of implantable radiotelemetry under such circumstances.

2.11. Postoperative Care, Pain Management and Recovery

Failure to support the animals during postoperative care can result in perioperative complications and death. Generally, small rodents (mice and rats) are placed back into their home cages after implantation. We normally place the animals on a synthetic fleece or a small, sterile blanket or towel on the wood shavings or sawdust as cage bedding. Special care has to be taken to avoid wound healing being impaired by the "blankets".

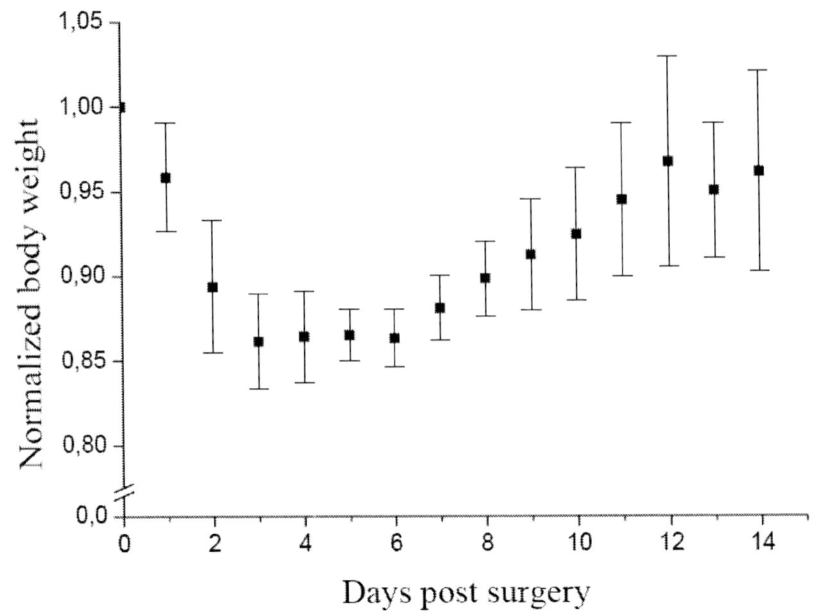

Reprinted from Brain Research Protocols, 14, Weiergräber M, Henry M, Hescheler J, Smyth N, Schneider T, Electrocorticographic and deep intracerebral EEG recording in mice using a telemetry system, p154-164, 2005, with permission from Elsevier.

Figure 5. Postoperative development of body weight. Postsurgical monitoring reveals that maximum decrease in body weight occurs 4 - 5 days after intraperitoneal implantation (displayed as mean ± SEM). The results presented here were obtained from intraperitoneal ECG implantation. Our own test measurements for EEG implantation showed the same trend in weight loss, no matter whether the transmitter was implanted into the belly or subcutaneously. According to behavioral criteria however, recovery following subcutaneous implantation is much faster.

As animals easily exhibit hypothermia, the ambient temperature should be maintained at 26 – 33 °C in a quite environment. This can be achieved by using warming lamps or heating pads within the first 3 – 4 days post-surgery. External temperature support should not be removed until the animals become able to maintain their physiological body temperature themselves. Sometimes administration of oxygen supply is also helpful due the initial respiratory depression following anesthesia and this can be applied by using a face mask. As chewing is likely to be impaired after longitudinal incision of the scalp, animals are fed with moistened food pellets.

Care should also be taken to avoid the animals being injured during recovery from anesthesia, as ataxia or transient agitation may occur at the early stage of recovery. Animals should be carefully observed early after implantation, particularly for alterations in heart rate, respiration, potential aspiration of gastric contents, vomiting or regurgitation.

Systemic administration or local application of antibiotics is often recommended in mice and rats but may not be necessary. Broad-spectrum prophylactic antibiotics can be used in both mice and rats. Infections usually occur within the first 8 h following surgery, so it is reasonable to administer antibiotics prior to implantation. It is essential to choose an appropriate agent in relation to its activity and possible unwanted side effects. Some experimenters topically apply antibiotics, e.g. penicillin, in suspension, whereas others systemically inject those, like cefazolin (10 mg/kg im.) or sulfadoxin (20 mg/kg) and trimethoprim (5 mg/kg). Generally, the list of useful antibiotics is extensive. However, when the experimental procedures are carried out in highly sterile conditions, the administration of antibiotics is not necessary. In fact, we generally do not give antibiotics in our studied animals and there have been no signs of peri- or postsurgical infection, such as meningitis, encephalitis or other local or generalized infection detected at postmortem examination.

The postoperative pain management is principally carried out using the non-steroidal anti-inflammatory drugs (NSAID), such as acetominophen, aspririn, carprofen, flunixin, ibuprofen, indomethacin, ketoprofen, piroxocam or phenylbutazone. Opiate analgesics, such as tramadol, buprenorphine, butorphanol, fentanyl, meperidine, morphine, nalbuphine, oxymorphone or pentazocine can also be used. One of the NSAID often being used is metamizole at dosages of 100 mg/kg ip. for four days post-surgery. Due to its short-half life (2 – 3 hrs) several injections per day are necessary. Another NSAID drug, carprofen (5 mg/kg sc.) needs to be applied only once a day for three to four days due to its long half-life (12 hrs). Alternatively, tramadolhydrochloride can be applied directly on the tongue or be added to the drinking water. However, the latter approach may not be sufficient as the animals have difficulty to drink water, particularly in the early phase of recovery. For any analgesic drug used, we highly recommend that it should be dissolved in 0.3 - 0.5 ml of 5 % glucose in Ringer solution or 0.9 % NaCl as water uptake within the first few hours after implantation is essentially impaired.

Postoperative recovery after intraperitoneal or subcutaneous pouch implantation of the transmitter can be assessed by checking the animal's body weight. A sharp drop in body weight is generally observed around 4 - 5 days post surgery, followed by a slight, but steady increase of weight during a 10 - 14 day recovery period, and the body weight remains constant afterwards (Figure 5) (Kramer and Kinter, 2003; Leon et al., 2004; Weiergräber et al., 2005a). The time pattern of recovery is consistent with the postoperative development of body weight after intraperitoneal or subcutaneous implantation for ECG recording (Weiergräber et al., 2005b) thus the electrode localization itself does not seem to influence the time course of body weight regain. By using the automated Laboratory Animal Behavior System (LABORAS, Metris System Engeneering, Hoofddorp, The Netherlands), investigators are capable of monitoring climbing, locomotion, immobility, grooming, food and water uptake as well as defecation and urination (Van de Weerd et al., 2001). Studies on implanted animals reveal that in the early phase after implantation, climbing, locomotion, and food uptake are decreased, whereas grooming and immobility are increased. Around 2 weeks after surgery, no difference in physiological parameters between transmitter implanted and sham-operated animals are detected. Due to this pattern of recovery time course, a minimum period of 10 - 14 days of recovery is therefore recommended in order to obtain valid EEG recordings (Kramer and Kinter, 2003).

2.12. Signal-to-noise Ratio

EEG recording systems of a *tethered* type require special connectors to attach the sensing leads to the implanted wires. This can introduce significant noise to the system. Radiotelemetry systems using screw electrodes or isolated stainless steel wires for implantation can similarly be susceptible to noise generation in the recording. Since the EEG signal is small compared to both EMG and ECG signals, it is important to eliminate or maximally reduce the influence of these biopotentials from the EEG. To avoid introduction of noise, inert materials should be used in the sensing leads and implanted electrodes, and both should be attached mechanically, e.g. by clipping, but not by soldering. For surface recordings, the sensing leads can simply be bent at 90° and implanted to the desired stereotaxic coordinates and that helps reduce the background noise within a range of 100 nV – 1 μV in our recordings after implantation.

2.13. Epidural Surface and Deep Intracerebral EEG Recordings – Examples and Practical Approaches

In the following sections, we will use examples of various physiological and pathophysiological characteristics and alterations of the murine EEG to illustrate electroencephalographic variability. During spontaneous EEG recordings, it is essential that the animals are left undisturbed, unstressed and not being exposed to sensory stimuli, unless these are part of the experimental regime. When performing pharmacological injection experiments in implanted animals, the route of application can often be a challenging task. The single-handed restraint approach is not applicable in implanted animals due to the subcutaneous lead placement and the risk of tearing off the recordings electrodes. Therefore, it is necessary to use restraining devices, such as translucent plexiglas tubings or tiny cages made of steel mesh and to perform subcutaneous or intraperitoneal administration using small diameter needles (26 – 30 G).

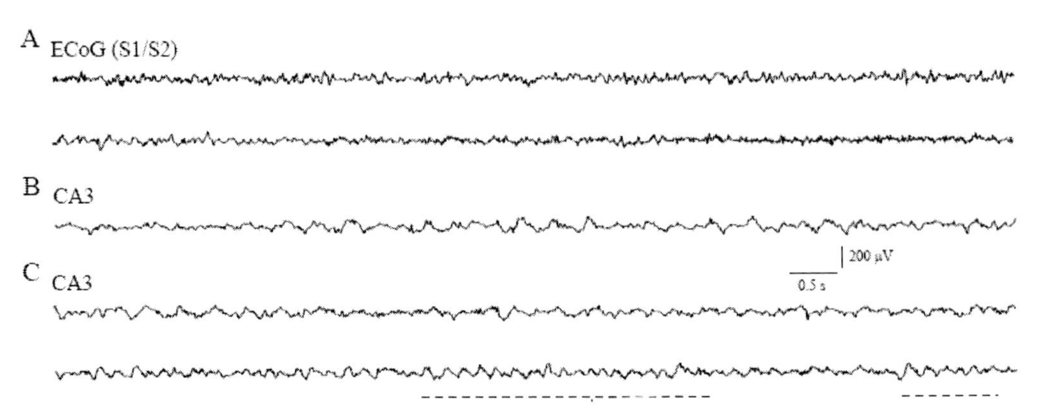

Modified from Weiergräber M, Henry M, Krieger A, Kamp M, Radhakrishnan K, Hescheler J, Schneider T, Altered seizure susceptibility in mice lacking the $Ca_v2.3$ E-type Ca^{2+} channel. Epilepsia, 2006 May;47(5):839-50, Wiley Blackwell.

Figure 6. Electrocorticographic (ECoG, A) and deep intracerebral recordings (B, C) from C57Bl/6 mice. A) Representative surface recordings obtained from the somatosensory cortex (S1/S2) of a control mouse during the awake state. **B**) Spontaneous EEG activity following deep intrahippocampal CA3 implantation. After induction of exploratory behavior by changing cages, the hippocampus exhibits typical theta-wave activity (**C**, dashed lines). If leads are implanted properly, artifacts are rarely seen. Isolated single spike events due to implantation related brain damage may occur, underlining the necessity to carefully analyze and compare this phenomenon when neurologically phenotyping transgenic mice.

2.13.1. Spontaneous Electrocorticographic Recordings from the Somatosensory Cortex (S1/S2) and Spontaneous Intracerebral, Deep EEG Recordings from the Hippocampus

An example of a spontaneous electrocorticographic recording from the somatosensory cortex of a C57Bl/6 mouse is given in Figure 6A. This approach can be used, for example to analyze thalamocortical rhythmicity under various physiological and pathophysiological states, e.g. different stages of consciousness like the awake state or slow-wave sleep. Surface recordings from the motor cortex can be very helpful in characterizing motor seizures. The primary and secondary motor cortex (M1/M2) is a good region to start with when observing tonic, clonic or tonic-clonic phenomena, or repetitive, stereotyped behavioral exacerbations in transgenic mouse models. If only part of the fore- or hind-legs, head, or truck is affected, it is likely to be a partial seizure. However, if the whole body is involved in seizure-like activity, it is likely to be a primary or secondarily generalized tonic-clonic seizure. As the number of EEG channels available in EEG radiotelemetry is still limited in mice (maximum of two), its necessary to test various surface und deep implantation regions to get an overall view of the seizure initiation sites and potential seizure propagation pathways.

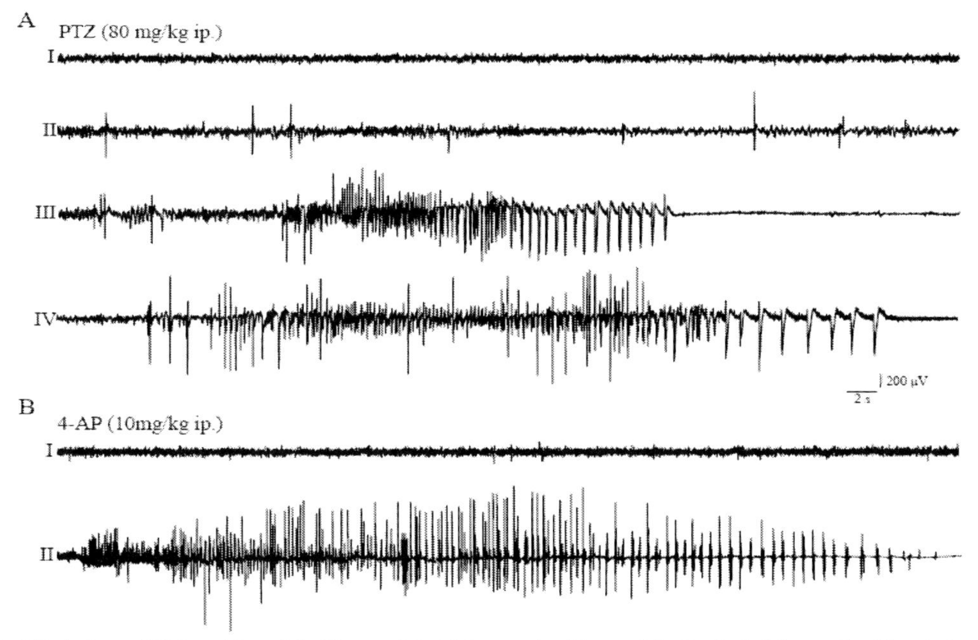

Modified from Weiergräber M, Henry M, Krieger A, Kamp M, Radhakrishnan K, Hescheler J, Schneider T, Altered seizure susceptibility in mice lacking the $Ca_v2.3$ E-type Ca^{2+} channel. Epilepsia, 2006 May;47(5):839-50, Wiley Blackwell.

Figure 7. Electrocorticographic (ECoG) characteristics following administration of pentylenetetrazol (PTZ, 80 mg/kg ip., A_I-$_{IV}$) and 4-aminopyridine (4-AP, 10 mg/kg ip., $B_{I,II}$) in C57Bl/6 mice. No epileptiform discharges are recorded during the hypoactive state (A_I). Stage 2 is characterized by repetitive occurrence of single spikes (A_{II}) that correlate behaviorally with whole-body twitchings as has been confirmed by simultaneous video EEG-EMG monitoring (not shown). Finally, spiking gains frequency evolving into a generalized myoclonus (stage 3, A_{III}) characterized by continuous trains of spikes and / or spike-waves. If seizure activity exhibits maximal spread, a generalized tonic-clonic event occurs (stage 4, A_{IV}), most likely causing death due to respiratory insufficiency. Although PTZ- and 4-AP-treated mice display similar seizure stages, the ECoG does not show prominent spike activity in stage 1-3 (B_I), but only during maximal generalized clonic-tonic seizure stages (stage 4, B_{II}).

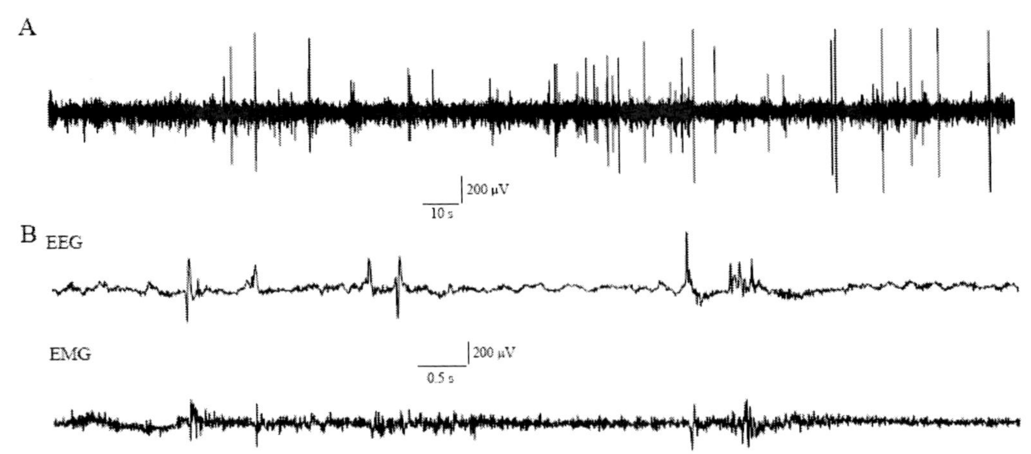

Modified from Weiergräber M, Henry M, Krieger A, Kamp M, Radhakrishnan K, Hescheler J, Schneider T, Altered seizure susceptibility in mice lacking the $Ca_v2.3$ E-type Ca^{2+} channel. Epilepsia, 2006 May;47(5):839-50, Wiley Blackwell.

Figure 8. Interictal spike activity in electrocorticograms from C57Bl/6 mice after PTZ (80 mg/kg ip.) administration. A) ECoG from the interictal stage. B) Simultaneous video-EEG-EMG monitoring using a two channel biopotential radiotransmitter (F20EET, DSI) demonstrates that single-spike activity in surface EEG recordings (upper trace) correlates with whole-body twitching (see EMG, lower trace).

Recordings from the CA3 hippocampal region of a C57Bl/6 mouse are depicted as an example of deep, intracerebral EEG recordings (Figure 6B,C). The hippocampus is of special interest in investigating limbic seizures, but also in theta-wave activity related to exploration and memory formation.

2.13.2. Electrocorticographic Recordings from the Somatosensory Cortex during Generalized Tonic-clonic Seizures

Control recordings of at least one hour were performed prior to injection. Systemic administration of pentylenetetrazole (PTZ, e.g. (80 mg/kg ip.) or 4-aminopyridine (4-AP, e.g. 10 mg/kg ip.) in mice and rats provokes generalized tonic-clonic seizures (Figure 7). The animals develop a typical progressive sequence of seizure activity of increasing severity. Shortly after the injection, mice are hypoactive (Figure $7A_I$), followed by a mild (partial) myoclonus (displaying clonic seizure activity) affecting the face (vibrissal twitching), the head and / or forelimbs (Figure $7A_{II}$). This state evolves into a generalized clonus characterized by loss of upright posture (Figure $7A_{III}$), whole body clonus involving all four limbs, jumping, wild running and finally, a tonic extension of the hindlimbs (Figure $7A_{IV}$).

Figure 7B illustrates a typical recording after 4-AP administration (10 mg/kg ip.). At the early stages of seizure development (myoclonus of the head, face and forelimbs) the EEG exhibits only marginal EMG contamination (Figure $7B_I$). Following sporadic spike activity the generalized clonus starts with a characteristic deflection and a subsequent episode of continuous spike activity (Figure $7B_{II}$). Although this period is characterized by massive muscle activity due to whole body clonus, spike activity of the brain can clearly be determined and EMG contamination is unexpectedly low, indicating that the implantation procedure was capable of picking up electroencephalographic signals selectively even under generalized seizure conditions, when EEG signals might expected to be masked by EMG artifacts. After a generalized clonus we often observed an EEG (postictal) depression before initiation of the next clonus. Interestingly, implantable EEG radiotelemetry can also unmask isolated spike activity during the interictal stage (Figure 8).

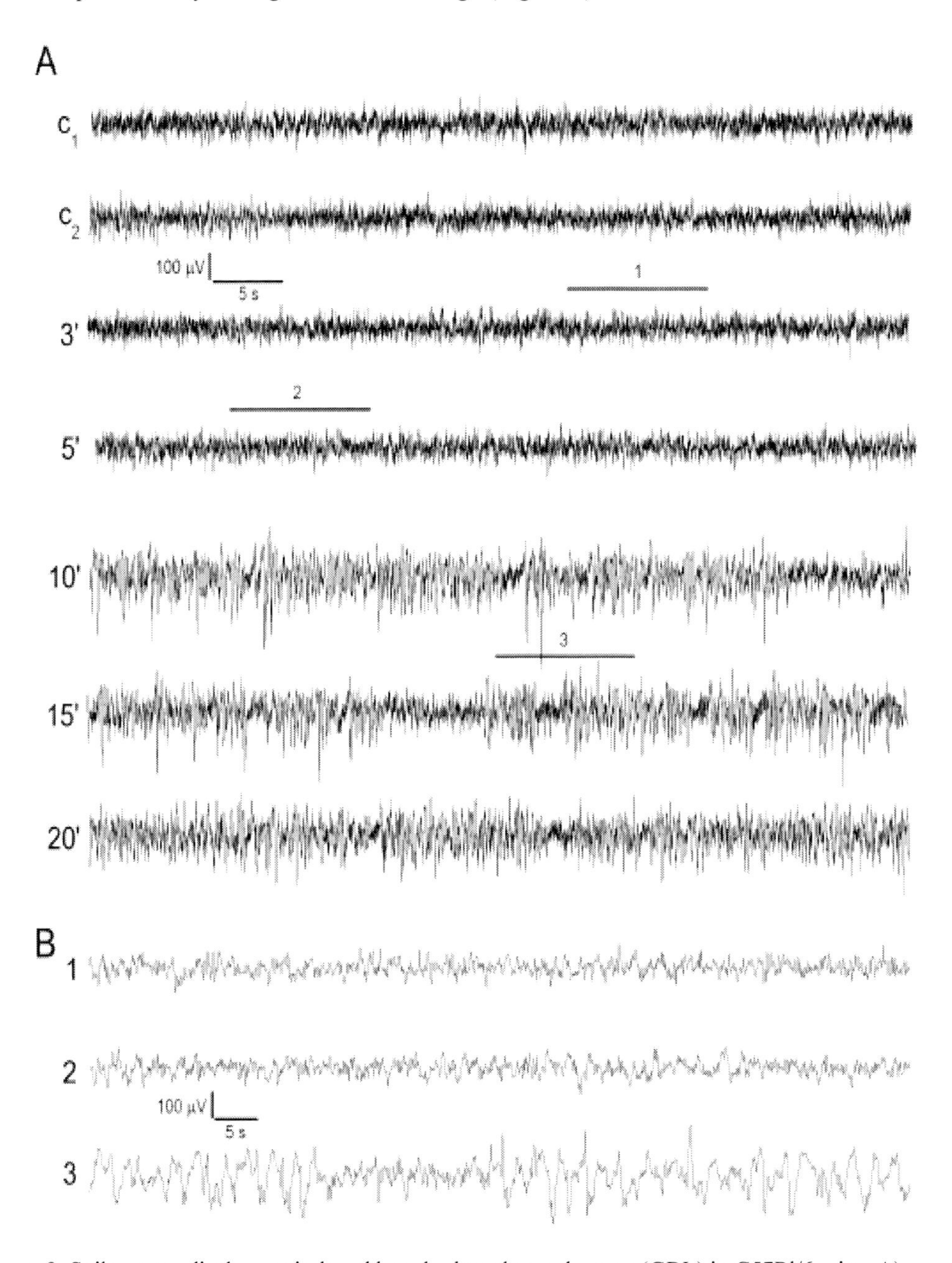

Figure 9. Spike-wave discharges induced by γ-hydroxybutyrolactone (GBL) in C57Bl/6 mice. A) Representative 1 min radiotelemetric ECoG recordings before (c1,c2) and after administration of GBL (70 mg/kg ip.) at different time points as indicated (3', 5', 10', 15' and 20'). Horizontal bars (labelled 1 – 3) represent EEG segments that are displayed with an expanded time scale in B1-3. Electrodes are positioned on the somatosensory cortex (S1/S2) and the EEG exhibits typical bilateral synchronous SWDs based on thalamocortical hyperoscillation.

2.13.3. Electrocorticographic Recordings from the Somatosensory Cortex during Absence Seizure Activity

Pharmacological injection experiments can be used to provoke various seizure types. In the following sections, we illustrate how to provoke and test for absence seizures in small rodents as an example. Absence epilepsy belongs to the group of *petit mal* seizures, that are generalized and thus associated with loss of consciousness. Behaviorally, they are often characterized by hypolocomotion or cessation of activity. However, there are absence seizure subtypes that are also associated with stereotyped, perseverative movement or vegetative phenomena (Manning et al., 2003). Childhood absence epilepsy (pycnolepsy) in humans is a very pure form of absence seizures with little to no motoric components and characterized electroencephalographically by highly organized, bilaterally synchronized, 3 Hz spike-wave discharges (SWD), that are actually pathognomonic for this type of seizure (Khosravani and Zamponi, 2006). The absence specific SWD frequency is species-specific. In cats, the mean frequency is 4.5 Hz, in GAERS it varies from 7 to 11 Hz (Danober et al., 1998) and in mice between 3 - 5 Hz (Kim et al., 2001). In contrast to the highly organized SWD pattern in human childhood absence epilepsy, the spontaneous or pharmacologically induced SWDs in rodents are often less regular (Figure 9). Pathophysiologically, SWD activity results from hyperoscillation within the thalamocortical system, a neuronal circuitry that is also involved in slow-wave sleep.

Spike-wave discharge activity originating from the thalamocortical-corticothalamic circuitry can be pharmacologically induced by γ-hydroxybutyrolactone (GBL, 70 mg/kg), R/S-baclofen (20 mg/kg) or bicucullinemethobromide (BMB) at 10 mg/kg. Both substances, baclofen (a $GABA_B$ receptor agonist) and bicuculline (a $GABA_A$ receptor antagonist), are capable of provoking spikes / spike-waves in the ECoG based on hyperoscillation within the thalamocortical circuitry. GBL generally serves as a weak $GABA_B$ receptor agonist and exerts strong agonistic effects on the newly characterized GBL receptors (Andriamampandry et al., 2007). It is interesting to note that both GBL and baclofen provoke SWDs typical of absence seizures predominantly on the thalamic level, whereas bicuculline induced SWDs seem to originate from the cortex. Figure 9 illustrates exemplary bipolar surface EEG recordings after the i.p. administration of GBL (70 mg/kg ip.).

2.13.4. Deep Intrahippocampal Recordings (CA3, CA1) under Control Conditions and following Kainic Acid (KA) Injection

The hippocampus is one of the most intensively studied areas of the brains. It is functionally relevant as an interface between the short-term and long-term memory and thus is essential for the consolidation of memory engrams. Hyperexcitability of the hippocampus leading to hippocampal seizures are an intensely studied field in epilepsy research as about 40 % of seizures in humans are temporal lobe seizures which are strongly related to hippocampal dysfunction. Hippocampal seizures are complex partial seizures characterized by stereotyped, automatistic behaviour, affecting the face, neck and forelimbs. In rodents, perseverative chewing and eye blinking can sometimes be hard to determine. However, excessive, bizarre grooming, scratching and forelimb clonus might be the first indication of hippocampal seizure activity. To evaluate whether the seizure activity is originating from or involving the hippocampus, deep EEG recordings, e.g. from CA1, CA3 or the dentate gyrus can be performed.

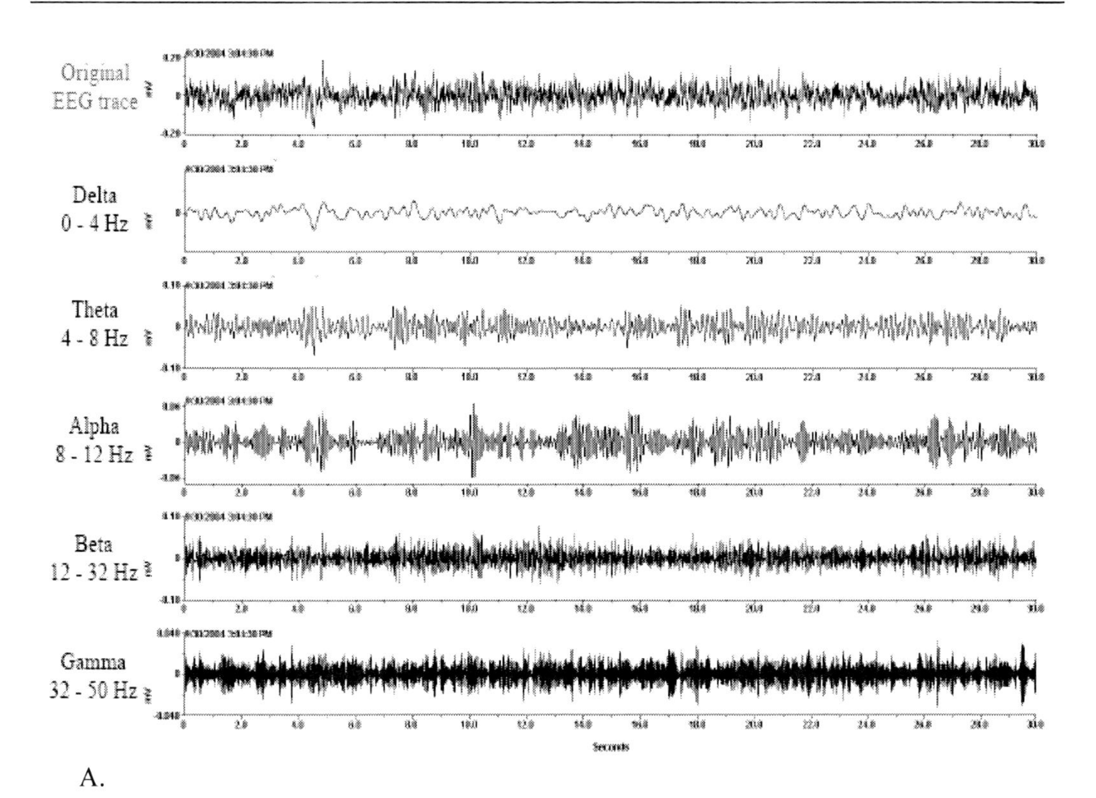
Original
EEG trace
Delta
0 - 4 Hz
Theta
4 - 8 Hz
Alpha
8 - 12 Hz
Beta
12 - 32 Hz
Gamma
32 - 50 Hz
Seconds

A.

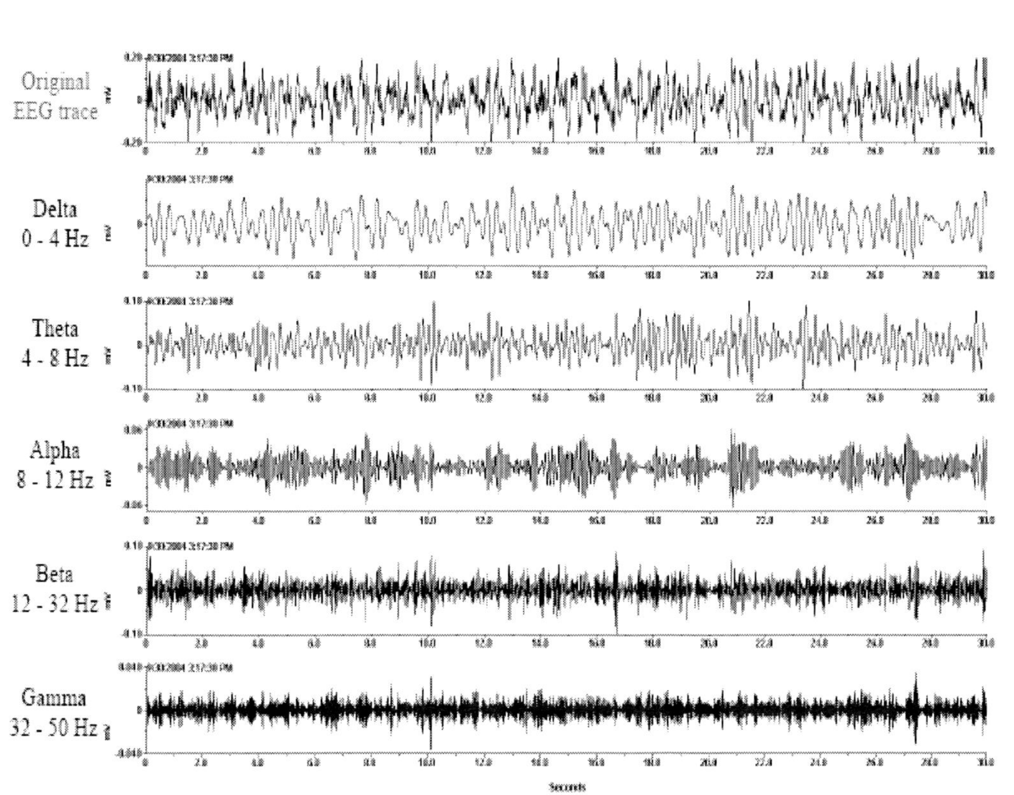
Original
EEG trace
Delta
0 - 4 Hz
Theta
4 - 8 Hz
Alpha
8 - 12 Hz
Beta
12 - 32 Hz
Gamma
32 - 50 Hz
Seconds

B.

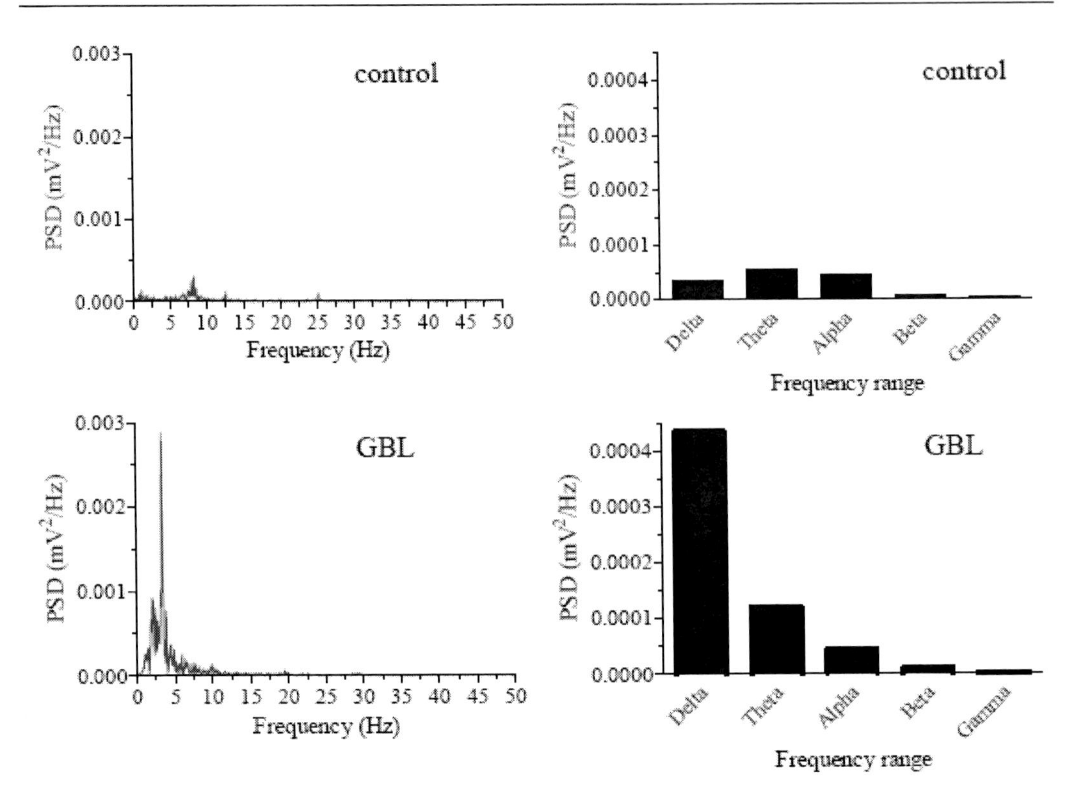

Figure 10. Qualitative and quantitative analysis of EEG recordings. A) The original EEG trace displayed at the top was obtained from a surface EEG (S1/S2) recording of a C57Bl/6 mouse during the awake state. This 30 sec ECoG episode was band-pass filtered for the different frequency ranges, delta, 0 - 4 Hz, theta, 4 - 8 Hz, alpha, 4 - 8 Hz, beta, 12 - 32 Hz and gamma, 32 - 50 Hz. B) The upper trace illustrates a 30 sec ECoG episode after administration of GBL (70 mg/kg ip.) in the same animal. Compared to the previous original trace in A, the EEG pattern has markedly changed: the EEG has gained amplitude and exhibits lower frequency. This initial impression is validated by band-pass filtering, eliciting that delta and theta wave activity is strongly increased, whereas the other frequency entities seem to be unchanged. C) In order to quantify these apparent effects, one can perform a continuous (left) or discontinuous (right) power spectrum density (PSD) analysis that provides quantitative data on the frequency distribution within the original EEG trace. As became obvious by visual inspection of the EEG, GBL injection results in SWD activity, predominantly in the delta and theta wave range. The analysis was carried out using the Dataquest ART™ 4.1 (DSI).

Many transgenic mouse models do not exhibit spontaneous hippocampal seizures, but display differences in hippocampal seizure susceptibility (Weiergräber et al., 2007). Apart from seizures induced by NMDA, the ictal activity induced by kainic acid (KA) is one of the most well-established models of hippocampal seizures. Depending on the KA dosage, animals can exhibit a characteristic sequence of seizure stages (Baran et al., 1994): stage 1, no behavioural change; stage 2, facial clonus; stage 3, forlimb clonus; stage 4, rearing; stage 5, falling; stage 6, status epilepticus; stage 7, jumping, tonic seizures; stage 8, maximum generalized seizure activity, respiratory arrest, and death. Deep recordings from the hippocampus illustrate typical ictal discharges composed of spike, poly-spike and spike-wave activity during hippocampal seizure (Figure 11B,C). However, from the EEG it is hardly impossible to infer a particular seizure stage. As in humans, exacerbating seizure activity is often followed by so called postictal depression, a phenomenon of neuronal silencing after excessive hyperexcitability (Figure 11B). Interestingly, due to KA related excitotoxic effects in the hippocampus, epileptiform activity though slightly declining still persists several days after KA administration (Figure 11D,E). With higher KA dosages administered, animals finally tend to exhibit continuous seizure activity, a hippocampal status epilepticus (Figure 12).

2.13.5. Deep Intrahippocampal Recordings from the CA1 Region: Spontaneous and Pharmacologically Induced Theta Oscillations

As outlined above, the hippocampus plays an intriguing role in the contextual integration of information perceived by an organism (Buzsaki, 2002). This includes the comparison of new environmental information with short- and long-term memory information. The typical extracellular field potential correlated to such activity is known as theta oscillations. Within the hippocampus, theta-wave activity appears strikingly regular, especially when the mouse is engaged in exploration, and during rapid-eye movement (REM) sleep. Theta activity can be provoked by changing the experimental environmental, e.g., just simply placing the mouse into a new cage and the dominant theta activity associated with the exploratory behavior can be confirmed quantitatively by power spectrum density analysis. In addition, theta oscillations can be induced pharmacologically, such as by systemic administration of urethane (800 mg/kg ip.) (Gillies et al., 2002; Sakata, 2006). Figure 13 illustrates representative traces of urethane induced, atropine sensitive theta activity in C57Bl/6 mice. Furthermore, theta oscillations can be induced by sensory stimuli, such as tail pinching (Figure 13).

2.14. Combination of Implantable EEG Radiotelemetry with other Techniques

Implantable EEG radiotelemetry can be effectively combined with various other techniques, e.g. video monitoring. Simultaneous EEG-video monitoring is essential in characterizing seizure-like activity by excluding those EEG graphoelements that mimic ictal discharges (spikes, polyspikes or spike-waves) and in identifying the activity which originates from the brain and not from behavioural movement, such as grooming, scratching, chewing etc. Video recording systems can be either analog and digital ones. For analog approaches, cameras are connected to videotape recorder, e.g. VHS or super VHS and the analog signal can then be converted into digital signals. Direct recordings by digital camera are more preferable as they can be stored in different formats of different resolutions. In most application, black and white recordings seem to be sufficient.

In principal, it is possible to combine radiotelemetric EEG recordings with other type of telemetric approach, such as ECG or EMG. A combined surface EEG recording and EMG monitoring from the trapezoid muscle together with video monitoring is an ideal approach to perform sleep analysis in small rodents. If it is necessary to acutely apply drugs to the brain or specific brain regions, EEG radiotelemetry is also compatible with microdialysis procedures (Vogel et al., 2002). For intracerebral or intracerebroventricular long-term administration of drugs, the use of osmotic pumps by brain infusion kits is recommended. Furthermore, radiotelemetry in small rodents can also be effectively combined with other imaging techniques such as X-ray, computer tomography (CT), magnetic resonance imaging (MRT) and positron emission computer tomography (PET) (Antier et al., 1998).

2.15. Electromyographic and Electrocardiographic Contamination of the EEG

The implantation techniques described above provide a reliable way of recording EEG with a high signal-to-noise ratio in both mice and rats. However, as the EEG signal is weak both electrocardiographic (Figure 14B) and electromyographic contamination (Figure 14C) can evolve into severe problems. Contamination often occurs when the sensing lead is improperly secured or not isolated from extracranial tissues by glas ionomer cement, or when the silicone insulation of the sensing leads has been damaged during the implantation procedure.

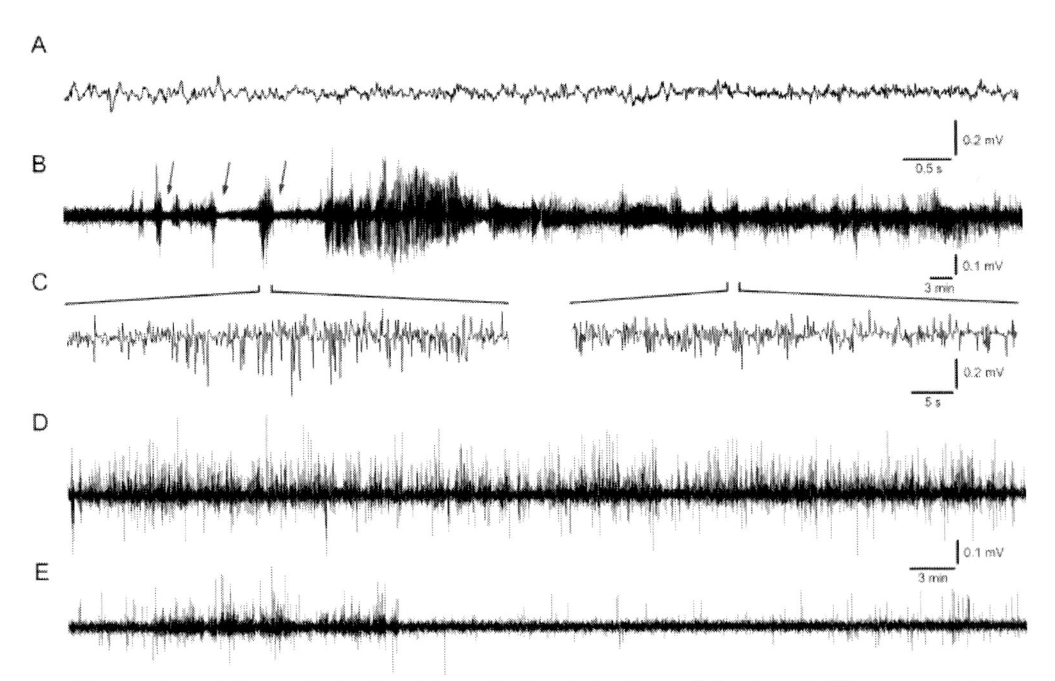

from Weiergräber M, Henry M, Radhakrishnan K, Hescheler J, and Schneider T: Hippocampal Seizure Resistance and Reduced Neuronal Excitotoxicity in Mice Lacking the $Ca_v2.3$ E/R-Type Voltage-Gated Calcium Channel; J Neurophysiol 97: 3660-3669, 2007; modified with permission from The American Physiological Society (APS).

Figure 11. Intrahippocampal EEG recordings after intraperitoneal administration of kanic acid (30 mg/kg). A) Representative spontaneous EEG traces (10 s) obtained from the CA1 region of the hippocampus from a C57Bl/6 mouse before KA injection. B) Deep intrahippocampal CA1 recording from a C57Bl/6 mouse for 2 h immediately after KA administration. Exacerbating hippocampal seizure activity is occasionally interrupted by postictal depression (red arrows). As displayed in the insets below (C), ictal discharges are characterized by spike and / or spike-wave activity in the delta and theta frequency range. At day 1 (D) and day 5 (E) post injection 1-h CA1 EEG recordings exhibit slightly declining but still continuous ictiform activity related to excitotoxic neurodegeneration.

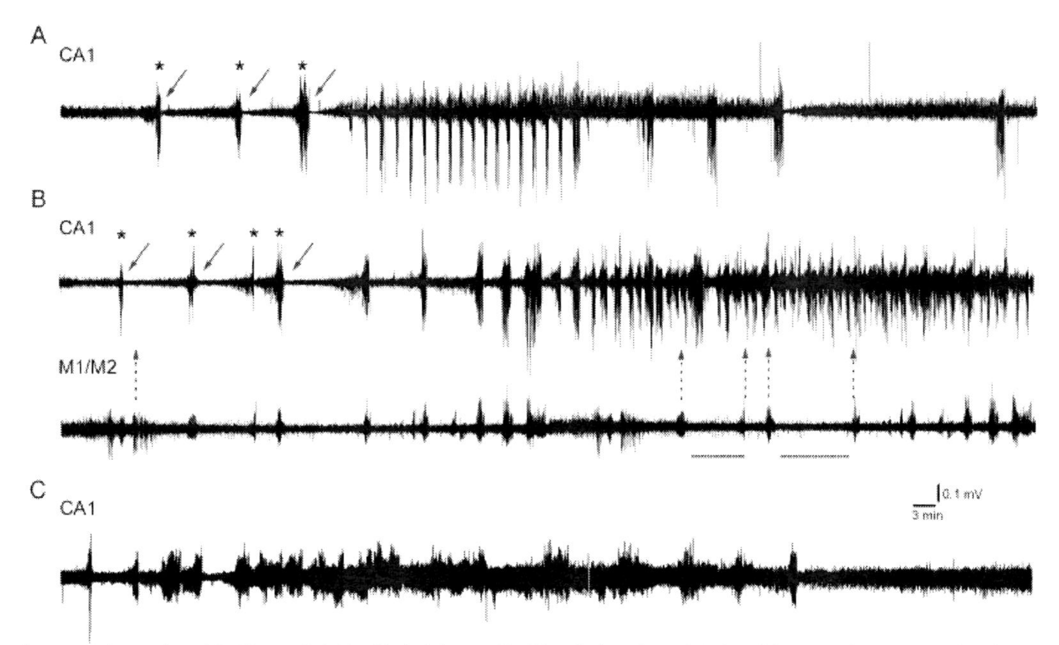

from Weiergräber M, Henry M, Radhakrishnan K, Hescheler J, and Schneider T: Hippocampal Seizure Resistance and Reduced Neuronal Excitotoxicity in Mice Lacking the $Ca_v2.3$ E/R-Type Voltage-Gated Calcium Channel; J Neurophysiol 97: 3660-3669, 2007; modified with permission from The American Physiological Society (APS).

Figure 12. Intrahippocampal CA1 EEG recordings after administration of KA (30 mg/kg A, B and 10 mg/kg, C). A) Deep CA1 EEG recordings from C57Bl/6 mice for 2 h. In some animals seizure activity is predominantly characterized by repetitive and delimitable high-amplitude exacerbation of spike and/or spike-wave activity (*) usually followed by postictal depression (red arrows), whereas in others this activity merges resulting in continuous status-like seizure activity. B) Simultaneous deep CA1 and surface EEG recordings from the motor cortex (M1/M2) reveal a partially overlapping but also distinct pattern of neuronal hyperexcitability in the different brain regions (see dashed red arrows and dashed lines), pointing out the regional specificity of the EEG recordings. C) Even 10 mg/kg KA turned out to be far beyond hippocampal seizure threshold in these mice.

Six to eight weeks after lead implantation for EEG recording, we occasionally observed proliferation of connective tissue as well as ossification in the burr holes. These tend to disconnect the electrodes from the underlying dura mater or lift up the deeply implanted electrodes out of the brain, resulting in both EMG and ECG contamination. The latter one can easily identified by its typical regular R-spike pattern of 7 – 9 Hz (according to an averaged heart rate of 420 – 540 bpm in Figure 14B) whereas EMG contamination often exhibits higher frequencies and augments if the animal is active.

2.16. Data Acquisition and Analysis

Special attention has to paid to the digitizing frequency embedded in the EEG recordings. The radiotelemetric EEG recordings are digitized EEG signals converted from the original analog EEG signal by an A/D converter. Current commercially available radiotelemetry systems allow the sampling frequency adjusted to a maximum of, e.g. 1000 Hz. For many experimental applications, digitizing frequencies of up to 300 Hz are sufficient but analysis of gamma oscillations above 200 Hz may require higher digitizing frequency. However, the higher the sampling frequency, the larger the storage size is required, and in many cases also the higher susceptibility to environmental noise.

3. Critical Appraisal of EEG Radiotelemtry

Though the EEG radiotelemetry has its high potential in various aspects, it is highly essential to beware of its potential pitfalls. In the following sections, we will discuss in detail of all these aspects.

The body weight, circadian rhythmicity and motor activity of animals are important parameters for consideration. These parameters from radiotransmitter implanted animals should not differ significantly from the non-implanted control groups. The regain in body weight and motor activity indicate general recovery from implantation. Such recovery minimally takes 10-14 days. Understandably, the competitive scientific society demands for rapider data acquisition, analysis and prompt publication. However, data acquisition obtained at day 1 or a few days following radiotransmitter implantation (Leon et al., 2004) could be problematic. When this is not being aware, the results are often highly questionable, even in those well-published articles on mouse models of human diseases.

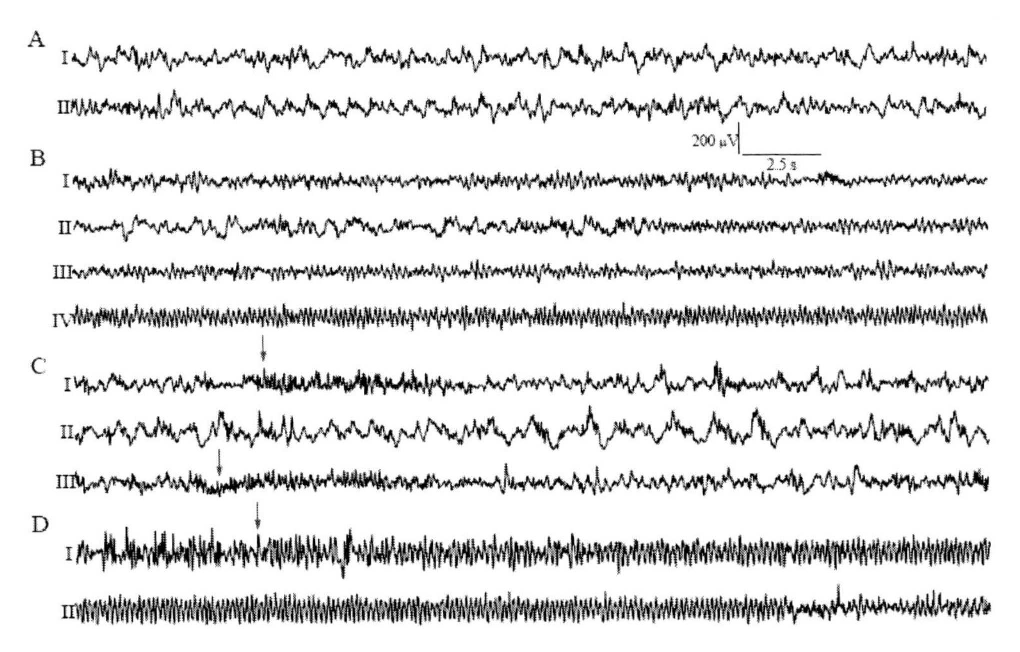

Figure 13. Deep intrahippocampal EEG recordings from the CA1 region in control mice following administration of urethane (800 mg/kg ip.) and atropine (50 mg/kg ip.). A) After injection of urethane, control mice exhibit typical large-amplitude irregular activity (LIA, $A_{I,II}$) interrupted by spontaneous fragmented or continuous segments of highly synchronized urethane-induced theta episodes (B_{I-IV}). The urethane induced synchronized activity can still be seen shortly after atropine administration (C_I), but is totally abolished not later than one hour after atropine injection with predominance of LIA activity (C_{II}). With atropine being metabolized, synchronized theta activity is slowly reoccurring (C_{III}). D) Following urethane injection (800 mg/kg ip.) sensory stimuli, such as tail pinching (see arrow) can induce robust theta oscillation in the CA1 hippocampal EEG recordings.

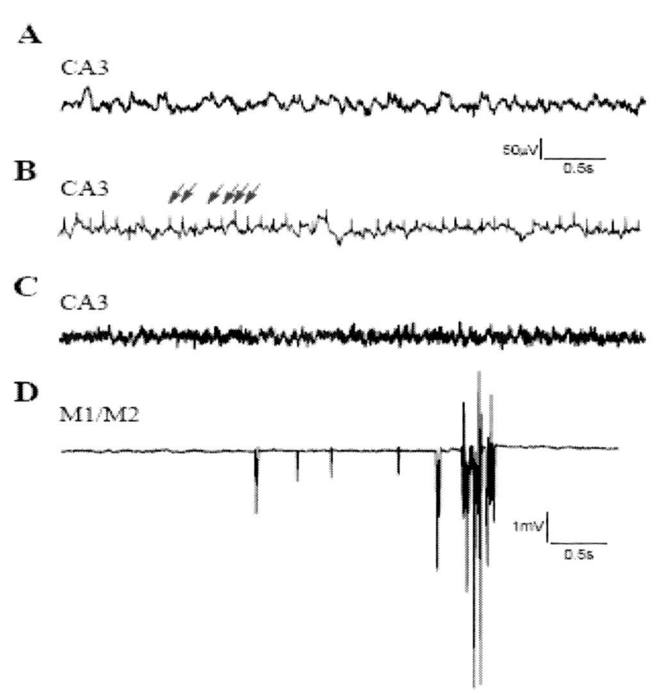

Reprinted from Brain Research Protocols, 14, Weiergräber M, Henry M, Hescheler J, Smyth N, Schneider T, Electrocorticographic and deep intracerebral EEG recording in mice using a telemetry system, p154-164, 2005, with permission from Elsevier.

Figure 14. EEG artifacts. Though deeply implanted, the strength of brain related extracellular field potentials is low compared to ECG and EMG signals. A representative intrahippocampal (CA3) recording from a C57Bl/6 mouse is displayed in A. Lesions of the silicon insulation as well as ossification around the burred holes tearing out the electrodes can result in ECG and EMG contamination of the EEG (B). As ECG R-spike contamination is often most regular and within the range of normal mouse heart rate (see arrows), it is easy to identify. The implantation procedure described in this chapter was proven to be effective in minimizing potential ECG related artifacts. Generally, EMG-contamination can have similar sources (C). Performing simultaneous video-EEG monitoring, the experimenter should pay specific attention to whether high-frequency, EMG related EEG activity coincides with locomotion, eating, scratching or grooming, as this can be misinterpreted as seizure-like activity. Apart from biopotentials themselves, another source of artifacts is external electrical noise picked up from the room lights or electrical devices close to the receiver as well as crosstalk between different receiver plates (D). These artifacts are easy to identify, as the morphology of such events is often unrelated to biological electrical phenomena and the amplitude is often unphysiologically high. Experimentally, the researcher should shield the receiver and home cage using a Faraday cage.

The rate of recovery depends on many physiological factors. The surgical procedure alone definitely affect the time for recovery. Butz and Davisson (2001) compared the subcutaneous pouch with the intraperitoneal implantation. They found the former route could result in faster recovery. As discussed earlier, any type of surgical intervention will cause a number of well-documented physiological alterations, affecting cardiovascular parameters (heart rate, blood pressure), respiration, sleep alterations and other CNS related effects, changes in body temperature, immunological reactions, etc. Laparotomy, for example, has been shown to cause a decrease in body weight and the ability to reach water bottle immediately after recovery from anesthetics is hindered by the presence of the transmitter and the pain associated with the surgical wound. Another factor influencing the body weight recovery is the anesthesia which largely affects the basal metabolism of mice, resulting in hypothermia and subsequent increased metabolism can lead to the loss of body weight. This is particularly relevant to mice due to their high surface area to body mass ratio and therefore they need to receive temperature support after implantation. In addition, any pre-, peri- or post-operative handling and restraint applied on the animals (mice or rats) can increase their blood pressure, heart rate, body and rectal temperature, and can cause elevated levels of norepinephrine and epinephrine together with decreased oxygen consumption and hence affecting the metabolism. These adverse effects are most critical in mice. In rats the adverse effects due to implantation are relatively transient, probably due to the higher body to transmitter mass ratio. Circadian rhythms in mice were shown to recover faster than body weight, whereas in rats the opposite was observed/found (Leon et al., 2004).

All the above documented changes after the implantation clearly indicate that recordings shortly after implantation do not represent the normal physiological state and strictly speaking they are prone to generate artifacts. More critical and standardized protocol has to be established and enforced in this research area in the future. Finally, EEG radiotelemetry data should not stand on its own, they should be combined with data obtained from less complex approaches, such as brain slice technology, for better assessment and evaluation.

4. Validation of Radiotelemetry and Good Radiotelemetric Practice

Validation of a telemetry system includes calibration and verification of the sensor functions so that the sensor signal is accurately received and collected within the data acquisition system. When multiple animals are recorded simultaneously, verification of the accuracy and integrity of each data channel has to be included. If commercially available telemetry systems are used, information on validation of transmitter function and data acquisition and analysis systems can normally be obtained from the manufacturer's manual.

It is necessary to validate telemetry transmitter signals for the accuracy over the anticipated physiological range and for the stability over the duration of intended use. One way of validating transmitters is to simultaneously measure in a single animal or in separate groups of animals and compare the telemetry signal to that obtained from one or more conventional techniques. When comparing telemetry and conventional techniques, small differences in the absolute values can also be anticipated because of differences in lead placement. Short-acting pharmacological agents and manipulations can also be used to induce acute changes in physiological parameters to test dynamic ranges. Kramer et al. (2003) have described a nice 5-step protocols for such type of internal validation. In contrast to commercially available telemetry systems, self-made approaches definitively require even stricter internal validation procedures of transmitter and data acquisition and analysis systems. In the first line, validation of EEG biopotentials is a qualitative issue and less a quantitative one, as the EEG predominantly focuses on the lege-artis morphology of waveforms, i.e. physiological or pathological ones, such as ictal-like discharges (spikes, polyspikes or spike-waves). Thus, delimitation of EMG or ECG artifacts that can mimic normal or abnormal EEG graphoelements is of central importance and further relevance also in quantitative EEG analysis.

5. Conclusion

Current biomedical research is strongly founded on transgenic technology which demands for sophisticated and reliable techniques in order to analyse animal models of human diseases. Implantable radiotelemetry represents the state-of-the art approach that conforms these requirements. We have demonstrated that EEG radiotelemetry is of particular relevance in investigating brain related disorders, such as epilepsies, sleep disturbances, analysis of neurotoxicity and neuropsychiatric disorders. Special attention has to be paid to the planning of biotelemetry studies, the transmitter implantation itself, postoperative treatment of the animals but also potential pitfalls. Unlike other approaches, radiotelemetry most strongly demands for the integrative (patho)physiological capabilities of the experimenter and if applied adequately, radiotelemetry will evolve more and more as a precious tool in animal research.

6. Acknowledgements

The authors are deeply grateful to Prof. Dr. Edward H. Bertram III. (Department of Neurology, University of Virginia, Charlottesville) for his helpful and fruitful discussions during the establishment of EEG radiotelemetry in our lab. This work was financially supported by the Center of Molecular Medicine Cologne (CMMC, to MW, JH and TS), Faculty of Medicine, University of Cologne. The authors further declare that there is no conflict of interests.

References

Andriamampandry, C, Taleb, O, Kemmel, V, Humbert, JP, Aunis, D, Maitre, M. Cloning and functional characterization of a gamma-hydroxybutyrate receptor identified in the human brain. *FASEB Journal* 2007 21, 885-895.

Antier, D, Zhang, BL, Mailliet, F, Akoka, S, Pourcelot, L, Sannajust, F. Effects of neonatal focal cerebral hypoxia-ischemia on sleep-waking pattern, ECoG power spectra and locomotor activity in the adult rat. *Brain Res.* 1998 807, 29-37.

Baran, H, Vass, K, Lassmann, H, Hornykiewicz, O. The cyclooxygenase and lipoxygenase inhibitor BW755C protects rats against kainic acid-induced seizures and neurotoxicity. *Brain Res.* 1994 646, 201-206.

Bastlund, JF, Jennum, P, Mohapel, P, Vogel, V, Watson, WP. Measurement of cortical and hippocampal epileptiform activity in freely moving rats by means of implantable radiotelemetry. *J. Neurosci. Methods* 2004 138, 65-72.

Bennet, MK, Scheller, RH. The molecular machinery for secretion is conserved from yeast to neurons. *Proceedings Of The National Academy Of Sciences Of The United States Of America* 1993 90, 2559-2563.

Bertram, EH. Functional Anatomy of Spontaneous Seizures in a Rat Model of Limbic Epilepsy. *Epilepsia* 1997a 38, 95-105.

Bertram, EH. Functional anatomy of spontaneous seizures in a rat model of limbic epilepsy. *Epilepsia* 1997b 38, 95-105.

Bertram, EH, Cornett, JF. The evolution of a rat model of chronic spontaneous limbic seizures. *Brain Res.* 1994 661, 157-162.

Bertram, EH, Lothman, EW. Ambulatory EEG cassette recorders for prolonged electroencephalographic monitoring in animals. *Electroencephalogr. Clin. Neurophysiol.* 1991 79, 510-512.

Bertram, EH, Williamson, JM, Cornett, JF, Spradlin, S, Chen, ZF. Design and construction of a long-term continuous video-EEG monitoring unit for simultaneous recording of multiple small animals. *Brain Res. Brain Res. Protoc.* 1997 2, 85-97.

Bertram, EH. Monitoring for seizures in rodents. In: Pitkänen A, Schwartzkroin P, Moshe S. *Models of seizures and epilepsy*. Academic Press; 2006; 569-582.

Boutros, NN. A review of indications for routine EEG in clinical psychiatry. *Hosp. Community Psychiatry* 1992 43, 716-719.

Brookes, ZL, Reilly, CS, Brown, NJ. Differential effects of propofol, ketamine, and thiopental anaesthesia on the skeletal muscle microcirculation of normotensive and hypertensive rats in vivo. *Br. J. Anaesth.* 2004 93, 249-256.

Butz, GM, Davisson, RL. Long-term telemetric measurement of cardiovascular parameters in awake mice: a physiological genomics tool. *Physiol Genomics* 2001 5, 89-97.

Buzsaki, G. Theta oscillations in the hippocampus. *Neuron* 2002 33, 325-340.

Colby, LA, Morenko, BJ. Clinical considerations in rodent bioimaging. *Comp. Med.* 2004 54, 623-630.

Conn, PM. *Sourcebook of Models for Biomedical Research.* 1st edition. Humana Press, 2008.

Cotugno, M, Mandile, P, D'Angiolillo, D, Montagnese, P, Giuditta, A. Implantation of an EEG telemetric transmitter in the rat. *Ital. J. Neurol. Sci.* 1996 17, 131-134.

Danober, L, Deransart, C, Depaulis, A, Vergnes, M, Marescaux, C. Pathophysiological mechanisms of genetic absence epilepsy in the rat. *Prog. Neurobiol.* 1998 55, 27-57.

Davis, JA. Mouse and rat anesthesia and analgesia. *Curr. Protoc. Neurosci.* 2008 Appendix 4, Appendix.

Flecknell, PA. *Laboratory Animal Anaesthesia.* 2nd edition. Academic Press, 1996.

Forsgren, I, Beghi, E, Ekman, M. Cost of epilepsy in Europe. *Eur. J. Neurol.* 2005a 12 Suppl 1, 54-58.

Forsgren, L, Beghi, E, Oun, A, Sillanpaa, M. The epidemiology of epilepsy in Europe - a systematic review. *Eur. J. Neurol.* 2005b 12, 245-253.

Forsgren, L, Hauser, WA, Olafsson, E, Sander, JW, Sillanpaa, M, Tomson, T. Mortality of epilepsy in developed countries: a review. *Epilepsia* 2005c 46 Suppl 11, 18-27.

Franklin KBJ and Paxinos G. *The Mouse Brain in Stereotaxic Coordinates.* 3rd edition. Academic Press, 2008.

Gehrmann, J, Hammer, PE, Maguire, CT, Wakimoto, H, Triedman, JK, Berul, CI. Phenotypic screening for heart rate variability in the mouse. *Am. J. Physiol Heart Circ. Physiol* 2000 279, H733-H740.

Gillies, MJ, Traub, RD, LeBeau, FE, Davies, CH, Gloveli, T, Buhl, EH, Whittington, MA. A model of atropine-resistant theta oscillations in rat hippocampal area CA1. *J. Physiol.* 2002 543, 779-793.

Gioanni, Y, Goyon, D, Prevost, J. Intracerebroventricular dermorphin, but not dermenkephalin, is epileptogenic in the rat. *Neuroreport* 1991 2, 49-52.

Grohrock, P, Hausler, U, Jurgens, U. Dual-channel telemetry system for recording vocalization-correlated neuronal activity in freely moving squirrel monkeys. *J. Neurosci. Methods* 1997 76, 7-13.

Hanusch, C, Hoeger, S, Beck, GC. Anaesthesia of small rodents during magnetic resonance imaging. *Methods* 2007 43, 68-78.

Hedrich, H. *The Laboratory Mouse.* 1st edition. London, San Diego: Academic Press; 2004.

Herzog, W, Stano, A, Leonard, TR. Telemetry system to record force and EMG from cat ankle extensor and tibialis anterior muscles. *J. Biomech.* 1993 26, 1463-1471.

Hoffmann, K, Lindner, M, Groticke, I, Stangel, M, Loscher, W. Epileptic seizures and hippocampal damage after cuprizone-induced demyelination in C57BL/6 mice. *Exp. Neurol.* 2008 210, 308-321.

Hunter, AJ, Nolan, PM, Brown, SD. Towards new models of disease and physiology in the neurosciences: the role of induced and naturally occurring mutations. *Hum. Mol. Genet.* 2000 9, 893-900.

Ishida, N, Kasamo, K, Nakamoto, Y, Suzuki, J. Epileptic seizure of El mouse initiates at the parietal cortex: depth EEG observation in freely moving condition using buffer amplifier. *Brain Res.* 1993 608, 52-57.

Kaplan, BJ, Seyfried, TN, Glaser, GH. Spontaneous polyspike discharges in an epileptic mutant mouse (tottering). *Exp. Neurol.* 1979 66, 577-586.

Khosravani, H, Zamponi, GW. Voltage-gated calcium channels and idiopathic generalized epilepsies. *Physiol. Rev.* 2006 86, 941-966.

Kim, D, Song, I, Keum, S, Lee, T, Jeong, MJ, Kim, SS, McEnery, MW, Shin, HS. Lack of the burst firing of thalamocortical relay neurons and resistance to absence seizures in mice lacking alpha(1G) T-type Ca^{2+} channels. *Neuron* 2001 31, 35-45.

Ko, WH, Neuman, MR. Implant biotelemetry and microelectronics. *Science* 1967 156, 351-360.

Kramer, K, Kinter, L, Brockway, BP, Voss, HP, Remie, R, Van Zutphen, BL. The use of radiotelemetry in small laboratory animals: recent advances. *Contemp. Top. Lab. Anim. Sci.* 2001b 40, 8-16.

Kramer, K, Kinter, L, Brockway, BP, Voss, HP, Remie, R, Van Zutphen, BL. The use of radiotelemetry in small laboratory animals: recent advances. *Contemp. Top. Lab. Anim. Sci.* 2001a 40, 8-16.

Kramer, K, Kinter, LB. Evaluation and applications of radiotelemetry in small laboratory animals. *Physiol. Genomics* 2003 13, 197-205.

Kramer, K, van Acker, SA, Voss, HP, Grimbergen, JA, van der Vijgh, WJ, Bast, A. Use of telemetry to record electrocardiogram and heart rate in freely moving mice. *J. Pharmacol. Toxicol. Methods* 1993 30, 209-215.

Kramer, K, Voss, HP, Grimbergen, JA, Mills, PA, Huetteman, D, Zwiers, L, Brockway, B. Telemetric monitoring of blood pressure in freely moving mice: a preliminary study. *Lab. Anim.* 2000 34, 272-280.

Krauss, GL, Kaplan, P, Fisher, RS. Parenteral magnesium sulfate fails to control electroshock and pentylenetetrazol seizures in mice. *Epilepsy Res.* 1989 4, 201-206.

Kullmann, DM. Genetics of epilepsy. *J. Neurol. Neurosurg. Psychiatry* 2002 73 Suppl 2, II32-II35.

Kullmann, DM, Hanna, MG. Neurological disorders caused by inherited ion-channel mutations. *Lancet Neurol.* 2002 1, 157-166.

Lapray, D, Bergeler, J, Dupont, E, Thews, O, Luhmann, HJ. A novel miniature telemetric system for recording EEG activity in freely moving rats. *J. Neurosci. Methods* 2008 168, 119-126.

Leon, LR, Walker, LD, DuBose, DA, Stephenson, LA. Biotelemetry transmitter implantation in rodents: impact on growth and circadian rhythms. *Am. J. Physiol. Regul. Integr. Comp Physiol* 2004 286, R967-R974.

Lewanowitsch, T, White, JM, Irvine, RJ. Use of radiotelemetry to evaluate respiratory depression produced by chronic methadone administration. *European Journal of Pharmacology* 2004 484, 303-310.

Manning, JP, Richards, DA, Bowery, NG. Pharmacology of absence epilepsy. *Trends Pharmacol. Sci.* 2003 24, 542-549.

Matthew, CB. Telemetry augments the validity of the rat as a model for heat acclimation. *Ann. N. Y. Acad. Sci.* 1997 813, 233-238.

Mejias-Aponte, CA, Jimenez-Rivera, CA, Segarra, AC. Sex differences in models of temporal lobe epilepsy: role of testosterone. *Brain Res.* 2002 944, 210-218.

Mills, PA, Huetteman, DA, Brockway, BP, Zwiers, LM, Gelsema, AJ, Schwartz, RS, Kramer, K. A new method for measurement of blood pressure, heart rate, and activity in the mouse by radiotelemetry. *J. Appl. Physiol* 2000 88, 1537-1544.

Morton, DB, Hawkins, P, Bevan, R, Heath, K, Kirkwood, J, Pearce, P, Scott, L, Whelan, G, Webb, A. Refinements in telemetry procedures. Seventh report of the BVAAWF/FRAME/RSPCA/UFAW Joint Working Group on Refinement, Part A. *Lab. Anim* 2003 37, 261-299.

Nakano, H, Saito, K, Suzuki, K. Chronic implantation technique for monopolar EEG monitoring of epileptic seizures in mice. *Brain Res. Bull.* 1994 35, 261-268.

Nesathurai, S, Andrew, GW, Edell, DJ, Rosene, DL, Mansfield, K, Sehgal, P, Magill, D, Sledge, JB. Electromyographic telemetry in the development of humane primate model of spinal cord injury. *J. Med. Primatol.* 2006 35, 397-400.

Noebels, JL, Qiao, X, Bronson, RT, Spencer, C, Davisson, MT. Stargazer: a new neurological mutant on chromosome 15 in the mouse with prolonged cortical seizures. *Epilepsy Res.* 1990 7, 129-135.

Noebels, JL, Sidman, RL. Inherited epilepsy: spike-wave and focal motor seizures in the mutant mouse tottering. *Science* 1979 204, 1334-1336.

Ondze, B, Espa, F, Dauvilliers, Y, Billiard, M, Besset, A. Sleep architecture, slow wave activity and sleep spindles in mild sleep disordered breathing. *Clin. Neurophysiol.* 2003 114, 867-874.

Paxinos G and Watson C. *The Rat Brain in Stereotaxic Coordinates.* 5th edition. Academic Press, 2004.

Pitsch, J, Schoch, S, Gueler, N, Flor, PJ, van Der, PH, Becker, AJ. Functional role of mGluR1 and mGluR4 in pilocarpine-induced temporal lobe epilepsy. *Neurobiol. Dis.* 2007 26, 623-633.

Poindron, P., Piguet, P. *New Animal Models of Human Neurological Diseases.* 1st edition. S Karger AG; 2008.

Printz, MP. Radiotelemetry comes of age--perhaps just in time. *Am. J. Physiol Regul. Integr. Comp Physiol* 2004 286, R818-R819.

Richardson, CA, Flecknell, PA. Anaesthesia and post-operative analgesia following experimental surgery in laboratory rodents: are we making progress? *Altern. Lab. Anim.* 2005 33, 119-127.

Rosen, I. Electroencephalography as a diagnostic tool in dementia. *Dement. Geriatr. Cogn Disord.* 1997 8, 110-116.

Sakata, S. Timing and hippocampal theta in animals. *Rev. Neurosci.* 2006 17, 157-162.

Snelderwaard, PC, van, G, V, Witte, F, Voss, HP, Kramer, K. Surgical procedure for implanting a radiotelemetry transmitter to monitor ECG, heart rate and body temperature in small Carassius auratus and Carassius auratus gibelio under laboratory conditions. *Lab. Anim.* 2006 40, 465-468.

Soltysik, S, Jackson, R, Jelen, P. Apparatus for studying behavior and learning in restrained rats. *Acta Neurobiol. Exp. (Wars.)* 1996 56, 697-701.

Spoelstra, EN, Ince, C, Koeman, A, Emons, VM, Brouwer, LA, van Luyn, MJ, Westerink, BH, Remie, R. A novel and simple method for endotracheal intubation of mice. *Lab Anim* 2007 41, 128-135.

Stephens, ML, Conlee, K, Alvino, G, Rowan, AN. Possibilities for refinement and reduction: future improvements within regulatory testing 1. *ILAR. J.* 2002 43 Suppl, S74-S79.

Sundstrom, LE, Sundstrom, KE, Mellanby, JH. A new protocol for the transmission of physiological signals by digital telemetry. *J. Neurosci. Methods* 1997 77, 55-60.

Swoap, SJ, Overton, JM, Garber, G. Effect of ambient temperature on cardiovascular parameters in rats and mice: a comparative approach. *Am. J. Physiol Regul. Integr. Comp Physiol* 2004 287, R391-R396.

Tang, X, Sanford, LD. Telemetric recording of sleep and home cage activity in mice. *Sleep* 2002 25, 691-699.

Terrell, RC. The invention and development of enflurane, isoflurane, sevoflurane, and desflurane. *Anesthesiology* 2008 108, 531-533.

Turnbull, J, Lohi, H, Kearney, JA, Rouleau, GA, Delgado-Escueta, AV, Meisler, MH, Cossette, P, Minassian, BA. Sacred Disease Secrets Revealed: The Genetics of Human Epilepsy. *Hum. Mol. Genet.* 2005.

Valatx, JL. [Study of sleep in C57BR-cd and CBA mice]. *J. Physiol (Paris)* 1970 62 Suppl 2, 325-326.

Valatx, JL. [Genetics of EEG, of certain forms of epilepsy, sleep and learning]. *Rev. Electroencephalogr. Neurophysiol. Clin.* 1975 5, 319-329.

Van de Weerd, HA, Bulthuis, RJ, Bergman, AF, Schlingmann, F, Tolboom, J, Van Loo, PL, Remie, R, Baumans, V, Van Zutphen, LF. Validation of a new system for the automatic registration of behaviour in mice and rats. *Behav. Processes* 2001 53, 11-20.

Vanhatalo, S, Voipio, J, Kaila, K. Full-band EEG (fbEEG): a new standard for clinical electroencephalography. *Clin. EEG. Neurosci.* 2005 36, 311-317.

Vergari, A, Gunnella, B, Rodola, F, Frassanito, L, Musumeci, M, Palazzesi, S, Casalinuovo, IA. A new method of orotracheal intubation in mice. *Eur. Rev. Med. Pharmacol. Sci.* 2004 8, 103-106.

Vogel, V, Sanchez, C, Jennum, P. EEG measurements by means of radiotelemetry after intracerebroventricular (ICV) cannulation in rodents. *J. Neurosci. Methods* 2002 118, 89-96.

Wang, C, Slikker, W, Jr. Strategies and experimental models for evaluating anesthetics: effects on the developing nervous system. *Anesth. Analg.* 2008 106, 1643-1658.

Weiergräber, M, Henry, M, Hescheler, J, Smyth, N, Schneider, T. Electrocorticographic and deep intracerebral EEG recording in mice using a telemetry system. *Brain Res. Brain Res. Protoc.* 2005a 14, 154-164.

Weiergräber, M, Henry, M, Krieger, A, Kamp, MA, Radhakrishnan, K, Hescheler, J, Schneider, T. Altered seizure susceptibility in mice lacking the $Ca_v2.3$ E-type Ca^{2+} channel. *Epilepsia* 2006a 47, 839-850.

Weiergräber, M, Henry, M, Radhakrishnan, K, Hescheler, J, Schneider, T. Hippocampal seizure resistance and reduced neuronal excitotoxicity in mice lacking the $Ca_v2.3$ E/R-type voltage-gated calcium channel. *J. Neurophysiology* 2007 97, 3660-3669.

Weiergräber, M, Henry, M, Südkamp, M, De Vivie, ER, Hescheler, J, Schneider, T. Ablation of $Ca_v2.3$ / E-type voltage-gated calcium channel results in cardiac arrhythmia and altered autonomic control within the murine cardiovascular system. *Basic Res. Cardiol.* 2005b 100, 1-13.

Weiergräber, M, Kamp, MA, Radhakrishnan, K, Hescheler, J, Schneider, T. The $Ca_v2.3$ Voltage-gated calcium channel in epileptogenesis. Shedding new light on an enigmatic channel. *Neurosci. Biobehav. Rev.* 2006b 30, 1122-1144.

Yang, Y, Frankel, WN. Genetic approaches to studying mouse models of human seizure disorders. *Adv. Exp. Med. Biol.* 2004 548, 1-11.

Zhao, X, Wu, N, Zhou, J, Yang, Y, Fang, Y, Cheng, W, Ma, R, Tian, Y, Huang, L. A technique for retrograde intubation in mice. *Lab Anim (NY)* 2006 35, 39-42.

Zornetzer, S, McGaugh, JL. Effects of frontal brain electroshock stimulation on EEG activity and memory in rats: relationship to ECS-produced retrograde amnesia. *J. Neurobiol.* 1970 1, 379-394.

In: Telemetry: Research, Technology and Applications ISBN 978-1-60692-509-6
Editors: Diana Barculo and Julia Daniels

Chapter 4

TELEMETRY: AN ECOLOGIST SWISS-ARMY KNIFE

Filipa Loureiro[1] and Luís Miguel Rosalino[2]
Universidade de Lisboa, Centro de Biologia Ambiental,
Faculdade de Ciências, Ed C2, Campo Grande, 1749-016 Lisboa, Portugal

ABSTRACT

Telemetry is nowadays one of the most important tools used by ecologists allowing a privileged view of animal life while contributing decisively to its effective conservation. From the smallest rodent to the big blue whale, this is a tool that currently can be used for almost any moving vertebrate, from terrestrial to marine or flying species. Since the early 1960s, when this technique began to be applied, numerous improvements have occurred, mostly at a technological level, but most importantly that allowed increasing the amount of information that might be collected when telemetry studies are implemented. In this chapter we will present a brief revision of how this technique has evolved throughout the years and its numerous applications. By providing the geographical location of animals, and sometimes concurrently allowing the registration of other parameters (e.g., activity, temperature, etc.), this technique is often used to study home-range size and shape, habitat selection, activity patterns, den use, migration routes, path tortuosity and many other aspects of an animal's ecology. However, telemetry studies can also be an important source of many other biological data. Because it is an intrusive technique, animals must be trapped live and consequently samples of blood, fur, parasites, as well as morphometric measurements might be acquired. Thanks to today's advances in microbiology and molecular biology, data on kinship, taxonomy, biogeography, presence of virus, bacteria and fungus, and much other biological information might also be derived from samples collected during radio-tracking studies, which can help to improve our knowledge of a population's natural history, ecology and biology. A case study using a Mediterranean population of Eurasian badgers (*Meles meles*) will be presented in order to illustrate the usefulness of this technique.

1 filipa_loureiro@fc.ul.pt.
2 lmrosalino@fc.ul.pt.

INTRODUCTION

Radio-tracking, or wildlife telemetry, is the technique that has most revolutionized wildlife research (Mech and Barber 2002). It allows remote monitoring of free-ranging animals while they pursue their normal movements and activities, and nowadays is one of the most widely-used techniques in ecological studies contributing to animal conservation. Moreover, since many wildlife species are secretive and difficult to observe, radio-telemetry has provided a valuable tool to learn more about their life-histories (Resources Inventory Committee, 1998).

Often, there is some confusion between "radio-tracking" and "telemetry". Nevertheless, radio-tracking might be defined as the technique itself, which allows determination of information about an animal through the use of radio signals from or to a device carried by the animal; while telemetry is the process of transmission of information through the atmosphere, usually by radio waves (Mech and Barber, 2002). So radio-tracking involves telemetry. Other terms used to refer to wildlife related telemetry are "radio tagging", or simply "tagging" or "tracking".

It is difficult to say exactly when and who started the use of radio tracking (Kimmich, 1980). The fact is that the idea of using a short-range transmitter in animals to track them occurred to several investigators in the late 1950s and the beginning of the 1960s in order to save time and resources. Before that, the studies of animal movements depended upon live-trapping individual animals and then either observing them in the wild or recovering them through recapture. A later refinement was the use of visual markers such as color-coded collars that allowed observers to identify individuals from afar (Mech and Barber, 2002). Another more indirect method used at that time, and still in use, was finding deceased remains or the presence signs such as footprints, scats, etc.

It was only in 1963 that the first complete workable system was designed and published by W.W. Cochran and R.D. Lord (Cochran and Lord, 1963). At that time, the first experiences with this technique involved only land mammals and birds. Since then, many improvements emerged in basic radio-tracking materials and field techniques for wildlife studying. These have included more efficient transmitters, more effective materials for protecting transmitters from the weather and animals, lighter weight batteries (thus allowing them to be used on smaller species), the higher miniaturization of electronic components, development of a high diversity of sensors, development of satellite and Global Positioning System (GPS) tracking, and probably the most important—the commercial availability of a readymade variety of transmitters (collars, harness, etc.) suitable for a wide diversity of species, including aquatic animals (Mech, 1983).

The progress in radio tracking since the 1960s has been amazing, and nowadays we might radio-track and follow in great detail the paths of almost any moving vertebrate, from the smallest rodent to the big blue whale, from terrestrial to marine or even flying species. Moreover, currently there is a variety of sensors that might be incorporated within the transmitters that allow us to collect multiple parameters and enhance enormously the acquired information.

Sensor development is one of the most important improvements in radio-tracking that helped to reduce intrusiveness by researchers and to acquire an extraordinary amount of extra information. When sensors first started to be applied, the additional information collected was still very limited. At that time, sensors in collars could detect whether an animal was active or not (Knowlton et al., 1968), if it had had died (Kolz et al., 1973; Stoddart, 1970), or even when it had urinated (Charles-Dominique, 1977). Presently, with adequate sensors, simultaneously with locomotor activity, it is possible to monitor cardiovascular function and collect data related to blood pressure, heart rate, respiratory rates, ECG, body temperatures, etc. (Kanwisher et al., 1978; Horii et al., 2002). In the marine environment, for example, acoustic tags with sensors have also been extensively applied to large fish (e.g., tunas, billfishes, sharks), and besides geographic location these have the capability to measure a wide variety of parameters, including depth, swimming speed, tailbeat frequency, internal body temperature, bottom depth, ocean currents, thermal profiles and even salinity (Boehlert, 1996).

Other important recent refinement in radio-tracking is the ability to program radio-collars to transmit only at certain times ("duty cycling") rather than continually. This can double or triple transmitter's life, thus reducing or eliminating the need to recapture an animal for replacing an expired collar (Mech and Barber, 2002). Moreover, these transmitters often have the option of "drop-off" available, leaving the animal free of the transmitter. This is especially important in animals that are difficult to capture and even more difficult to recapture, since this option allows the removal of the collar of an animal that otherwise would have it for the rest of its life.

Finally, probably one of the most important developments is the satellite and GPS transmitters. VHF radio-tracking was the standard technique in use since 1963; an animal could only be tracked by a person on the ground or in the air with a special receiver and directional antenna (Mech and Barber, 2002). Satellite and GPS transmitters allow biologists to study many species without having to venture into the field to determine each location, and more importantly have the advantage to track animals over long distances and in remote areas. The main difference between these two types of telemetry is the fact that, with satellite telemetry, the animal's transmitter sends information to the satellite receivers which relay this information to a recording center on Earth, whilst with GPS telemetry, a different set of satellites function as transmitters, while the animal's telemetry unit acts as a receiver (Mech and Barber, 2002). Beside the time and resources that might be saved, with these transmitters is possible to follow a higher number of animals simultaneously with a small effort and the locations are very precise.

Despite its popularity and its numerous improvements, radio-telemetry might be inappropriate under many circumstances. Thus, a researcher should cautiously consider a number of questions before embarking on a study involving radio-tracking, namely the objectives of the study, the type of data to be collected, the constrains on funding, field conditions, equipment limitations, and the species under study (White and Garrott, 1990). In fact, this is an expensive and time-consuming technique that has proven to be unsuitable for use in some species (due to the animal's size or life-history traits). Moreover, despite the high frequency with which radio collars and other transmitters are attached to research animals, surprisingly little is known about their effects on the behaviour and survivorship of the species in question because hardly a researcher can prove that the transmitter is not having an effect. Thus, common sense should be used in placing transmitters on animals, to minimize potential impacts on the animal and to maximize the chances of obtaining reliable data from the study (White and Garrott, 1990).

Briefly, although the fact that radio-tracking technique is a revolutionary technique and that there is no other wildlife research technique that comes close to approximating its many benefits (Mech, 1983), a circumspect consideration must occur before deciding to use it.

Direct Applications

If after carefully thought the decision is to go forward on a study involving telemetry, as we mentioned above, the potential of new information acquired with this technique is almost unlimited (Mech and Barber, 2002). Radio-tracking allows the precise identification of individual animals and potentially allows a researcher to locate each individual as often as desired (Mech, 1983); thus, the uses in which it may be applied are numerous.

A key step in the implementation of any radio-tracking study is it careful design and planning (White and Garrott, 1990). Depending on that, two important measures of sample size need to be defined: the number of individuals followed, and the number and timing of relocations of each individual. This distinction is very important since the number of animals tagged must be adequate for statistical tests and because it is the number of animals and not of locations that is the true sample size (Resources Inventory Committee, 1998). Thus, when properly applied and planned wildlife telemetry allows acquiring detailed data on many aspects of wildlife biology. Immediate data obtained with this technique is the location of an animal at a certain time and place. Having this, and depending on the rate at which locations are registered, several types of information about the individual can be learned (Mech, 1983).

The most common application of this technique is related to the study of the spatial ecology of a species or populations of a species, by the estimation of home-ranges (i.e., spatial extent used by an animal, or group of animals, during the course of the everyday activities), core-areas (i.e., area of heaviest use within a home range, Wray et al., 1992) and habitat utilization patterns. By registering several locations per day over a long period, a good approximation of the size and shape of an individual's home range can be achieved. Depending on the species, this period should last from weeks to several months, ideally for an entire year, in order to register information during all seasons. If locations are precise enough and if good habitat data is available, then habitat selection by species in a given area can also be studied (Mech, 1983) and priority habitats for the conservation of a species might be known.

On the other hand frequent locations, registered throughout the day or night, provide excellent insights on the circadian activity rhythms of an animal, its daily movement pattern and information about resting sites. It is therefore possible to know if an animal is mainly nocturnal (e.g., Bats, *Myotis grisescens* and *Myotis sodalist* - LaVal et al., 1977), diurnal (e.g., Golden Spiny Mouse, *Acomys russatus* - Shkolnik, 1971), or if switches between diurnal and nocturnal activity (e.g., Atlantic Salmon, *Salmo salar* - Fraser et al., 1993). Additionally, and by locating animals while sleeping, radio-tracking also provide information about types of resting sites, namely dens and nests, pattern of use and fidelity to certain sites. Is even possible to study dormancy and hibernation in various species, and to know when it begins, finish and its duration. This information is especially important to help delineate conservation plans for endangered species, since resting sites are essential to the survival of most animals. Concerning movement pattern, it precision increases with the rate at which the researcher gathers the telemetric data. When this rate is good enough, and almost continuous locations are collected, path tortuosity of individuals, populations or species, can be studied and animal movements during the activity period can be characterized (e.g., Loureiro et al., 2007b). Rarely animal movements occur randomly. Usually, they are determined by the spatial arrangement of habitat patches as well as by animal behaviour while searching for food, partner or scent marking, and often they have a straight or convoluted pattern. Because organism movement represents one manifestation of it interaction with their environment (McIntyre and Wiens, 1999), by understanding how it occurs enhances the knowledge on animal's ecology.

Beside daily movements, longer movements can also be registered. These usually affect population dynamics and are dispersion and migratory movements. Thus, radio-tracking provide excellent information about season, direction, distance and duration of dispersion, age of dispersers, new areas of settlement and mechanisms of colonization of new areas, and allows to study many aspects of migration, such as triggering conditions, migratory route, duration and distance (Mech, 1983).

Although not so obvious, demographic and population data might also be obtained with radio-tracking studies. From these studies, researchers are able to determine individual longevity, cause of mortality and the age structure of populations.

Herein we have mentioned some of the most direct and common applications of radio-tracking, which permit to know many aspects of an animal's ecology. Nevertheless, nowadays, telemetry applications in the fields of scientific research are constantly being developed and many other aspects of an animal's ecology can be studied.

Telemetry Studies as Data Source for Other Bio-Ecological Questions

Telemetry studies are also an important source of many other biological data, which usually are not the main aim of this kind of studies and are slightly different from those described earlier in the chapter. Because telemetry is an intrusive technique, animals must be live trapped, and an array of biological data can be obtained, maximizing the amount of information available to a researcher and enhancing the profitability of a wildlife project.

The first obvious data that an ecologist should look into, since animals generally must be anesthetised for collar (implant or harness) fitting, is morphology. Thus, body condition, body size and shape, gender, age and coat pattern might be some important parameters to be registered. Moreover, the collection of morphometric measurements can later help to detect sexual dimorphism that could justify possible differences in behavioural patterns (e.g., diet variation between males and females - Thom et al., 2004). Evolutionary mechanisms such as "character displacement" or "character release" (see Dayan and Simberloff, 1998 for details), perceived in the first place through the analysis of sex differences in animals size, can help researchers to understand regional and population ecological adaptations. Possible geographical variation in size (e.g., Bergman's rule - Rodríguez et al., 2008), and coat pattern (e.g., Veron et al., 2004), are also good surrogates of ecological traits that are important to better understand the role of a particular species in the ecosystem.

Another aspect of an animal's natural history is parasites burden: specifically ectoparasites (e.g., fleas, ticks, mites) and endoparasites (e.g., helminths). These wildlife players have the potential to decisively affect the structure and stability of natural communities (Hudson et al., 1998), and therefore are fundamental to perform a qualitative evaluation of the sanitary status of a population. Moreover, the importance of parasitological investigations is further enhanced by the role of some wildlife as reservoirs of parasites that can infest domestic animals (e.g., *Molineus patens*—dogs - Durette-Desset and Pesson, 1987), or even man (e.g., bovine tuberculose, *Mycobacterium bovis* – Cosivi et al., 1995).

The collection of ectoparasites is an easy and low cost task, which can produce very interesting results, especially if combined with body condition, sex, age, reproductive status and behavioural ecological data. Inversely, while ectoparasites collection and identification is usually not a very expensive assignment, since it is a direct method which requires low technology, the quantification of endoparasites (as well as other infectious agents) involves more elaborate techniques, which frequently imply the collection of biological material (e.g., faeces, blood, tissues) for further analysis. Biochemical approaches (e.g., formol-ether concentration method for helminths collected in faeces - Allen and Ridley, 1970), together with bacteriological (e.g., conventional identification of bacteria collected in faeces based on macro- and microscopic morphology – e.g., Oliveira et al., 2008) and genetic tools (e.g., viral DNA and RNA extraction from faeces – Oliveira et al., 2008; *Borrelia* spirochete isolation from ear clips – Oliver et al., 2000; bovine tuberculosis detection by the analyse of serum, plasma or blood – Twomey et al., 2007) give researchers the possibility to understand how pathologies shape a community. Fur or feathers samples, collected while animals are anesthetised, are also an interesting material to investigate fungal infestation on wildlife (e.g., Czeczuga et al., 2004).

Using the same sample's framework (e.g., fur, blood, tissues, etc.) researchers have now powerful molecular tools (such as allozymes, microsatellites and mitochondrial and nuclear DNA sequences analysis) that can also be used to estimate numerous biological features of interest, such as migration rates and routes, population size and structure, bottlenecks, kinship, population trends, biogeographical patterns, and much more (see Selkoe and Toonen, 2006). These methods are extremely useful, and their importance is enhanced when combined with telemetry studies, since they allow the interpretation of, for example, social structure, territory size and configuration (obtained from radiotracking data) in light of kinship (e.g., Janečka et al., 2006; Charif et al., 2005). Moreover, with this data is possible to assess potential effects of fragmentation by combining genetic differentiation between populations determined by genetic analysis (e.g., randomly amplified polymorphic DNA - RAPD - and microsatellite markers) with estimates of effective population size and dispersal distances, calculated from telemetry studies (e.g., Antolin et al., 2001).

As mentioned in the beginning of this section, since animals must be captured, anesthetised and manipulated during radio-tagging, we have an excellent and unique opportunity to take as much information as possible. Therefore, it is also possible to collect samples from radio-monitored animals to determined hormone levels in faeces or blood collected during collar-fitting procedures. Faecal hormone levels (as well those assessed through blood analysis) can be linked, for example, with diet quality and social rank (e.g., Taillon and Côté, 2008), stress (e.g., Blumstein et al., 2006), reproductive status (Rolland et al., 2005) or even be a useful accurate method for sex determination (e.g., Oates et al., 2002).

Finally, the effect of human activities and the population's resilience can also be studied using the collected blood samples. For example Skaare et al. (2001), in Svalbard (Norway) allied a telemetry study with the evaluation of pollutants concentration (e.g., Organochlorines), retinol and thyroid hormone in the polar bears (*Ursus maritimus*) population. This approach allowed them to assess the deleterious effect of that chemical compound on bear, since the decrease of plasma retinol and thyroid hormone levels due to PCB exposure may be caused by the displacement of the physiological compounds from their binding sites on plasma binding proteins by PCB metabolites (Skaare et al., 2001).

Thus, thanks to today's advances in microbiology and molecular biology, data on kinship, taxonomy, biogeography, presence of virus, bacteria and fungus, and many other biological information might also be derived from samples collected during radio-tracking studies. This data permits to cover a wide range of a population's bio-ecology, from kinships to sanitary status and can help us to improve our knowledge of a population's natural history, ecology and biology.

A Case Study Using a Mediterranean Population of Eurasian Badgers (*Meles meles*)

To illustrate what we have been describing in earlier sections, we will now present a study developed in SW Portugal, using Eurasian badgers (*Meles meles*) as the project's model.

1) Main Aims of the Project

Badgers (*Meles meles* Linnaeus, 1758) are medium-sized carnivores, whose distribution includes most Eurasia, and are one of the few Mustelid species that are gregarious, living in social groups or clans. Badgers dig out extensive underground burrows, called setts, which are often used for many years by several generations.

Until the implementation of our project, badgers' ecology was the focus of numerous studies throughout Western Europe, mainly Great Britain (see Neal and Cheeseman, 1996), but only scarce information was available on Mediterranean or low density populations. In this scenario, we aimed to study a badger population inhabiting a typical multiuse agro-forestry system (cork oak woodland or *montado*) in Mediterranean Portugal, to understand how this meso-carnivore is adapted to southern landscapes, shaped by environmental Mediterraneity (Virgós et al., 1999). Thus, some of our central questions were focused on:

- physical parameters;
- activity rhythms;
- social structure;
- spacial ecology;
- habitat selection;
- distribution.

To answer all this bio-ecological questions, telemetry (or radio-tracking) was a central piece on our methodological puzzle, from which all other technical approaches in our study design depended upon.

2) Study Area

This project was implemented in Southwest of Portugal (38° 07'N; 8° 36W, elevation range, 150 to 270m a.s.l., c.a. 66km^2, Figure 1), in Serra de Grândola. This is a Mediterranean landscape where the habitat is dominated by one of the last examples of traditional rural agro-silvo-pastoral systems in Europe – the *montado*, namely the cork oak (*Quercus suber) montado.*

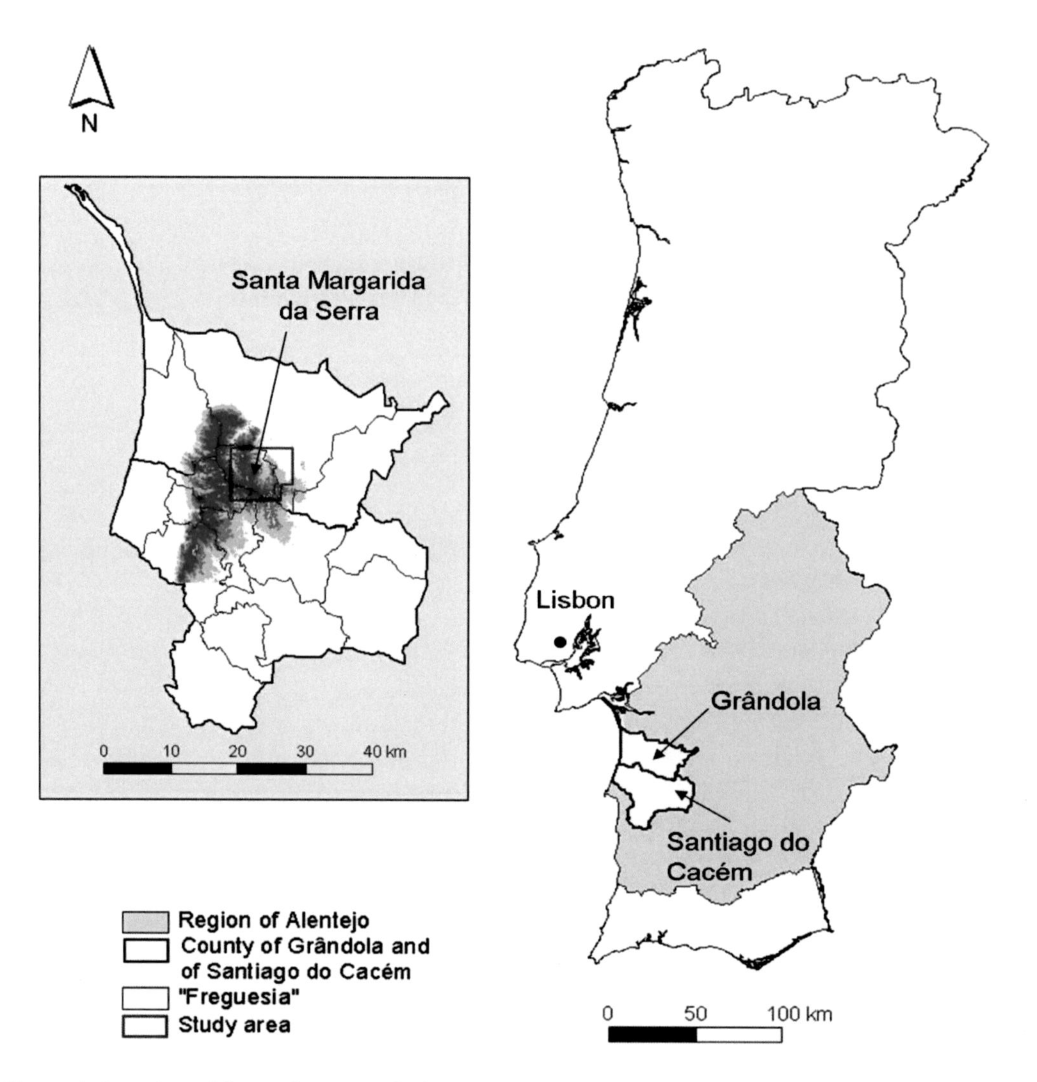

Figure 1. Location of the study area and of Serra de Grândola within Portugal; Serra de Grândola is in green (In Loureiro, 2008).

The matrix of this landscape is composed mainly by cork oak (*Quercus suber*) and holm (*Q. rotundifolia*) woodland, with and without understory. Interspersed within the *montado* there are several other patches of different land covers such as riparian vegetation, orchards, olive yards and, in smaller proportion, pine (*Pinus pinea* and *Pinus pinaster*) and eucalyptus (*Eucalyptus globosus*) stands (for more details see Rosalino et al., 2004). Corridors of riparian vegetation, occasionally very dense, are dominated by blackberry bushes (*Rubus* spp.), poplar (*Populus alba*) and ash (*Fraxinus angustifolia*).

The climate is Mediterranean, but humid due to the Atlantic influence, and clearly seasonal in terms of temperature and precipitation. Dry seasons are spring and summer, being spring less humid and hotter. Winters, although mild, are characterized by very cold temperatures and some precipitation. Fall is generally less cold but more rainy (for more details see Santos-Reis and Correia, 1999; Loureiro, 2008).

Human density is very low in Serra de Grândola (about 18.5 indivuals/km^2) since most farm houses are nowadays abandoned, and the study area is only crossed by two main roads and several trails. Trails are dirt roads used to travel between farm houses, but together with main roads currently carry only very light traffic. Human activities are nowadays restricted to agriculture, cork and wood extraction and cattle rearing (mainly sheep and black Iberian pigs).

3) Methods

The project started in 1997, with a first trapping campaign (with appropriate licence from the Portuguese Institute for Nature Conservation and Biodiversity - ICNB) aiming to characterize the meso-carnivore guild of Serra de Grândola, and more specifically to determine badgers' abundance and confirm the presence and permanent occupancy (through reproduction evidences: cub capture) of this mustelid in the study area. In this first approach a grid of box traps (Tomahawk Livetrap Co., Wisconsin, USA) and foot-hold-traps (Victor Soft Catch nº 2, Long Spring Woodsteam Corporation, Pennsylvania, USA) were used (3 traps per km^2) (Rosalino et al., 2005c). All captured animals were anesthetized with 0.1mL ketamine hydrochloride per kg (Imalgene 1000, Rhöne Mérieux, Lyon, France), individually marked by a combination of ear cuts, sexed and aged (tooth wear) (see Rosalino et al., 2005c and Rosalino et al., 2004 for details).

From 2000 to 2002 another trapping session was implemented focused only on badgers. Foot-hold traps (Victor Soft Catch No. 2, Long Spring Woodstream Corp., Lititz, Pennsylvania) and box traps (model 608, Tomahawk Live Trap Co., Tomahawk, Wisconsin) were set along their paths and near setts (burrows that are main resting sites of badgers). Animals' manipulation followed the procedures described above. Extra data samples were also collected: fur, faeces, blood, tissue (ear clip) and ectoparasites. Adult badgers were radio-tagged using Biotrack (model TW- 5, Biotrack, Wareham, UK) VHF (154 000 – 155 000 Hz) transmitters equipped with motion sensors, and monitored from 2000 to 2004.

Table 1. Identification, sex and radio-tracking period for each captured badger (Win = winter; Spr = spring; Sum = summer)

Ind.	2000			2001				2002				2003				2004	
	Spr	Sum	Fall	Win	Spr	Sum	Fall	Win	Spr	Sum	Fall	Win	Spr	Sum	Fall	Win	Spr
T7♀																	
T11♂																	
T13♀																	
T17♀																	
T18♂																	
T21♂																	
T24♂																	
T25♀																	
T30♂																	

Some preliminary 24 hour-periods tracking sessions confirmed badgers' nocturnal activity pattern (Rosalino et al. 2005c). Thereafter, badgers were regularly located using traditional telemetry techniques (triangulation and homing, Mech, 1983) every 30 min during their activity period (night period). Telemetry of activity periods commenced at least one hour before dusk (with the animal still in the sett) and ended one hour after sunrise (always after the animal stopped moving), and represented activity cycles.

Several days per month badgers were located during daytime, while inactive in their resting sites (see Rosalino et al., 2004; Loureiro et al., 2007a, for details). Habitat use was assessed by overlapping animals' positions and home-range/cores areas limits (determined by using Animal movement extension of ArcView 3.1) to a landscape cover map (see Rosalino et al., 2004 for details).

Movement paths were analysed trough path tortuosity using the fractal dimension. Fractal dimension (D) is a natural measure of tortuosity, which classifies paths according to their degree of convolution or straightness (for more details see Katz and George, 1985; Loureiro et al., 2007b). Paths were obtained from activity cycles that had continuous tracked of an animal following it entire movement and foraging route. We have excluded activity cycles for which no movement was registered throughout the entire night, and paths for which no locations were registered during 3 hours or more (because the animal eluded us, or bad weather conditions or technical problems foreshortened the tracking session).

4) Project Results

Trapping Success and Monitoring Scheme

With a trapping effort of 4530 trap-nights, 30 badger were live trapped (17 males and 13 females), mostly cubs/juvenile (almost 67%, Rosalino et al., 2005c) and a large set of samples related to different parameters was collected (e.g., morphometric measures, social organization data, etc.). Moreover, posterior analysis of the biological samples collected allowed the investigation of many other novel aspects this carnivore's population ecology.

From those, only nine animals, four adult females, four adult males and one juvenile (with more than 9 months), were radio tagged and tracked for several months (Table 1). These animals belong to 3 different social groups (later confirmed). Radio tracking monitoring varied between 5 months (T24 and T30) and 38 months (T21); those that were radio tracked only for a short period either died or lost their collar.

In the overall tracking period, a total of 61 night cycles were made and more than 1600 nocturnal and 1700 diurnal locations were registered for all individuals.

Morphometrics

In this project task, we aimed to describe and understand badger biometric characteristics and evaluate if this species is sexually dimorphic.

Badgers were smaller than their northern counterparts, but overlapped with other Mediterranean population, confirming the predicted pattern associated with the Bergman's rule: negative correlation between body size and environmental temperature and consequently latitude (Ashton et al., 2000). Moreover, animals living in less productive areas are more likely to be smaller, since lower energetic requirements associated with lower size counterbalance lesser productivity (Revilla et al., 1999). No significant sexual dimorphism was detected, although males have a tendency to be heavier (8.1kg vs 6.5kg), bigger (body and head length: 73.4cm vs 68.4cm) and taller (Hilt length: 32.2cm vs 31.9cm) than females (Rosalino et al., 2005c).

This kind of data is essential to understand the spatial structure as well as the feeding ecology of a predator, since size influences the animals' energetic needs, which together with food availability and dispersion, and mate accessibility usually determines home-range size and configuration, and constrains the type of prey a carnivore might consume. The absence of sexual dimorphism can be considered a clue to a non-segregative use of trophic resource among genders.

Population Structure

The use of capture data, a sub-product of the need to detained animals to radio-tag them, can also provide an important glimpse on how a population is structured. In our example, badgers' population was dominated by females, since the sex-ratio was 1 male:1.5 females, although when juveniles and cubs are included in the analyses the female-biased structure is attenuated. Moreover, the capture of lactating females and cubs in all the monitored years allow us to confirm reproduction in the area for the considered period (Rosalino et al., 2004). This structure is also indicative of a possible differential philopatric behaviour among genders, since cubs sex –ratio is male biased (1 male:0.7 females).

Although in high density populations females are more prone to disperse (Woodroffe et al., 1993), probably to achieve a successful reproduction, in low density areas female dispersion is null (Neal and Cheeseman, 1996). This can result in the detected adult female biased sex-ratio. The radio tracking data showed that there was a high overlap between yearly home ranges of several individual, which was a good predictor of some social behaviour of the studied population. Moreover, this kind of information (animals' location) also gave a good clue on the groups (or clan) constitution: groups comprised 3–4 adults plus 3–4 cubs of the year for a total of 6–8 badgers/group (mean = 7, SD = 1, n = 3) (see Rosalino et al., 2004 for details).

The complementary analysis of the trapping data, together with group size and home-range dimension (see below), lead us to believe that, in our study area, badgers where living in low density, varying between 0.36 and 0.48 individuals/km^2 (see Rosalino et al., 2004 for details).

Spatial Ecology

Contrarily to most of the above results, this is one of the direct aims of most radio-tracking study: to evaluate how a population uses the spatial environment, especially how it uses the available landscape units and what is the spatial arrangement and structure of home-ranges/territories or core areas.

In our study area, badgers showed home-ranges very similar to those described for Iberia, but larger than those described for other Western Europe populations (home-ranges: 4.46 km^2; core-areas: 0.38 km^2) (Figure 2).

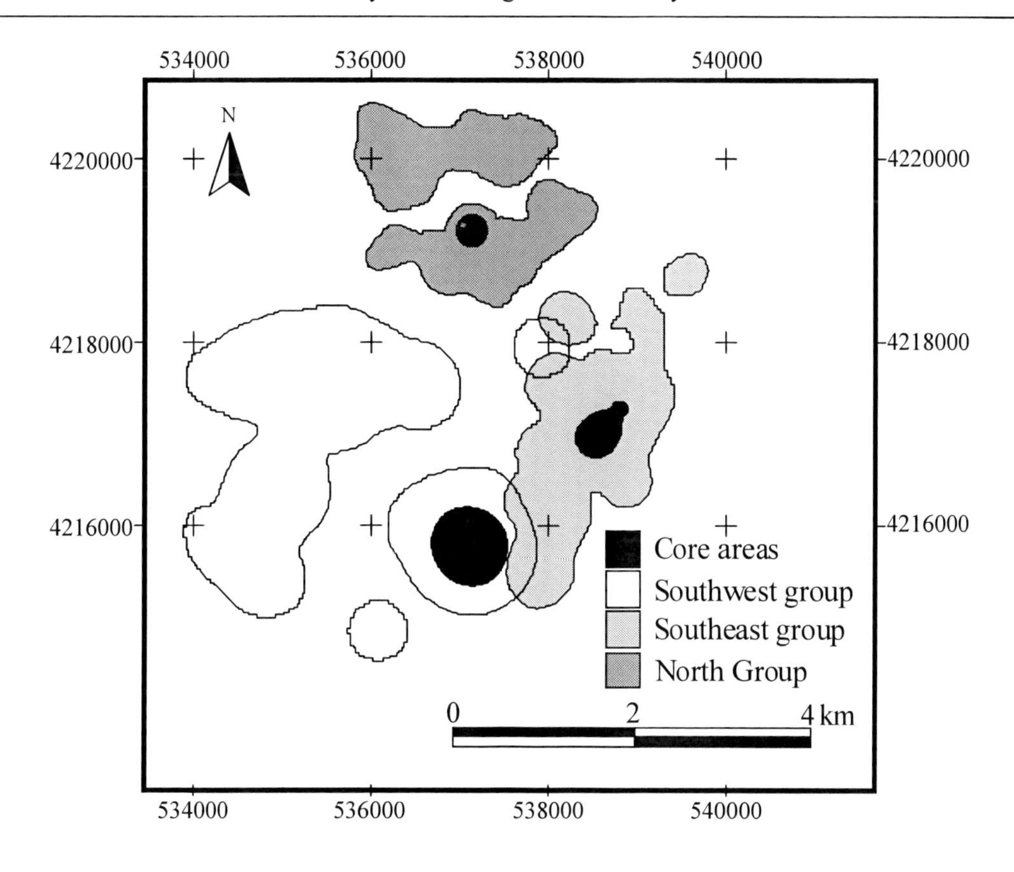

Figure 2. Group home ranges (and core areas) of the radio-tagged badgers using 95% Fixed Kernel estimator. (Adapted from Rosalino et al. 2004)

Home-range sizes were positively correlated with food-patch dispersion, namely orchards and olive yards (Rosalino et al., 2004). On the other hand the population seems to be positively selecting cork oak woodland with understory and riparian vegetation, for different purposes: 1) provide a safe travel routes between food patches and shelter for setts emplacements (Rosalino et al., 2005b); 2) both habitats are important sources of food (Rosalino et al., 2004). This analysis allowed us to infer that in south-western Portugal, badgers depend upon an environmental mosaic such as olive groves, orchards and vegetable gardens for food and cork oak woodlands for shelter and protection.

Circadian Activity Rhythms

As mentioned above, if radio-collars (implants or harness) are equipped with activity sensors, as was the case of our study, researchers can go beyond the spatial picture of an animal's life. Therefore, in this project we were able to study the activity rhythms of badgers. We manage to confirm that, as in much of it range, in southern regions of Europe this carnivore has crepuscular or nocturnal activity, generally starting after sunset and ending before sunrise, with no inactive intervals in between, being active on average for 8.26 hours per day (Figure 3 and 4).

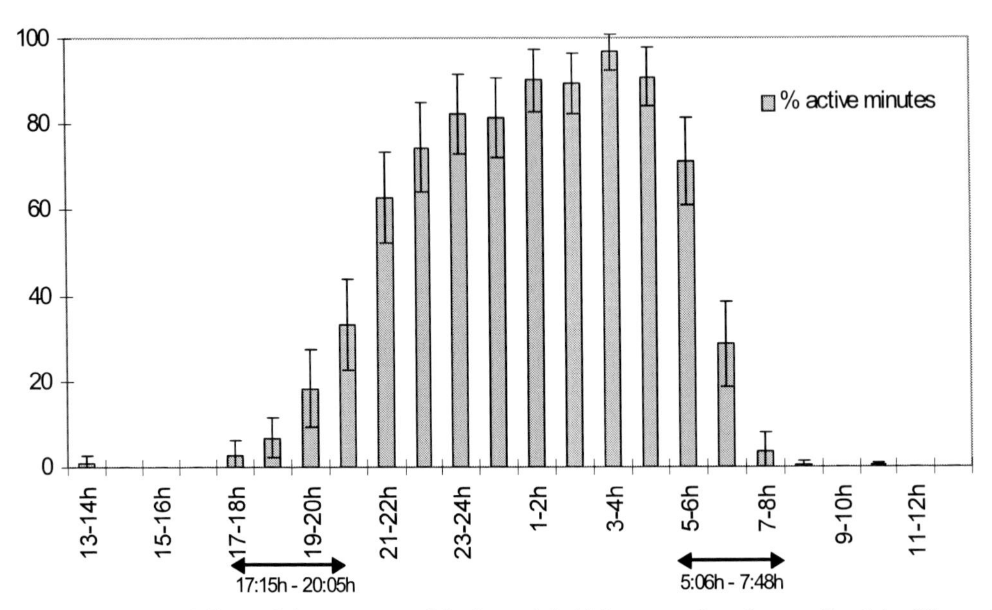

Figure 3. Average daily activity pattern of badgers inhabiting a cork oak woodland in "Serra de Grândola". Arrows indicate sunrise and sunset period (Adapted from Rosalino et al. 2005a)

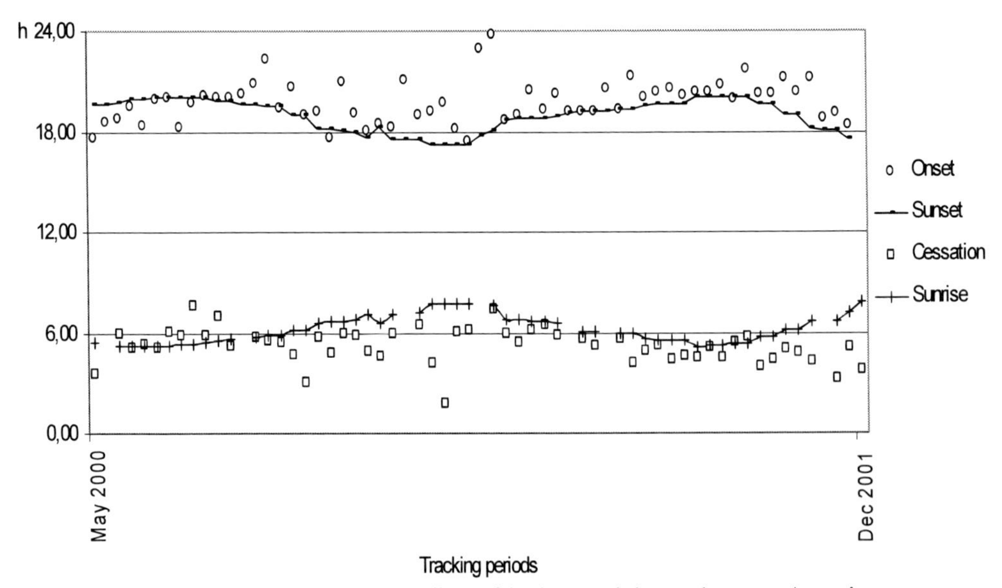

Figure 4. Relation between beginning / ending of badger activity and sunset / sunrise moments per activity cycle. (Adapted from Rosalino et al. 2005a)

No hibernation period was detected and no significant correlation was found between the onset of activity and sunset or between cessation of activity and sunrise (see Rosalino et al., 2005a).

Resting Sites

As previously referred, once a day badgers were located, while inactive, in their resting sites. These locations allowed registering badgers sleeping in 48 different diurnal resting sites (Loureiro et al. 2007a). When compared to other populations of Europe, Grândola's badgers used a large number and variety of resting sites (e.g., low-density areas: Norway, 12 – Brøseth et al., 1997; Poland, 9 – Kowalczyk et al., 2004; high-density areas: England, 5.3 – Roper et al., 2001; Ireland, 5.0 – Feore and Montgomery, 1999). This high number of shelters could be explained by the large home ranges of badgers in Serra de Grândola (4.46km^2 - Rosalino et al., 2004). As demonstrated by other authors (Revilla et al., 2001; Kowalczyk et al., 2004), there is a significant correlation between number of resting sites and territory size, suggesting that the larger the territory the higher the number of resting sites.

On the other hand, there is certainly a high diversity of structurally different sites available to Grândola's badgers. In fact, they used very different resting sites: burrows, shrubs (dominated by *Rubus* spp. and *Cistus* spp.), rocks, hollow trees and man-made structures (e.g., small culverts). Burrows were by far the most used type of resting site (91.77% of locations), followed by shrubs (4.8%). It was interesting to note that members of different social groups sometimes used exclusive types of resting sites; for example, although available, members of the northern group never used man-made structures, whereas animals from the southwest social group never slept in rocky shelters. The same happened for members of different sexes; males were never found denning in man-made structures. Although, females used almost twice as many shelters as did males, once again, for both sexes burrows were the most common resting site used (about 90%), followed by shrubs (about 5%).

Badgers' resting sites are not only different in terms of structure, often they have different functions and, therefore, might be classified as main setts (used as winter resting site by all members of a social group, and where reproduction occurs, Roper, 1992), secondary setts (used frequently by more than one individual of a social group) and occasional resting sites (used sporadically generally only by a single animal, for more details see Loureiro et al., 2007a).

All badgers from each social groups used one main, one secondary and several occasional resting sites. Nonetheless, not all males used the main sett nor did all females use the secondary sett, and overall females used more than twice as many occasional resting sites as did males. In fact, although occasional resting sites were the most numerous (14±7.55), they were only used in 11.25% of all locations. In spite of this variety of types of resting sites, is important to note that 62.5% of our observations were of badgers using main setts in burrows (comparable to figures reported by Rodríguez et al., 1996: 63.3%; Do Linh San, 2002: 75.0%).

This data, also revealed that the population inhabiting our study area had lower sett fidelity than reported for their northern counterparts, especially those of lowland England. In fact, only on 36% of occasions did they come back to sleep in the same sett, with females returning more often to the same den. Moreover, the mean number of consecutive days spent in a sett by our badgers was higher in spring and lower in summer (Rosalino et al., 2005a)

Besides having the number of different resting sites and it frequency of use, it is also possible to know the pattern of use of these refuges and how it varies throughout year. Thus, the pattern of use of badgers resting sites in Serra de Grândola varied seasonally, showing differences not only according to sex, but also among individuals of different social groups. Overall females used more than twice as many occasional resting sites as did males. Generally burrows, predominantly main setts, were most frequently used during winter and autumn; whilst non-burrows shelters, were preferred during spring and summer (Figure 5 and 6).

Seasonal differences in the use of dens occurred not only in what concerns to the category of resting sites but also to their structure, suggesting that the observed pattern of resting sites use of badgers in Serra de Grândola might be influenced by climate conditions. Moore and Roper (2003), working in the UK, concluded that there was no evidence that temperatures within main setts differed from that of other burrows. The same may not apply to Mediterranean areas where thermal range is much wider, and shelters other than burrows are widely used. Not much is know about the environmental conditions inside other types of shelters and we have no data for internal temperature of the different resting sites. However, in our Mediterranean habitats, there was a clear general association between warmer weather and occupancy of sites outside burrows, namely of main setts (see Loureiro et al. 2007a). Badgers only used Mediterranean scrublands, rocks and man-made structures during spring and summer, the hottest seasons; whilst during autumn and winter, when temperatures reach lower values, burrows were more used.

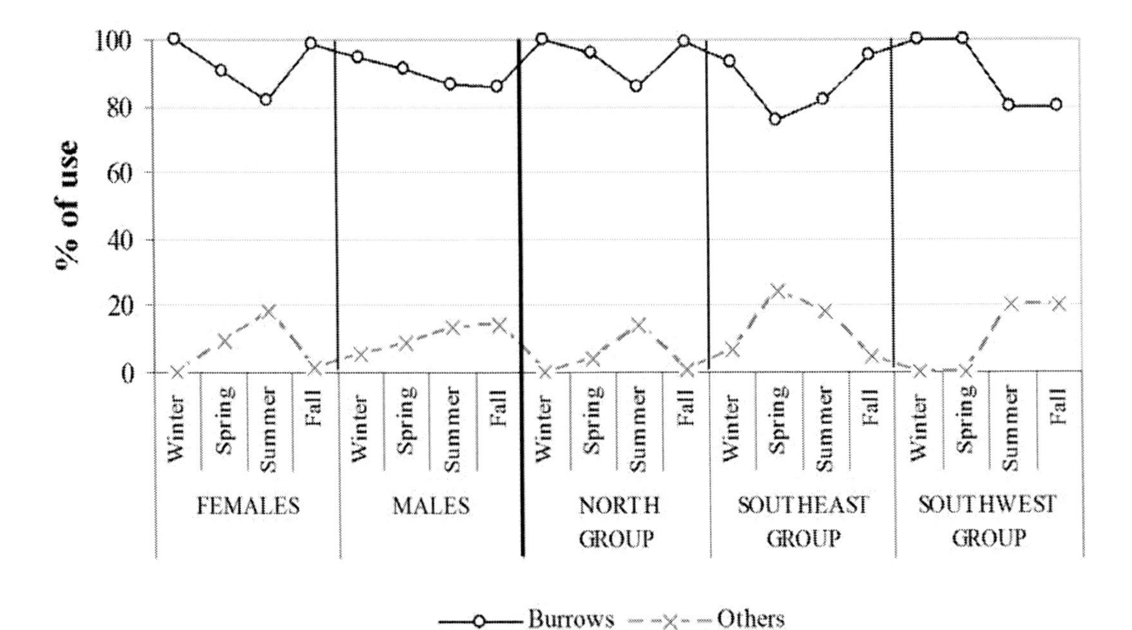

Figure 5. Seasonal changes in the use of the resting sites types by radio-collared badgers in Serra de Grândola, by gender: males (N= 5) and females (N=4), and by social group: animals from the North (N=3), Southeast (N=4) and Southwest (N=2). (Adapted from Loureiro et al. 2007)

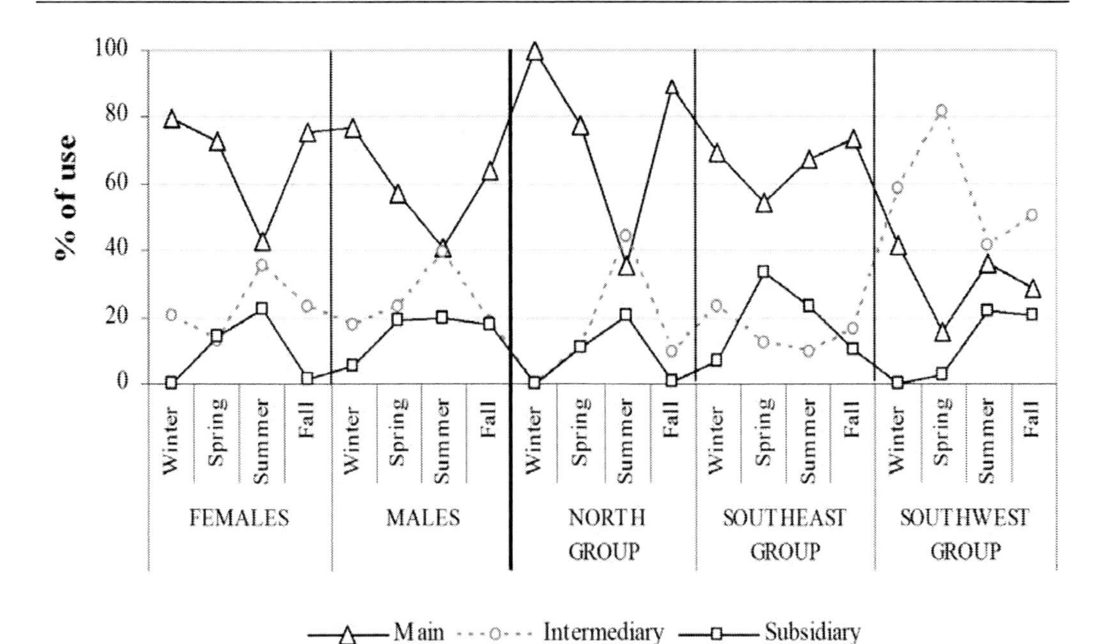

Figure 6. Seasonal changes in the use of the resting site categories by radio-collared badgers in Serra de Grândola by gender: male (N=5) and female (N=4), and by social group: animals from the North (N=3), Southeast (N=4) and Southwest (N=2). (Adapted from Loureiro et al. 2007)

If this seasonal association is causal, or merely a correlation with some other, for example sociological, reproductive or parasitological factors, is still unknown. Reproduction, for instance, clearly contributes to the observed pattern of use, since breeding females spend most of winter and early spring inside and close to the main setts, where cubs are born and raised (F. Loureiro, unpublished data). On the other hand, in cork oak forests, management activities such as shrub clear-cut for cork extraction, often lead to the destruction of badgers main setts during spring time, when cubs become to emerge from dens. Therefore, understanding the causal mechanisms of badgers' use of dens in this Mediterranean landscapes, might be very valuable, not only for determining the species ecological requirements, but also and more importantly, to help to create guidelines for cork oak managers to help badgers conservation.

Movement Pattern

Two years of data collection, and many continuous nocturnal locations, allow us to gather a total of 55 paths that were considered suitable for studying movement patterns of these animals.

Only six animals from two social groups provided suitable paths. Since females were monitored for a longer time, they provided most of the search paths (Table 2). On the other hand, during winter, badgers sometimes did not leave their resting sites throughout the entire night; thus fewer search paths were considered for this season.

Having suitable search paths it was possible to estimate total distances travelled by badgers in one night. Distances varied from 1.69 to 9.16 km and there were no significant differences among distances travelled between individuals of different social groups, and during different seasons, in opposition to the observed differences between sex (Table 2). On average, badgers travelled about 4 km m per night; males usually having more extensive movements than females, probably because maternal behaviour restricted female ranging (Rosalino et al., 2005a).

Table 2. Number of paths (N), average length and standard deviation (SD) travelled per sex, social group and season (km) and significance tests

		N	*Average ± SD (km)*	
Sex	*Males*	*38*	*5.44 ± 1.71*	*Z= -2.15, p< 0.05*
	Females	*17*	*4.19 ± 1.45*	
Social Group	*North Group*	*36*	*4.57 ± 1.72*	*t = -0.57, df = 53, p>0.05*
	South Group	*19*	*4.59 ± 1.46*	
Seasons	*Autumn*	*13*	*4.54 ± 1.90*	*F = 0.069, df =54, p>0.05*
	Spring	*19*	*4.70 ± 1.60*	
	Summer	*17*	*4.53 ± 1.80*	
	Winter	*6*	*4.39 ± 0.60*	

Concerning animals' path tortuosity during their foraging activity, individual search paths showed the three possible categories of tortuosity: straight, constrained and random walk (Figure 7, for more details see Katz and George 1985, Loureiro et al 2007b).

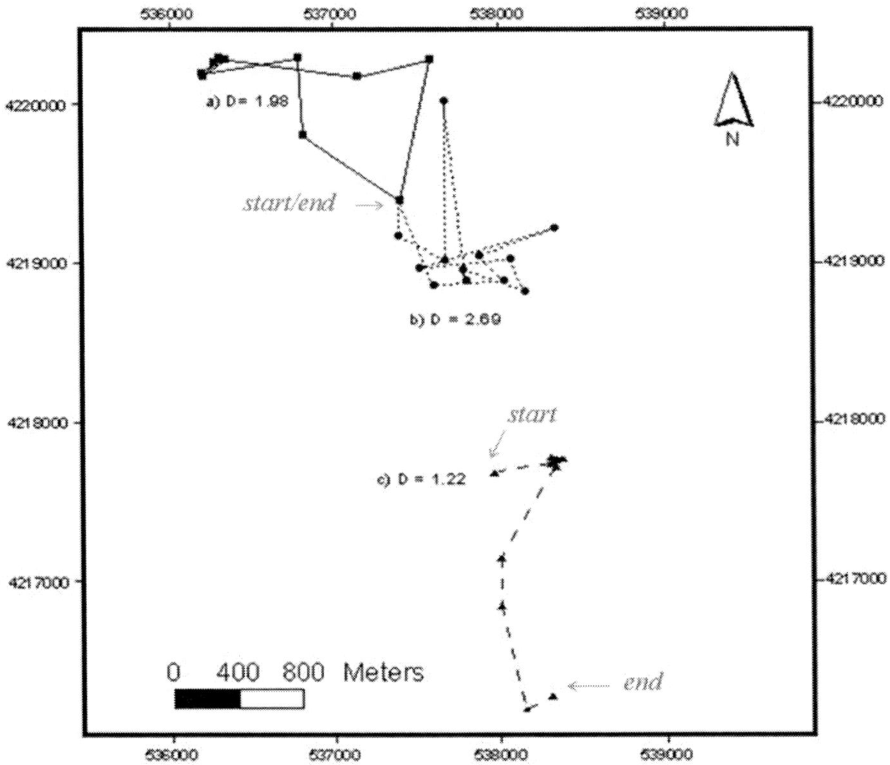

Figure 7. Examples of badgers' search paths in Serra de Grândola and their tortuosity (D = fractal dimension), a) random walk, female T7; b) constrained walk, female T7; c) straight walk, female T17. Outermost numbers are the coordinates in the transverse Mercator grid – UTM (European Datum 1950). (In Loureiro et al. 2007b)

In Serra de Grândola, nightly movement patterns of badgers were not random; although badgers generally undertook convoluted movements, this was not so in summer or for males – in which cases movement paths tended to be straighter. The convoluted movement pattern generally observed might be an adaptation to the clumped distribution of their food in the study area, once it allows an increased use of the area and a more successful foraging (Nams and Bourgeois, 2004). The straighter pattern observed during summer and for males can also be explained in light of the animal's ecology. As Rosalino et al. (2005a) showed, on 63.16 % of days badgers return to sleep in the same sett on consecutive days, contributing to the tortuosity reported here. Nonetheless, in summer badgers change resting sites more often (Rosalino et al. 2005a), resulting in a lower mean number of consecutive days spent in a den (Kowalczyk et al. 2004; Rosalino et al. 2005a) and therefore a straighter trajectory to their nightly movements. Females show greater fidelity to daytime resting sites than do males (Rosalino et al. 2005a).Thus, the lower fractal dimension of males' movements is probably a reflection of their lower fidelity to dens together with greater involvement in scent marking at latrines (Neal and Cheeseman, 1996; Revilla and Palomares, 2002; Rosalino et al., 2005b) and searching for females.

Modelling Badgers' Distribution

All of the information that a researcher can obtain from a telemetry methodological approach is important, per se, although he/she should move beyond the simple description of some aspects of an animal's natural history. More importantly, any researcher working in conservation biology should try to use its data to formulate management options or to generate models that could help understand the species ecology, not only on a local scale, but in a broader perspective, embracing regional or even national levels. In the present example, we try to modulate badgers' distribution, aiming to present an array of variables that could be constraining the species distribution, starting from a local scale to a regional one.

Assuming that their population density was related to home range sizes and that this, in turn, was influenced by food and water availability and the existence of substrate suitable for sett construction, we explored the relationship among these parameters. Sites with surface water in summer (the driest and more limiting season), as well as food patches, seem sufficient to support more badger groups than existed, leading us to believe that these factors were not limiting badger density. Inversely, four sett location predictors were fundamental for identifying areas where badgers were present: the existence of a geological fault/ discontinuity, ridges, valleys and the distance to abandoned farm houses. Hence, our results suggest that in the Mediterranean woodlands of Serra de Grândola, the main factor limiting badgers' density, and distribution, is the availability of suitable sites for setts emplacement. However, in areas where suitable sites for burrows existed, but food patches were absent, badgers were not found. This could indicate that the presence of both factors was necessary for badgers, although in this area sites suitable for digging setts appeared to be the primary limiting factor (see Rosalino at al., 2005b). With this data it was possible to create a distribution model for badgers in South western Portugal.

Other Data

Although not yet published, other data was collected from the radio-collared animals. For example, the sanitary status of the population and its integration with group constitution, density and distribution is being assessed by analysing faeces (endoparasites, helminths;

infectious agents), blood (infectious agents in plasma), tissue (ear tips—infectious agents) and fur (fungus) samples.

To understand the role of kinship on the spatial and social structure of our studied population, as well as to assess how the real sociability degree of the population constrains individual ecological options, tissue and blood sample are being processed by molecular techniques to integrate those ecological data with the genetic resemblance among individuals. This kinship relation will also help us comprehend the parasites' and infectious agents' persistence and dispersal in the wild population and contribute to a more effective wildlife management.

REFERENCES

Allen, AVH; Ridley, DS. Further observations on the formol-ether concentration technique for faecal parasite. *Journal of Clinical Pathology*, 1970 23, 545–546.

Antolin MF; Horne, BV; Holloway, AK; Roach, JL; Berger Jr., MD; Weeks Jr., RD. Effective population size and genetic structure of a Piute ground squirrel (*Spermophilus mollis*) population. *Canadian Journal of Zoology*, 2001 79(1), 26-34.

Ashton, KG; Tracy, MC; De Queiroz, A. Is Bergmann's rule valid for mammals? *American Naturalist*, 2000 156, 390-415.

Blumstein, DT; Patton, ML; Saltzman, W. Faecal glucocorticoid metabolites and alarm calling in free-living yellow-bellied marmots. *Biology Letters*, 2006 2, 29–32.

Boehlert, G. Application of acoustic and archival tags to assess estuarine, nearshore, and offshore habitat utilization by salmonids: Introduction and objectives of the workshop. In Proceedings of a workshop 'Application of acoustic and archival tags to assess estuarine, nearshore, and offshore habitat utilization and movement by salmonid', Seattle, Washington; 1996.

Brøseth, H; Bevanger, K; Knutsen, B. Function of multiple badger *Meles meles* setts: distribution and utilisation. *Wildlife Biology*, 1997 3, 89–96.

Charif, RA; Ramey II, RR; Langbauer Jr, WR; Payne, KB; Martin, RB; Brown, LM. Spatial relationships and matrilineal kinship in African savanna elephant (*Loxodonta africana*) clans. *Behavioral Ecology and Sociobiology*, 2005 57, 327–338.

Charles-Dominique, P. Urine marking and territoriality in *Galago alleni* (Waterhouse, 1837 – Lorisoidea, Primates): a field study by radio-telemetry. *Zeitschrift für Tierpsychology,* 1977 43, 113-138.

Cochran, WW; Lord Jr., RD. A radio-tracking system for wild animals. *Journal of Wildlife Management*, 1963 27,9-24.

Cosivi, O; Meslin, FX; Daborn, CJ; Grange, JM. Epidemiology of *Mycobacterium bovis* infection in animals and humans, with particular reference to Africa. *Revue Scientifique Et Technique*, 1995 14, 733-746.

Czeczuga, B; Godlewska, A; Kiziewicz, B. Aquatic fungi growing on feathers of wild and domestic bird species in limnologically different water bodies. *Polish Journal of Environmental Studies*, 2004 13(1), 21-31.

Dayan, T; Simberloff. D. Size patterns among competitors: ecological character displacement and character release in mammals, with special reference to island populations. *Mammal Review*, 1998 28, 99–124.

Do Linh San, E. Utilisation des terriers par le blaireau (*Meles meles*) et le renard (*Vulpes vulpes*) dans la Broye Vaudoise et Fribourgeoise. *Bulletin de la Société Fribourgeoise Sciences Naturelles,* 2002 91, 101-124.

Durette-Desset, MC; Pesson, B. *Molineus patens* (Dujardin, 1845) (Nematoda, Trichostrongyloidea) et autres espèces décrites sous ce nom. *Annales de Parasitologie Humaine et Comparée*, 1987 62, 326-344.

Feore, S; Montgomery, WI. Habitat effects on the spatial ecology of the European badger (*Meles meles*). *Journal of Zoology, London*, 1999 247, 537–549.

Fraser, NHC; Metcalfe, NB; Thorpe, JE .Temperature-dependent switch between diurnal and nocturnal foraging in salmon. *Proceedings: Biological Sciences*, 1993 252, 135-139.

Horii, K; Kito, G; Hamada, T; Jikuzono, T; Kobayashi, K; Hashimoto, K. Development of telemetry system in the common marmoset – cardiovascular effects of astemizole and nicardipine. *The Journal of Toxicological Sciences*, 2002 27,123-130.

Hudson, PJ; Dobson, AP; Newborn, D. Prevention of population cycles by parasite removal. *Science*, 1998 282, 2256-2258.

Janečka, JE; Blankenship, TL; Hirth, DH; Tewes, ME; Kilpatrick, CW; Grassman, LI. Kinship and social structure of bobcats (*Lynx rufus*) inferred from microsatellite and radio-telemetry data. *Journal of Zoology, London*, 2006 269(4), 494-501.

Kanwisher, JW; Williams, TC; Teal JM; Lawson Jr. KO. Radiotelemetry of heart rates from free-ranging Gulls. *The Auk*, 1978 95, 288-293.

Katz, MJ; George, EB. Fractals and the analysis of growth paths. *Bulletin of Mathematical Biology*, 1985 47, 273-286.

Kimmich, HP. *Artifact free measurement of biological parameters: Biotelemetry, a historical review and layout of a modern developments.* In CJ Amlaner Jr. and Macdonald DW, editors. *A Handbook of Biotelemetry and Radio Tracking*, Pergamon Press, Oxford; 1980.

Kolz, AL; Corner, GW; Johnson RE. A multiple-use wildlife transmitter. U.S. Fish Services Special Scientific Report on Wildlife; 1973.

Kowalczyk, R; Zalewski, A; Jedrzejejewska, B. Seasonal and spatial pattern of shelter use by badgers *Meles meles* in Bialowieza Primeval Forest (Poland). *Acta Theriologica*, 2004 49, 75–92.

Knowlton, FF; Martin, PE; Haug, JC. A telemetric monitor for determining animal activity. *Journal of Wildlife Management*, 1968 32, 943-948.

LaVal, RK; Clawson, RL; LaVal, ML; Caire W. Foraging behavior and nocturnal activity patterns of Missouri bats, with emphasis on the endangered species *Myotis grisescens* and *Myotis sodalis*. *Journal of Mammalogy*, 1977 58, 592-599.

Loureiro, F. Time and space use of key resources by the Eurasian badger *(Meles meles) in* a Mediterranean cork oak woodland: conservation implications. *Ph. D Thesis, University of Lisbon, Lisbon; 2008.*

Loureiro, F; Rosalino, LM; Macdonald, DW; Santos-Reis, M. Use of multiple den sites by Eurasian badgers, *Meles meles*, in a Mediterranean habitat. *Zoological Science*, 2007a 24, 978-985.

Loureiro, F; Rosalino, LM; Macdonald, DW; Santos-Reis, M. Path tortuosity of Eurasian badgers (*Meles meles*) in a heterogeneous Mediterranean landscape. *Ecological Research*, 2007b 22, 837-844.

McIntyre, NE; Wiens, JA. Interactions between habitat abundance and configuration: experimental validation of some predictions from percolation theory. *Oikos*, 1999 86, 129-137

Mech, LD. *Handbook of animal radiotracking*. Minneapolis: University of Minnesota; 1983.

Mech, LD; Barber SM. *A critique of wildlife radio-tracking and its use in National Parks: a report to the U.S.* National Park Services. Report; 2002.

Moore, JAH; Roper, TJ. Temperature and humidity in badger *Meles meles* setts. *Mammal Review*, 2003 33, 308–313.

Nams, VO; Bourgeois, M. Fractal analysis measures habitat use at different spatial scales: an example with American marten. *Canadian Journal of Zoology*, 2004 82, 1738–1747.

Neal, E; Cheeseman, C. *Badgers*. London: T and A Poyser Ltd; 1996.

Oates, JE; Bradshaw, FJ; Bradshaw, SD; Lonsdale, RA. Sex identification and evidence of gonadal activity in the short-beaked echidna (*Tachyglossus aculeatus*) (Monotremata : Tachyglossidae): non-invasive analysis of faecal sex steroids. *Australian Journal of Zoology*, 2002 50(4), 395-406.

Oliveira, M; Sales-Luís, T; Duarte, A; Nunes, SF; Carneiro, C; Tenreiro, T; Tenreiro, R; Santos-Reis, M; Tavares, L; Vilela, CL. First assessment of microbial diversity in faecal microflora of Eurasian otter (*Lutra lutra* Linnaeus, 1758) in Portugal. *European Journal of Wildlife Research*, 2008 54, 245–252.

Oliver, JHJ; Clarck, KL; Chandler, FWJ; Tao, L; James, AM; Banks, CW; Huey, LO; Banks, AR; Williams, DC; Durden, LA. Isolation, cultivation, and characterization of *Borrelia burgdorferi* from rodents and ticks in the charleston area of South Carolina. *Journal of Clinical Microbiology*, 2000 38, 120-124.

Resources Inventory Committee. *Wildlife Radio-telemetry Standards for Components of British Columbia's Biodiversity No. 5*, Prepared by Ministry of Environment, Lands and Parks Resources Inventory Branch for the Terrestrial Ecosystems Task Force; 1998.

Revilla, E; Delibes, M; Travaini, A; Palomares, F. Physical and population parameters of the Eurasian badgers (*Meles meles* L.) from Mediterranean Spain. *Zeitschrift für Säugetierkunde*, 1999 64, 269-276.

Revilla, E; Palomares, F; Fernández, N. Characteristics, location and selection of diurnal resting dens by Eurasian badgers (*Meles meles*) in a low density area. *Journal of Zoology, London,* 2001 255, 291–299.

Revilla, E; Palomares, F. Spatial organization, group living and ecological correlates in low-density populations of Eurasian badgers, *Meles meles*. *Journal of Animal Ecology,* 2002 71, 497–512.

Rodríguez, A; Martin, R; Delibes, M. Space use and activity in a Mediterranean population of badgers *Meles meles*. *Acta Theriologica*, 1996 41, 59-72.

Rodríguez, MÁ; Olalla-Tárraga, MÁ; Hawkins, BA. Bergmann's rule and the geography of mammal body size in the Western Hemisphere. *Global Ecology and Biogeography*, 2008 17, 274-283.

Rolland, RM; Hunt, KE; Kraus, SD; Wasser, SK. Assessing reproductive status of right whales (*Eubalaena glacialis*) using fecal hormone metabolites. *General and Comparative Endocrinology*, 2005 142, 308-317.

Roper, TJ. Badger *Meles meles* setts - architecture, internal environment and function. *Mammal Review*, 1992 22, 43-53.

Roper, TJ; Ostler, JR; Schmid, TK; Christian, SF. Sett use in European badgers *Meles meles*. *Behaviour*, 2001 138, 173-187.

Rosalino, LM; Macdonald, DW; Santos-Reis M. Spatial structure and land-cover use in a lowdensity Mediterranean population of Eurasian badgers. *Canadian Journal of Zoology*, 2004 82, 1493–1502.

Rosalino, LM; Macdonald, DW; Santos-Reis, M. Activity rhythms, movements and patterns of sett use by badgers, *Meles meles*, in a Mediterranean woodland. *Mammalia*, 2005a 69, 395-408.

Rosalino, LM; Macdonald, DW; Santos-Reis, M. Resource dispersion and badger population density in Mediterranean woodlands: is food, water or geology the limiting factor? *Oikos*, 2005b 110, 441-452.

Rosalino, LM; Santos, MJ; Domingos, S; Rodrigues, M; Santos-Reis, M. Population structure and body size of sympatric carnivores in a Mediterranean landscape of SW Portugal. *Revista de Biologia (Lisboa)*, 2005c 23, 135-146.

Santos-Reis, M; Correia, AI. Caracterização da flora e fauna do montado da Herdade da Ribeira Abaixo (Grândola - Baixo Alentejo), CBA, Lisboa; 1999.

Selkoe, KA; Toonen, RJ. Microsatellites for ecologists: a practical guide to using and evaluating microsatellite markers. *Ecology Letters*, 2006 9, 615–629.

Shkolnik, A. Diurnal activity in a small desert rodent. *International Journal of Biometeorology*, 1971 15, 115-120.

Skaare, JU; Bernhoft, A; Wiig, Ø; Norum, KR; Haug, E; Eide, DM; Derocher, AE. Relationships between plasma levels of organochlorines, retinol and thyroid hormones from polar bears (*Ursus maritimus*) at Svalbard. *Journal of Toxicology and Environmental Health, Part A*, 2001 62, 227-241.

Stoddart, LC. A telemetric method for detecting jackrabbit mortality. *Journal of Wildlife Management*, 1970 34, 501-507.

Taillon, J; Côté, SD. Are faecal hormone levels linked to winter progression, diet quality and social rank in young ungulates? An experiment with white-tailed deer (*Odocoileus virginianus*) fawns. *Behaviour Ecology and Sociobiology*, 2008 62, 1591-1600.

Thom, MD; Harrington, LA; Macdonald, DW. Why are American mink sexually dimorphic? A role for niche separation. *Oikos*, 2004 105, 525-535.

Twomey, DF; Crawshaw, TR; Anscombe, JE; Farrant, L; Evans, LJ; McElligott, WS; Higgins, RJ; Dean, G; Vordermeier, M; Jahans, K; de la Rua-Domenech, R. TB in llamas caused by *Mycobacterium bovis*. *Veterinary Record*, 2007 160, 170.

Veron, G; Laidlaw, R; Rosenthal, SH; Streicher, U; Roberton, S. Coat colour variation in the banded palm civet *Hemigalus derbyanus* and in Owston's civet *Chrotogale owstoni*. *Mammal Review*, 2004 34(4), 307–310.

Virgós, E; Llorente, M; Cortés, Y. Geographical variation in genet (*Genetta genetta*) diet: a literature review. *Mammal Review*, 1999 29, 117-126.

White, GC; Garrott, RA. *Analysis of wildlife radio-tracking data*. Academic Press. New York; 1990.

Woodroffe, R; Macdonald, DW; Silva, J. Dispersal and philopatry in the european badger, *Meles meles*. *Journal of Zoology, London*, 1993 237, 227-239.

Wray, S; Cresswell, WJ; Rogers, D. What, if anything, is a core area? An analysis of the problems of describing internal range configurations. In: Priede, JI and Swift, SM, editors. *Wildlife telemetry: remote monitoring and tracking of animals*, Ellis Horwood, New York; 1992.

In: Telemetry: Research, Technology and Applications
Editors: Diana Barculo and Julia Daniels
ISBN 978-1-60692-509-6

Chapter 5

BIOTELEMETRY NET FOR NEUROCHEMICAL BIOSENSOR AND MICROSENSOR APPLICATIONS: DESIGN, CONSTRUCTION AND VALIDATION

Pier Andrea Serra,[1] Martin Hebel,[2] Gaia Rocchitta,[1] and Ralph F. Tate[2]

1. Department of Neuroscience, Medical School,
University of Sassari, Viale S. Pietro 43/b, 07100 Sassari, Italy
2. Electronic Systems Technologies,
Southern Illinois University Carbondale, Carbondale, USA

ABSTRACT

Medical telemetry may be defined as "the measurement and recording of physiological parameters and other patient-related information via radiated bi- or unidirectional electromagnetic signals". Telemetric devices can be coupled with biosensors and microsensors that generate electrical signals related to electrochemical processes. Therefore, latter techniques involve the wireless *in-situ* detection of biologically-active molecules, including nitric oxide (NO), dopamine (DA), glucose, glutamate and lactate, in brain extracellular fluid (ECF), using implanted biosensors and microsensors. NO is a water-soluble free radical that readily diffuses through membranes and its actions in the central nervous system are largely studied. While low concentrations of NO modulate normal synaptic transmission, excess levels of NO may be neurotoxic. The constant-potential oxidation of NO occurs on Nafion poly-ortho-phenylenediamine (p-OPD)-coated carbon fibers at +865 mV vs Ag/AgCl reference electrode (RE) with good selectivity against electrochemically oxidizable anions. Dopamine is a catechol-like neurotransmitter having many functions in the brain including important roles in motor activity, reward and motivation but also in learning and attention. DA is oxidized to the corresponding orthoquinone (DA-OQ) on carbon fibers using different techniques including constant potential amperometry, chronoamperometry and fast scan cyclic voltammetry. However a biosensor is needed when it is not possible to directly oxidize molecules under standard conditions. Glucose, for example, is oxidized by means of the enzyme glucose oxidase (GOx), an oxidoreductase with a covalently-linked flavin adenine dinucleotide (FAD) cofactor. The reconversion of $FADH_2$ to FAD produces H_2O_2 in the presence of O_2. The application of a potential of +700mV to a platinum (Pt) working electrode, relative to RE, causes the generation of a measurable current that is proportional to the H_2O_2 produced. The high specificity and stability of GOx makes this enzyme suitable for biosensor construction when immobilized on the surface of Pt electrodes with p-OPD. The substitution of the GOx with glutamate or lactate oxydases, and the modification of biosensor design, allow detecting, respectively, glutamate and lactate, normally present in the ECF. While glucose and lactate are involved in the neural energetic metabolism, glutamate (GLUT) is considered the widespread excitatory neurotransmitter in the brain. Contrary to expectations, long term potentiation in the glutamatergic transmission determines an increasing of GLUT and NO in ECF with excitotoxic effects. Several biotelemetry systems for electrochemical sensor applications have been developed; in particular, a new parallel-computing embedded biotelemetry system is described in this chapter. The device, capable of driving two independent sensors, consists of a single-supply bipotentiostat-I/V converter, a parallel microcontroller unit (pMCU), a signal transmitter, and a stabilized power supply. The introduction of parallel computing programming techniques, could permit the development of low-cost devices, reduce complex multitasking firmware development, and offer the possibility of expanding the system simply and quickly. The use of CMOS components and careful electronics design will make miniaturization possible and permit implantable devices suitable for *in-vivo* applications.

INTRODUCTION

Medical telemetry is the measurement of a biomedical parameter at a distance [1]. The most important aim of biotelemetry is the rapid recognition of significant changes in physiological parameters, such therapeutic treatment that occur early [2]. This technique allows, for example, real-time reading of glucose levels in diabetic patients [3], critical care [4] and brain injuries [5]. In recent years sophisticated biotelemetry systems have been developed for research purposes for analyzing neurochemical data such as brain variations of dopamine in freely moving animals [6]. Telemetric devices can be coupled with microsensors or biosensors capable of generating electrical signals related to electrochemical processes [7].

Several biologically oxidable molecules can be directly detected on the surface of amperometric sensors connected to a potentiostat [8]. Figure 1 illustrates a typical carbon-based amperometric microsensor on the surface of which several molecules can be oxidized such as dopamine (DA) and nitric oxide (NO).

DA is a catechol-like neurotransmitter that has been implicated in cognitive functions [9], in reward pathways [10] and in Parkinson's disease [11-13]. A constant-potential oxidation of DA to the corresponding orthoquinone (DA-OQ) is possible on Nafion®-coated carbon fibers with a good selectivity against electrochemically oxidizable anions [14]:

$$DA\ (+500\ mV) \longrightarrow DA\text{-}OQ + 2e^- + 2H^+$$

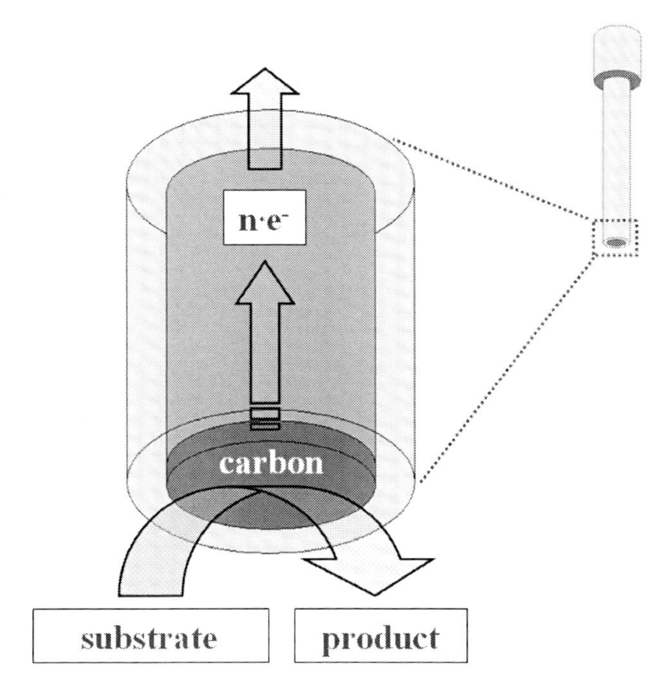

Figure 1. Schematic of a generic amperometric sensor used for neurochemical monitoring. The sensor is a simple Teflon-insulated platinum wire coated with a carbon-epoxy layer. When a positive potential (V_{App}) is applied to carbon surface, several biological molecules can be oxidized, transferring n electrons ($n{\cdot}e^-$) to input electronics and the subsequent generation of an anodic current. A further miniaturization of the sensor is possible using carbon fibers. The surface treatment of the carbon with Nafion® or/and p-OPD increases sensor selectivity to dopamine or nitric oxide.

NO is an essential molecule in biological systems involved in important roles such as blood pressure regulation, neuromodulation [15] and immune reaction [16]. High concentrations of NO have been reported to be present in different neurodegenerative conditions such as Alzheimer's and Parkinson's diseases [11, 12]. The electro-oxidation of NO occurs on carbon fiber at high working potential (+865 mV vs. Ag/AgCl): the sensor surface needs a permselective polymer to prevent interferences from other oxidizable biological compounds [17, 18]. The electrochemical oxidation of NO at the carbon surface occurs through a one-electron transfer process from NO to the working electrode/sensor:

$$NO\ (+865\ mV) \longrightarrow NO^+ + e^-$$

NO^+ subsequently leads to NO_2^- and NO_3^-.

When direct electrochemical oxidation of interesting biological compounds is not possible in standard experimental conditions, a biosensor may then be used instead [19]. A biosensor is a device that combines a transducer with a biologically selective and sensitive element [20]. A schematic representation of an oxidase-based biosensor is shown in Figure 2.

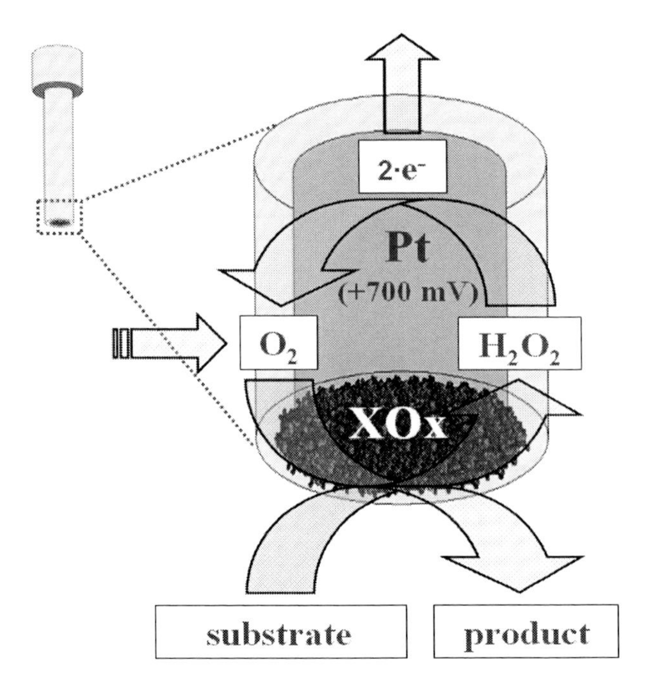

Figure 2. Schematic of an amperometric biosensor used for detecting non-oxidizable molecules. The biosensor is a simple Teflon-insulated platinum wire coated with an oxidase enzyme (XOx) immobilized by a non-conducting polymer (p-OPD). A neurochemical substrate (glutamate, lactate or glucose) is oxidized by the enzyme with the by-production of hydrogen peroxide in the presence of molecular oxygen. When a positive potential (V_{App} = +700 mV) is applied to the platinum surface, H_2O_2 is oxidized thus transferring two electrons ($2 \cdot e^-$) to input electronics and the subsequent generation of an anodic current. Biosensor selectivity to glutamate, lactate or glucose is related to the specific enzyme immobilized on platinum surface.

Glucose is the most important substrate for the production of ATP and its extracellular concentration [21] plays a key role in the energetic metabolism [22, 23]. Glucose is oxidized by the enzyme glucose oxidase (GOx), an oxidoreductase with a covalently linked flavin adenine dinucleotide (FAD) cofactor [24]:

$$\beta\text{-D-GLUCOSE} + \text{FAD+-oxidase} \rightarrow \text{D-GLUCONO-}\delta\text{-LACTONE} + \text{FADH}_2\text{-oxidase}$$

The oxidation of $FADH_2$ to FAD^+ produces H_2O_2 in the presence of O_2, as follows:

$$\text{FADH}_2\text{-oxidase} + O_2 \longrightarrow \text{FAD-oxidase} + H_2O_2$$

The application of a potential of +700 mV to the Pt working electrode, relative to an Ag/AgCl reference electrode, causes the following reaction:

$$H_2O_2 \longrightarrow O_2 + 2e^- + 2H^+$$

generating a current proportional to the concentration of the H_2O_2. The high stability and specificity of GOx makes this enzyme suitable for biosensor construction [25] when immobilized on the surface of platinum electrodes using poly-*o*-phenylenediamine [21]. Replacing GOx with glutamate or lactate oxydase, and modifying the biosensor design, allows detecting, respectively, of glutamate [26-31] and lactate [26, 32], normally present in the ECF:

L-GLUTAMATE + FAD+-oxidase → 2-OXOGLUTARATE + $FADH_2$-oxidase

L-LACTATE + FAD+-oxidase → L-PYRUVATE + $FADH_2$-oxidase

The back-oxidation of $FADH_2$ to FAD^+ produces H_2O_2 as previously described.

In this chapter we describe a fully automated multi-channel biotelemetry system, derived from previously published devices [33-35], which can be used with micro-biosensors for the measurement of DA, NO, glutamate and other electrochemically oxidizable molecules *in vitro* and potentially *in vivo*. The simultaneous implantation of several microsensors or biosensors or the development of a multisensor allows the study of complex biochemical pathways as illustrated in Figure 3.

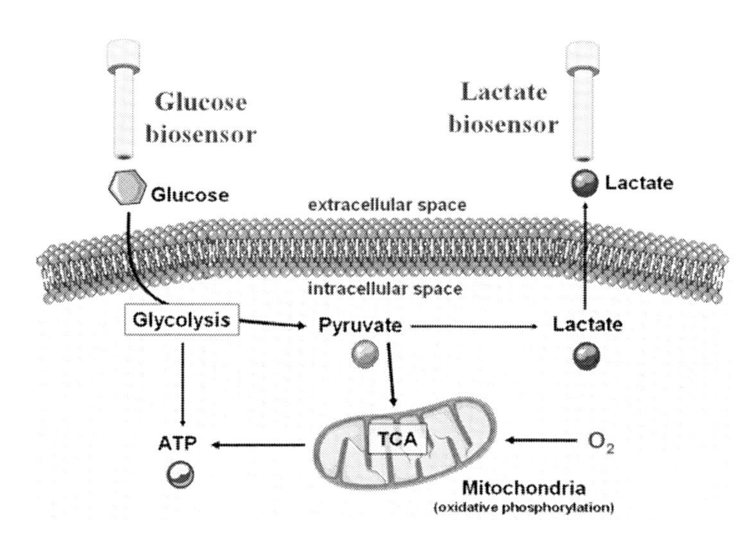

Figure 3. Schematic representation of brain energetic metabolism. Glucose is used for energy production (ATP) both in presence (tricarboxylic acids cycle, TCA) and in absence of molecular oxygen. During brain ischemia glucose is oxidized with the production of lactate. The simultaneous implantation of several microsensors or biosensors (glucose and lactate) or the development of a multisensor allows the real-time study of complex biochemical pathways of brain energetic metabolism.

THE POTENTIOSTAT

A potentiostat is an electronic device that controls the voltage difference between a reference electrode (RE) and a working electrode (WE) independently from the current (i_{redox}) flowing in the WE. The potentiostat fixes the oxidation potential (V_{App}) to allow optimal conditions for transferring electrons from oxidizable molecules to the WE. Both RE and WE are immersed in a conductive solution contained in an electrochemical cell. The potentiostat implements this control through the feedback circuit illustrated in Figure 4 by injecting current into the cell through an auxiliary, or counter, electrode (Aux/CE).

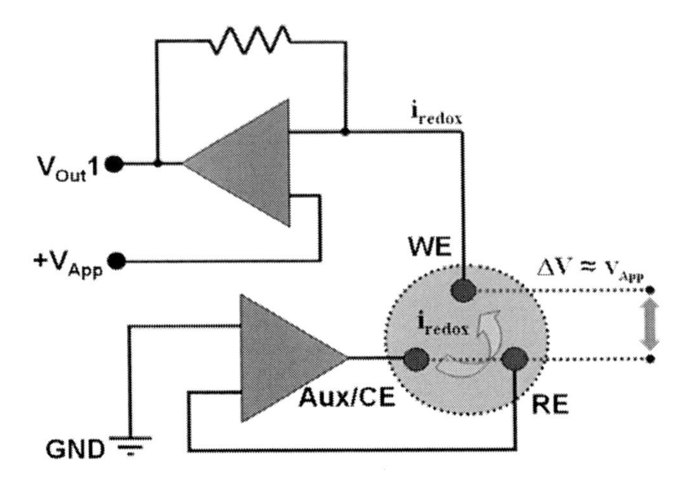

Figure 4. Three-electrode potentiostat schematic. The circuit function is to maintain a stable potential ($\Delta V \approx V_{App}$) between RE and WE while a redox current (i_{redox}) flows in the WE as the result of the electrochemical reaction on the WE surface. The potentiostat implements this control through a feedback circuit by injecting current into the cell through an auxiliary, or counter, electrode (Aux/CE). The current i_{redox} is converted to a corresponding voltage ($V_{Out}1$) by a current-to-voltage converter illustrated in detail in Figure 7.

The aim of this circuit is to maintain a stable redox potential between RE and WE while an oxidation current flows in the WE as the result of the electrochemical reaction on the WE surface. The role of Aux/CE is to compensate the current flowing in WE. The practical implementation of this circuit is realized using two operational amplifiers as illustrated in Figure 5.

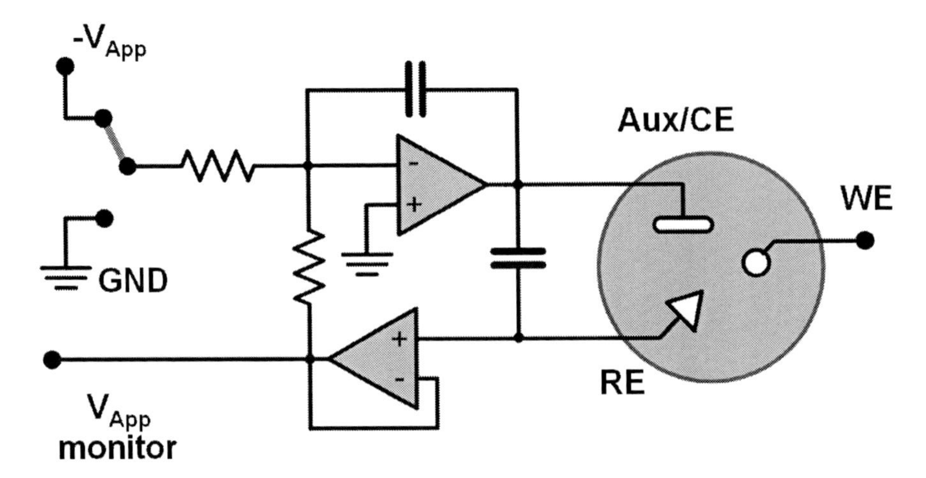

Figure 5. Practical implementation of a three-electrode potentiostat realized using two operational amplifiers. Two possible configurations of the circuit are possible depending on the potential (V_{App}) of WE relative to RE (see text). The applied potential can be monitored (V_{App} monitor) using an analog-to-digital converter (ADC) or a simple voltmeter.

Two possible configurations of the potentiostat circuit are possible depending on the potential of WE relative to RE: connecting the potentiostat to $-V_{App}$ and WE to GND or connecting the potentiostat to GND and WE to V_{App}. In both cases the oxidation potential drives the electrochemical reaction on the WE. The applied potential ($-V_{App}$) is fixed to the desired value using a potentiometer and a voltage follower as showed in Figure 6.

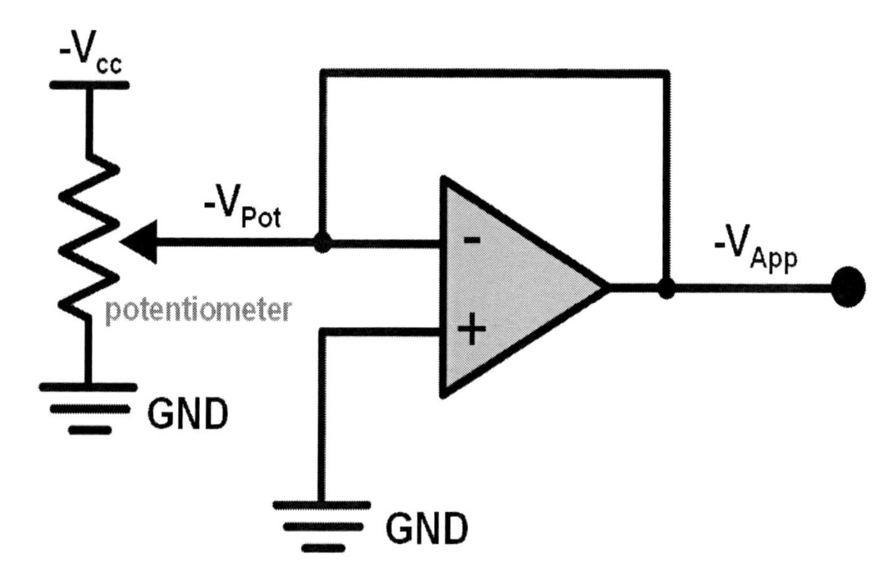

Figure 6. Voltage follower circuit used for stabilizing the applied potential (V_{App}). The sign of V_{App} depends on potentiostat and current-to-voltage converter configurations (see text). The applied potential is fixed to the desired value using a potentiometer.

When the potentiostat is grounded, the voltage follower circuit is used to polarize the WE to $+V_{App}$. In this case the potentiometer is connected to $+V_{cc}$. Indeed, the current-to-voltage converter circuitry is dependent on the potentiostat electronics, as will be discussed in the next paragraph. The characteristics of the potentiostat depend on the right OPA selection: very high input impedance (GigaΩ or TeraΩ) and low output resistance are fundamental to guarantee no current flowing in the RE and maximal current flowing from Aux/CE. To obtain the highest potentiostatic performance of the described circuits, it is necessary to select the right RE and Aux/CE. The RE is used for fixing and quantifying the WE potential. An ideal RE should have a constant electrochemical potential and it is necessary that no current flows through it. The most used REs are the saturated calomel electrode and the silver/silver chloride (Ag/AgCl) electrode, even if sometimes a pseudo-reference electrode is used (a piece of silver wire). The Ag/AgCl RE, for example, works as a redox electrode and the reaction occurs between the silver metal (Ag^0) and its corresponding salt (AgCl):

$$Ag^0 + Cl^- \leftrightarrow AgCl + e^-$$

no current flows in the RE because of this surface reaction. The standard electrode potential E^0 against the hydrogen electrode is about 0.230V [36].

In a three electrode system, Aux/CE allows that current does not pass through the RE and its surface area, much larger than that of the WE, ensures that the reactions occurring on the

WE are not surface area limited by the Aux/CE. As a consequence of an electrochemical reaction on WE surface, the Aux/CE polarity goes in the opposite direction. The most common Aux/CEs are large surface area (compared to WE) platinum wires.

The Current-to-voltage Converter

An operational amplifier can be used to easily convert the signal from a sensor that produces an output current, such as an amperometric micro-biosensor, into a voltage [37]. This is implemented with a single resistor and an optional capacitor in the feedback loop of the amplifier as shown in Figure 7.

The first electronic circuit connected to the WE is the current-to-voltage (I/V) converter. This circuit has the dual role of biasing the WE and converting the redox currents generated on the WE surface to a voltage easily readable by a voltmeter or an analog-to-digital converter (ADC). If an FET or CMOS OPA is used, the high input impedance of the OPA lets the current flow from the WE to the path of lower resistance R_f. The I/V converter is realized by an integrated (inverting) transimpedance amplifier composed by a single operational amplifier (OPA), a resistor (R_f) and a capacitor (C_f). The WE is directly connected to the inverting input of the OPA while the non-inverting input is connected to V_{App} or grounded (V_{App} = 0V) as previously described. Because the two OPA's inputs are at the same potential, there are two possible configurations of I/V converter related to the external connection of non-inverting input and the potential applied to it: this potential determines the potential of the WE, while the amplitude of the electric field between WE and RE is related to the potentiostat design. The gain of the I/V converter is fixed selecting the right value of R_f following the transfer function:

$$V_{Out}1 = -(i_{redox}R_f) + V_{App} \qquad (1)$$

where i_{redox} is the redox current flowing through the WE, R_f is the feedback resistor and V_{App} is the potential applied to the non-inverting input of the OPA. The capacitor C_f, in parallel with the resistor R_f, completes a low pass filter having a predefined cut-off frequency (F_{c-o}) and a value (C_f), calculated in Farads, according with the equation:

$$C_f = (F_{c-o}2\pi R_f)^{-1} \qquad (2)$$

The precision of the I/V conversion depends on different factors, including right OPA selection and R_f and C_f characteristics. BicMOS™ and DiFET™ technologies guarantee high input resistance (GigaΩ or TeraΩ), ultra-low input bias currents (a few femto-amperes), low input offset current and voltage and low power consumption. Modern CMOS fabrication processes allow operating from a single supply, having output swing extending to each supply rail (rail-to-rail output) and an input voltage range very close to power supply rails (rail-to-rail input) including ground in single supply configuration [38]. R_f should be a precision metal oxide thick film resistor with very high ohmic value (1MΩ-1GΩ), low power (125-250 mW) and small tolerance (0.1%). C_f should be a multilayer ceramic capacitor (NP0-type). It is generally recognized that any circuit which must operate with less than 1 nA of leakage current requires special layout of the PC board (PCB) [39]. Because of the ultra-low bias current of the OPA, typically in the fA current range, it is essential to have an excellent PCB layout. Under conditions of high humidity, contamination or dust, the surface leakage of the PCB will be appreciable. To minimize the effect of any surface leakage, a "guard ring" should be integrated in the design completely surrounding the OPA's inputs and the terminals of R_f and C_f connected to the op-amp's inputs, as illustrated in Figure 7.

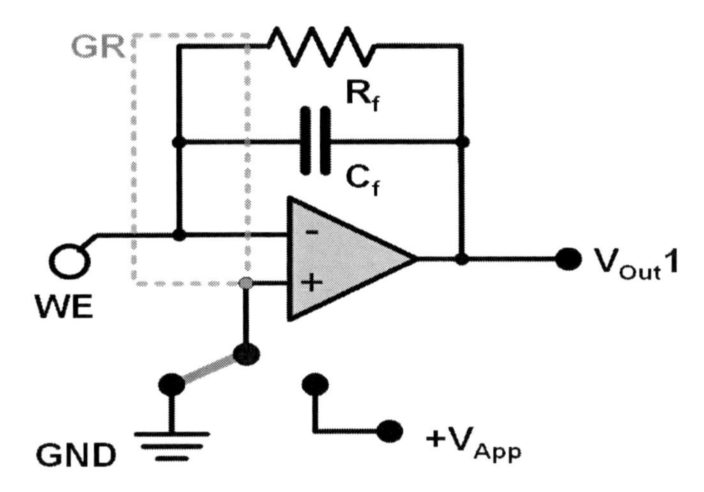

Figure 7. Current-to-voltage (I/V) converter. This circuit has the dual role of biasing the WE and converting the redox currents generated on the WE surface in a voltage (V_{Out1}) easily readable by an analog-to-digital converter (ADC) or a voltmeter. This is implemented with a single resistor (R_f) and an optional capacitor (C_f) in the feedback loop of the amplifier. Two configurations of the circuit are possible depending on the potential (V_{App}) of WE relative to RE (see text). To minimize the effect of any surface leakage, a "guard ring" (GR) should be integrated in the design.

To have a significant effect, "guard rings" should be placed on both the top and bottom of the PCB then connected at the same voltage as the amplifier inputs (V_{App} or ground) since no leakage current can flow between two points at the same potential. A simple alternative to "guard rings" is the use of point-to-point "up-in-the-air wiring" as shown in Figure 8.

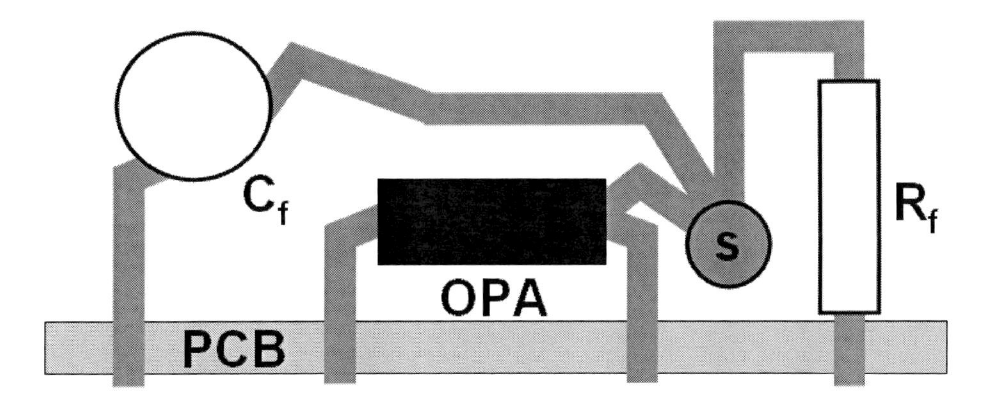

Figure 8. Noise reduction trick using air wiring of through-hole components in the current-to-voltage converter circuit. Components' pins are soldered to PCB except OPA's input pins that are bent and lifted out of the PCB, then directly soldered (S) to Rf and Cf terminals.

In this design, the OPA's input pins are not inserted into the PCB but are bent up in the air and directly soldered to R_f and C_f terminals using only air as an insulator which is an excellent insulator. While the "guard ring" solution is achievable both on through-hole and surface-mount components, the "up-in-the-air wiring" alternative is easy realizable only using through-hole components.

The Second Amplification Stage

The design of the second stage of amplification depends on the circuitry of the potentiostat and I/V converter. When the potentiostat inverting input is connected to $-V_{App}$ and the WE is grounded, the second stage of amplification is constituted by a simple inverting amplifier while a differentiator is needed when the potentiostat is grounded and the WE potential is fixed to V_{App}. The transfer function of the inverting amplifier (Figure 5) is:

$$V_{Out2} = - V_{In} (R_2/R_1) \quad (3)$$

The main role of the inverting amplifier is to invert the phase of V_{In} (changing the sign of the input current) offering the possibility of a further amplification of the WE signal increasing the value of R_2. The global transfer function, obtained from combining the first (1) and the second stage of amplification (3) and remembering that WE is grounded [V_{App} of (1) equals to 0V] is:

$$V_{Out2} = (i_{redox}R_f) (R_2/R_1) \quad (4)$$

The circuit schematic of the inverting amplifier is illustrated in Figure 9.

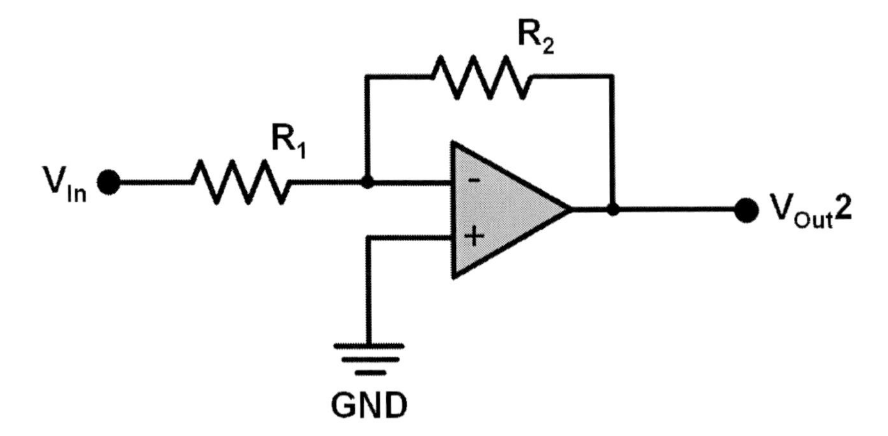

Figure 9. Inverting amplifier used in the second stage of amplification of the sensor signal. The main role of this circuit is to change the sign of the input current (V_{In}) offering the possibility of a further amplification of the WE signal increasing the value of R_2 and obtaining an output voltage (V_{Out2}) readable by an analog-to-digital converter (ADC) or a voltmeter.

The difference amplifier circuit is used when V_{App} is applied on the WE and has the dual role of subtracting V_{App} and amplifying the input signal. The transfer function of the differentiator is:

$$V_{Out2} = (V_{In}-V_{App})\,(R_2/R_1) + V_{Ref}(R_2/R_1) \quad (5)$$

where: V_{In} and V_{App} are the signals applied to the inputs and V_{Ref} is usually equal to 0V because the R_2 resistor, connected to OPA's non-inverting input, is grounded, then:

$$V_{Out2} = (V_{In}-V_{App})\,(R_2/R_1) \quad (6)$$

Figure 10 shows the difference amplifier previously described.

The combination of the two stages of amplification [(1) and (6)] results in the following equation:

$$V_{Out2} = -(i_{redox}R_f)\,(R_2/R_1) \quad (7)$$

The transfer function is identical to (4) with the exception of the sign: the differentiator (5, 6), indeed, does not invert the signal V_{In} applied to its inverting input.

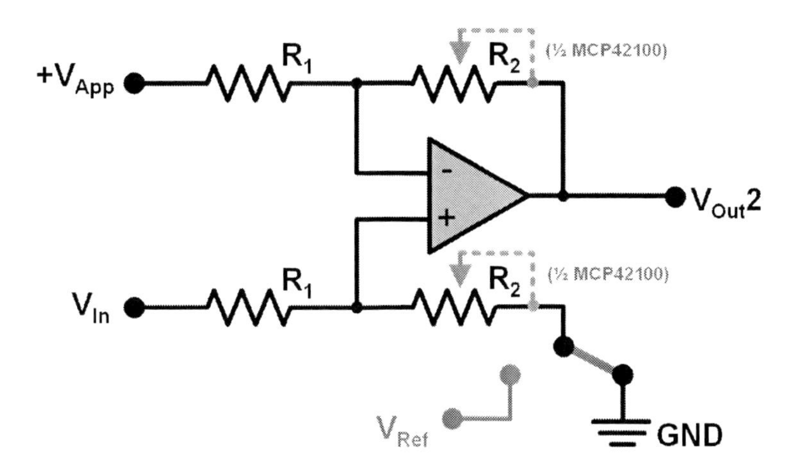

Figure 10. Difference amplifier (differentiator) implemented as second stage of amplification in alternative to the inverting amplifier previously described (see text). This circuit is used when V_{App} is applied on WE and has the dual role of subtracting V_{App} and amplifying the input signal (V_{In}) obtaining an output voltage (V_{Out2}) readable by an analog-to-digital converter (ADC) or a voltmeter. The gain of the differentiator can be digitally controlled using a dual digital potentiometer (MCP42100).

In both circuits (inverting amplifier and differentiator) the resistors R_1 and R_2 typically have the same value (10 KΩ – 100 KΩ), fixing the gain of the second stage of amplification to one, and are selected for having high thermal stability and small tolerance (0.1%). In this manner the transfer error is reduced and the amplification of the entire circuit, including I/V converter, is determined exclusively by R_f. The difference amplifier combines the inverting amplifier and non-inverting amplifier circuits into a signal block that subtracts two signals [37]. This circuit configuration will take the difference of two input signals as long as the signal source impedances are low. Errors can occur if the two signal source impedances are mismatched. Additionally, there will be a signal loss due to the voltage divider action between the source and the input resistors to the difference amplifier, if the signal source impedances are high, with respect to R_1 [37]. The real limitation of the differentiator circuit is mainly related to errors derived from R_1 and R_2 mismatches. To try to minimize this error and control the gain of the differentiator, a dual digital potentiometer (DPOT) can be used in the rheostat configuration with a very high degree of precision [33]. Using high precision DPOTs such as Microchip MCP42100, the nominal resistances between the two internal potentiometers, in a single MCP42100 part, are ratio-matched to a very small percentage (<0.2%). Common-mode rejection (CMR) error attributed to resistor mismatches has been demonstrated with a 12-bit, 5V system, with a gain of 100 V/V, the common-mode voltage (CMV) ranges 0–5V and the R_2 matching error is ±0.2%. The error of this type of common-mode excursion is equal to 0.2mV: five times less than 1 LSB [40]. Using a dual DPOT the value of R_2 could be digitally defined as [41]:

$$R_2 = (100\ \text{k}\Omega\ D_N)\ 2^{-N} + R_w \tag{8}$$

in which D_N is the 8-bit digital code (in decimal form) that is used to program the DPOT (0–255), N is the number of digital potentiometer bits (8 bits) and R_w is the parasitic resistance through the wiper of the DPOT (125Ω). The low tolerance resistors (0.1%) and the laser trimming process of production of the MCP42100 (mismatches between two internal potentiometers <0.2%) reduce the amplification error in the differentiator circuit [40, 42]. A V_{Ref} potential could be applied to the non-inverting input of the differentiator through R_2 (Figure 6) to introduce an offset in $V_{Out}2$ or to perform more complex single-supply designs based on supply splitters [37]. The introduction of monolithic differentiators such as the Maxim® MAX4199 or the Analog Devices® AD626 reduces the error sources previously described, allowing automated gain control and low-pass filtering of input signals [43].

Biotelemetry Unit Description

The biotelemetry unit comprises three different parts consisting of the power section, the amperometric and the digital sections. The power section of the device is stabilized using Linear® (LM1086 -5.0 and -3-3) voltage regulators. The two channel amperometric section is built around the integrated circuit (IC) MCP6042, a dual single-supply operational amplifier (OPA) from Arizona Microchip®. The following description concerns a single analog channel. The MAX5156 from Dallas-Maxim semiconductors® is the dual channel DAC that produces the voltage necessary to polarize the sensors between 0 and +2V according with the following equation:

$$V_{App} = (V_{Ref} D_N) / 2^N \tag{9}$$

where V_{Ref} is the reference voltage generated by the MAX6126, D_N the numeric value of the MAX5156 binary input code (0–4095) and N is the number of DAC bits (12 bits). A single-supply dual OPA stabilizes oxidation. The current-to-voltage (I/V) converter is a single-supply adaptation of a classic transimpedance amplifier [34] with a fixed gain depending on the value of R_f. The difference amplifier is built around ½ of an MCP6042 and has the dual role of subtracting the potential applied to the WE [33] and amplifying the resulting signal. The gain control in the differentiator has been implemented using a MCP42100 dual digital potentiometer (Arizona Microchip®) configured as illustrated in Figure 10. An alternative and more precise solution has been obtained using the AD626 from Analog Devices®. Moreover, this single supply, high precision differentiator, allows a pin selectable gain of 10X or 100X. Fixing R_f to 10 MΩ, the combination of the two stages of amplification provides selectable full scale ranging from 1 nA/V up to 100 nA/V. The Propeller™ microcontroller is the core of the digital module (Figure 2). This is a 32-bit multi-core CMOS IC from Parallax® with low-power features consisting of eight 32-bit independent processors called COGs. The pMCU, working at 80 MHz, controls the ADC, the DAC and all other connected peripherals as illustrated in Figure 11.

The 2.4 GHz transceiver, soldered on the biotelemetry unit, is an XBee PRO™ from Digi International®. It is a radio system-on-a-chip (SoC) module that, in conjunction with the pMCU, allows the realization of bi-directional data transfer devices, ranging up to 1000 meters and working at the maximum data rate of 250 Kbit. On the PC side, a USB/XBee dongle from Surveyor Corporation interfaces the software with an XBee PRO™ module through the CP2102 serial-to-USB converter (Silicon Labs®). A picture of the biotelemetry unit prototype is showed in Figure 12.

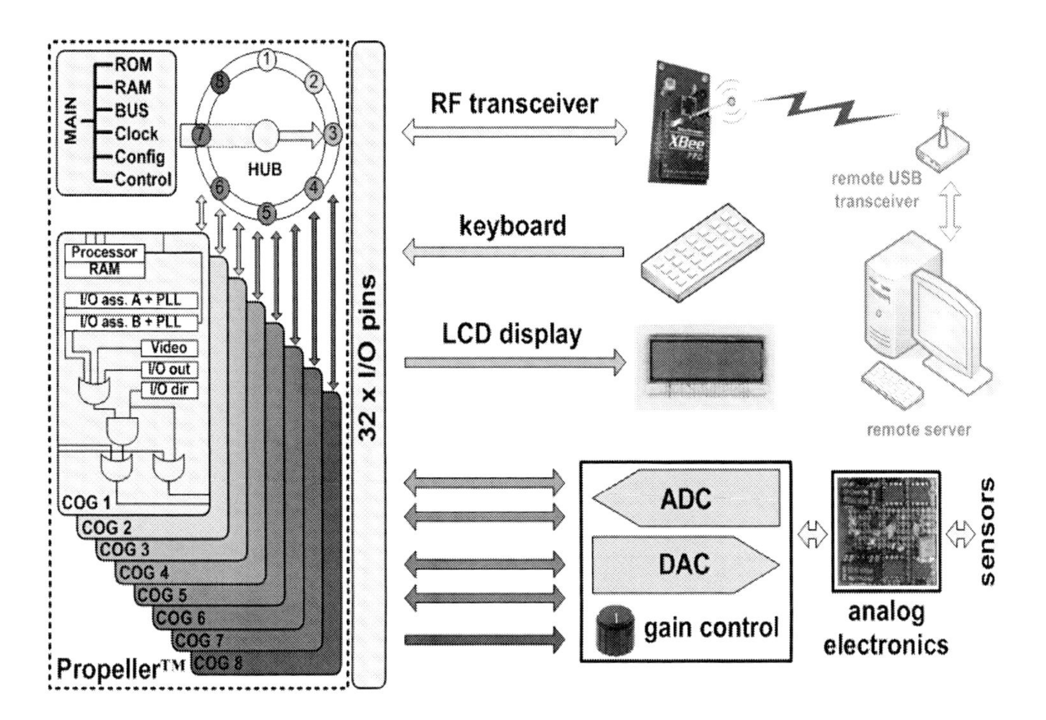

Figure 11. Schematic representation of the biotelemetry unit and peripherals connected to it. The Propeller™ pMCU contains eight identical 32 bit processors called COGs. The COGs perform tasks simultaneously either independently or with coordination from other COGs through main RAM.

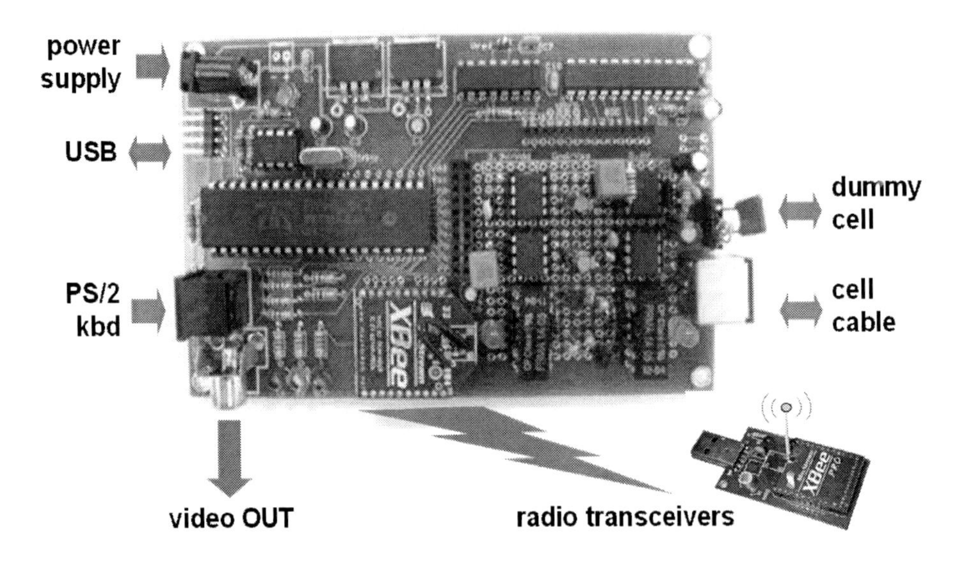

Figure 12. Picture of the biotelemetry unit prototype showing external connections. All pMCU peripherals have independent data and control lines to prevent hardware collisions during parallel-computing.

The biotelemetry unit can be interfaced with several peripherals and directly controlled by a keyboard. Data acquisition sample rate, gain and waveform generation can be controlled locally or remotely (from PC) for each channel. As illustrated in Figure 13, it is also possible to show and plot data locally using a LCD display driven by a single COG of the Propeller chip.

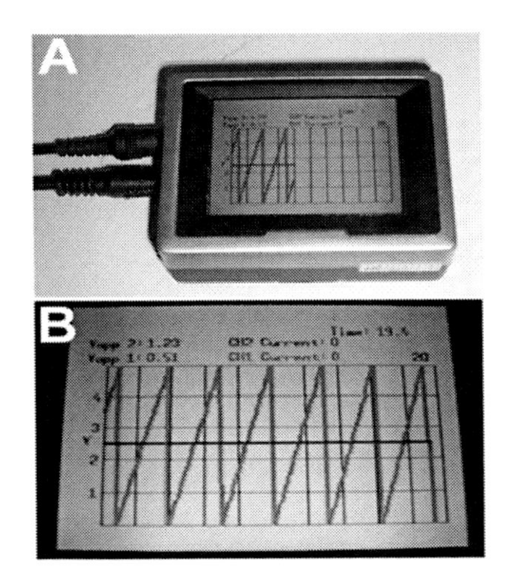

Figure 13. The biotelemetry unit can be controlled locally (using a PS/2 keyboard) or remotely (from PC). It is also possible to show and plot data locally using a LCD display driven by a single COG. Panel A: video composite signal interfaced with a 2.5' LCD display. Panel B: linear sweep waveform generated by the unit's DAC and plotted on LCD display.

THE DISTRIBUTED SENSOR NET

The distributed sensor NET has been developed using the XBee® platform [35]. The Digi International® XBee PRO™ module is a wireless transceiver using the IEEE 802.15.4 Low-Rate Wireless Personal Area Network protocol (LR-WPAN) for Wireless Sensor Networks (WSN). The devices use clear channel assessment (CCA) on a CSMA/CA network which helps ensure devices do not talk-over one another. The main features of XBee PRO™ can be summarized as follows:

- 2.4 GHz DSSS (Direct Sequence Spread Spectrum).
- Speed up to 250,000 bits per second.
- Error check, Acknowledgement and retry.
- Addressable with more than 65,000 addresses available.
- Point-to-Point and Point-to-Multipoint (broadcast) messaging.
- Channel and Network ID selectable.
- Fully configurable via serial commands.
- Transparent transmission and reception
- Flow Control.
- Receiver Signal Strength Indication (RSSI).

In addition, the transmission range of the modules reaches 1500 meters (outdoor with sufficient antenna height), the power consumption is not dramatically high in transmission mode while it is very low in power saving mode. Freely available configuration software (X-CTU™) allows full configuration of the module or it may be configured by the pMCU using a transparent serial protocol. Flow control is ensured through RTS and CTS serial lines allowing a transparent and direct interfacing with microcontrollers and USB-to-serial devices as illustrated in Figure 14.

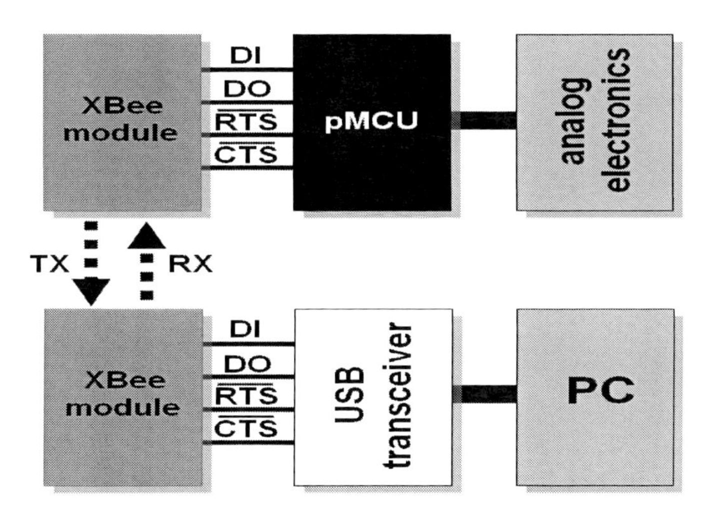

Figure 14. Schematic of the connections of a XBee PRO™ module with the pMCU and the PC. The flow control is ensured through RTS and CTS serial lines allowing a transparent and direct interfacing with microcontrollers and USB. DI: data in; DO: data out; RTS: request to send; CTS: clear to send.

The IEEE 802.15.4 protocol provides error checking and retries during transmission providing the user error free data. This alleviates the need for user developed low-level error checking (CRC or checksum algorithms) needed in complex TX-RX packet transmissions. The XBee PRO™ nodes allow addressable communications between modules. Data may be sent to individual modules (point-to-point), or to all modules in range (point-to-multipoint) using a broadcast address as illustrated in Figure 15.

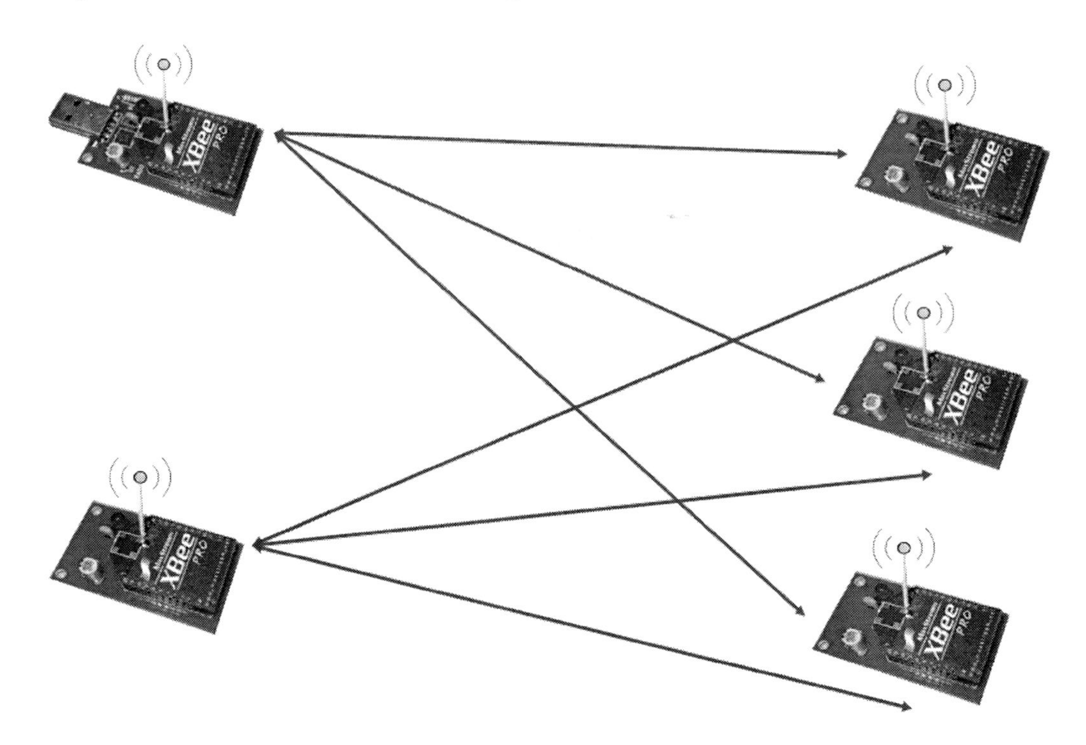

Figure 15. Distributed sensor NET developed using the ZigBee® Platform. The XBee PRO™ nodes allow addressable communications between modules. Data may be sent to individual modules (point-to-point), or to all modules in range (point-to-multipoint), using a broadcast address (see text).

In point-to-point communications data delivery is ensured by automated error checking, acknowledgements and retries providing a reliable data path. The XBee PRO™ module has a small socket for connecting a larger dipole antenna or is available with a small chip antenna directly soldered on the SoC. All transmission experiments have been made indoor and no communication problems were observed modifying the transmission distance in a linear range between 0 and 50 meters. The bidirectional telemetry, in particular, allowed the implementation of a distributed channel-sharing NET with error-free data transmission. The 25 Hz sample rate, used in the experiments, is sufficient to detect sub-second variations of oxidation current compared with previous studies [6, 44, 45].

FIRMWARE AND SOFTWARE

The firmware to drive the pMCU is developed in Spin™, a specialized interpreted language, freely available from www.parallax.com, and then transferred to the Propeller™ USB as illustrated in Figure 16.

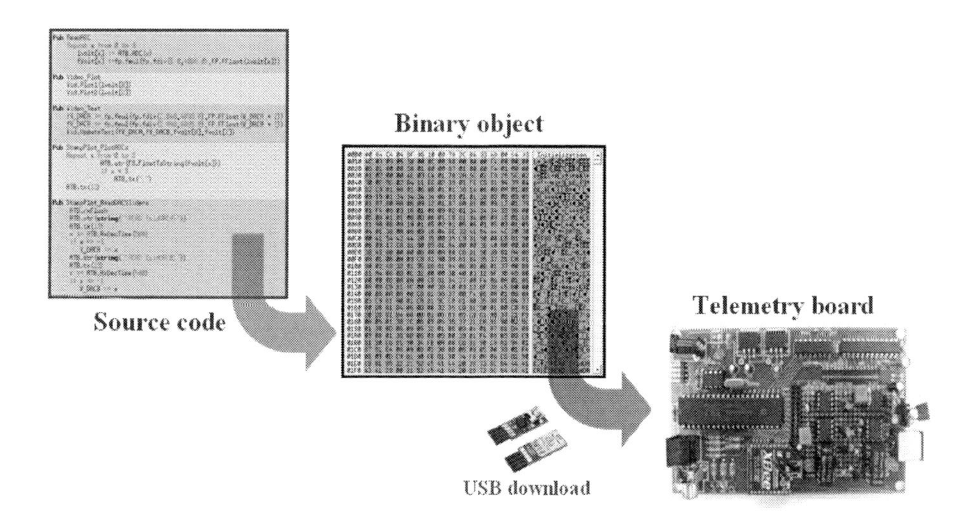

Figure 16. A specialized interpreted language called Spin™, running on a Windows®-based PC, allows the development of the firmware to drive the pMCU. The compiled program is easily transferred to the Propeller™ through the Universal Serial Bus (USB).

The compiled firmware consists of a main routine that is divided in two sections of code.

The first section initializes the communication port and all the other modules of the unit while the second section loads and starts specialized routines in the COGs' RAM. The sensor biasing waveforms are generated using a two-channel digital-to-analog converter (DAC) independently driven by two independent COGs. The resulting oxidation currents are converted into digital values using two more COGs interfaced to a multi-channel analog-to-digital converter (ADC). Six analog signals, V_{Out2}, V_{App} and V_{Bat} (+8.4 V and +5 V) are acquired "on demand" and sent to the transmitter module. Only for the V_{Out2} signals, digitized with a sample period of ~40 microseconds, the hardware ADC resolution (12 bit) is improved following the oversampling and averaging method [33]:

$$F_{os} = 4w \bullet F_s \qquad (10)$$

where w is the number of additional bits of resolution (4), F_s is the chosen sampling frequency (50 Hz) and F_{os} is the derived oversampling frequency (12.8 KHz). In accordance with the Nyquist's theorem F_{max}, the highest frequency of the input signal, is calculated as follows:

$$F_{max} = \frac{1}{2} \bullet F_s \qquad (11)$$

where F_{max} is equal to 25 Hz. To do that, the Propeller™ acquires and accumulates 256 consecutive samples in 20 ms (78.125 μS per sample) then right-shifted the sum by one bit (divided by 2). This technique increases the ADC resolution from 12 to 16 bits. Depending on the amplitude of the input signal, the sampling resolution was increased by reducing the MAX1270 full scale from +10 V to +5 V and then adjusting differentiator gain.

The remaining COGs interface a keyboard, generate a video composite signal (for monitoring sensor currents using a local LCD screen) and drive an XBee PRO™ radio transceiver (for remote monitoring).

DAC, ADC and all pMCU peripherals have independent data and control lines to prevent hardware collisions during parallel-computing. No hardware-access collisions or analog channel cross-interferences are observed during the implementation and testing of parallel-computing algorithms. A completely customizable data acquisition software package (StampPlot™ Pro by SelmaWare Solutions) allows recording, plotting and displaying the received data on a PC. The communication starts with a request directed to a specific unit and ends with the corresponding response. A software alarm is generated when a data reception time-out occurs or the battery level in the biotelemetry unit is too low (< 7.2 V). The application can interface up to 65535 biotelemetry units and connect them with a printer via USB and a Local Area Network (LAN) and Internet via TCP/IP. As shown in Figure 17, using desktop, laptop or PPC computers, several configurations of the biotelemetry NET are possible.

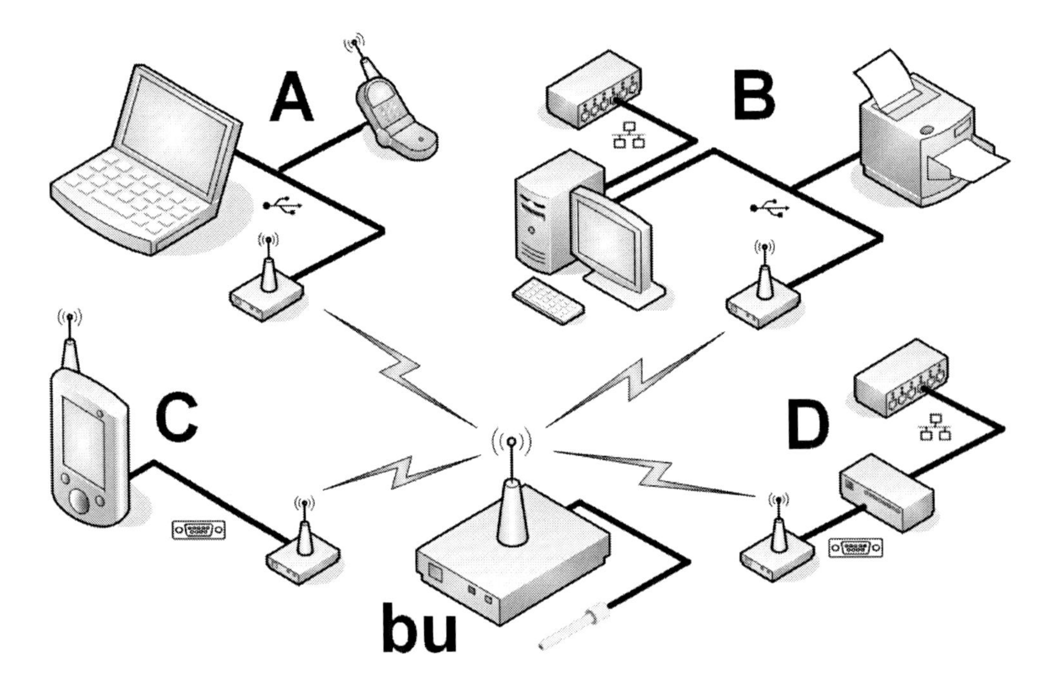

Figure 17. Schematic representation of the channel-sharing NET proposed in this chapter. Basically a biotelemetry unit (bu) connects to a PC using a Wireless Sensor Network protocol. Several configurations of the biotelemetry NET are possible by means of laptop (A), desktop (B), and PPC (C) computers or wiring a XBee PRO™ module to a serial-to-TCP converter (D).

The software, running on the Pocket PC (Windows CE), for example, has been developed using Embedded Visual Tools™ 3.0 from Microsoft and communicates with the biotelemetry units by using the integrated serial port wired to an XBee PRO™ module.

ANALOG ELECTRONICS CALIBRATION

For testing the amperometric module, consisting of the potentiostat, the I/V converter and the differentiator, an electrochemical dummy cell has been made as previously published [34]. The aim of the design shown in Figure 18 is to devise an electronic cell that would reproduce the constant amperometric response of a true electrochemical cell.

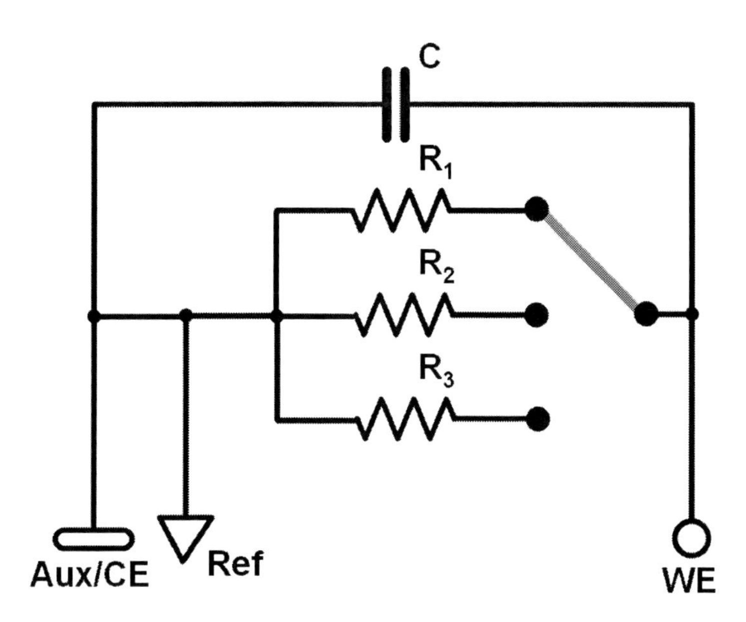

Figure 18. Electrochemical dummy cell used as Thevenin current source for testing and calibrating the amperometric electronics of the biotelemetry unit (see text).

The dummy cell was used as a Thevenin current source connected to the amperometric module. The voltage applied to the dummy cell was generated between the WE and the AE/RE electrodes and was equal to the voltage applied to the WE:

$$i_{dummy} = V_{app} / R_{dummy} \qquad (12)$$

The values of R_{dummy} (R_1, R_2 and R_3) were selected in order to obtain a full range of testing currents ranging from 1 nA/V up to 100 nA/V and were equal to 1, 10 and 100 MΩ respectively. The filter capacitor was selected having a value of 0.5 μF. The resulting anodic current (i_{dummy}) was read by the WE and converted to the resulting voltage [Eq. (7), $V_{out}2$] by the amperometric module. Selecting the right R_{dummy} value (1, 10 or 100 MΩ) and applying a known voltage (V_{App}), a 5-point calibration was made indoors for each of the amplification ranges with a linear distance between the biotelemetry unit and the PC of ~25 m. The logarithmic graph in Figure 19 represents the calibration currents plotted vs. the output voltage of the differentiator (V_{Out2}): from the same plot it is possible to extrapolate the values of R_{dummy} and V_{app} used for calibrating.

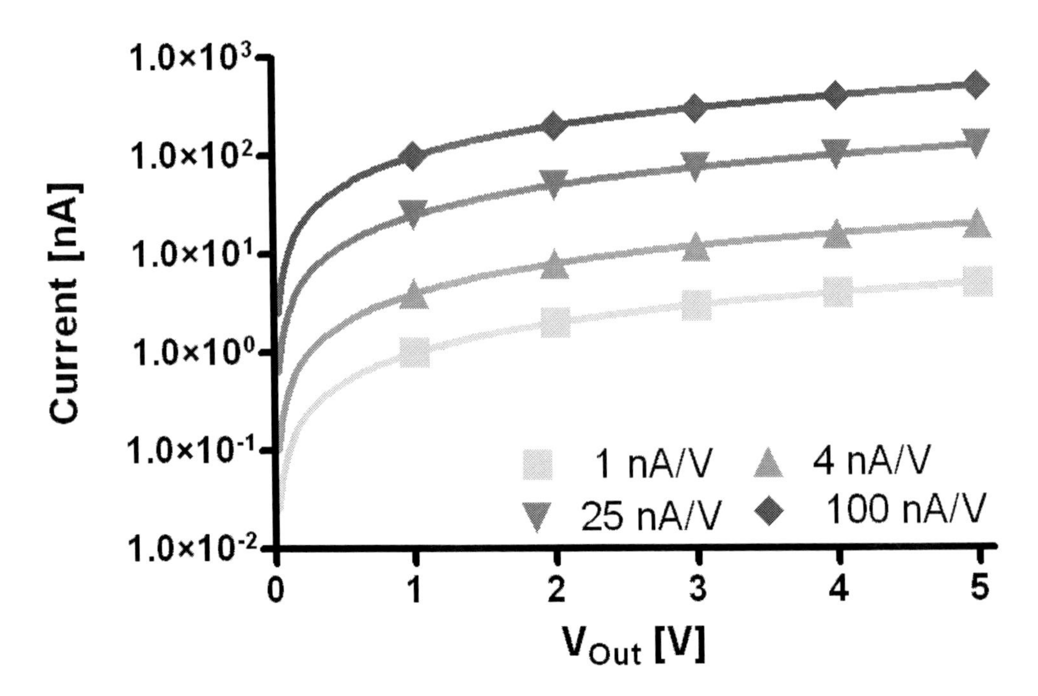

Figure 19. Calibration of the biotelemetry unit preformed by connecting the dummy cell and using it as a Thevenin current source (see text). The logarithmic graph represents the calibration currents plotted vs. the output voltage (V_{Out}) of the differentiator ($V_{Out}2$). From this graph it is possible to extrapolate the values of R_{dummy} and V_{app} used for calibrating.

Four ranges of amplification were obtained selecting an R_f value in (7) equal to 10 MΩ, R_1 equal to 1KΩ (7) and changing the value of R_2 according with (7) and (8). The value of R_2 was digitally programmed (9) to adjust the gain of the differentiator in the range between 1 V/V and 100 V/V and is empirically calculated as follows:

$$R_2 = V_{Out2} / (i_{dummy} \cdot R_f) \times 1000 \qquad (13)$$

where $V_{Out}2$ was fixed to 1V and i_{dummy} and R_f were respectively expressed in nA and MΩ. For example, a value of R_2 equal to 4 kΩ adjusts the amplification to 25 nA/V. The four ranges of amplification used in the electronics calibration are: 1 kΩ (100 nA/V), 4kΩ (25 nA/V), 25 kΩ (4 nA/V) and 100 kΩ (1 nA/V). After the calibration described above, a 2.5 nA current was generated though the dummy cell, setting the R_{dummy} to 100MΩ, selecting a gain of 1 nA/V and fixing V_{app} to + 250mV. The system was left under these conditions until the next calibration, which was performed the following day. A maximum V_{app} shift of 10mV was observed overnight while the current of the baseline noise was around 10 pA as illustrated in Figure 20.

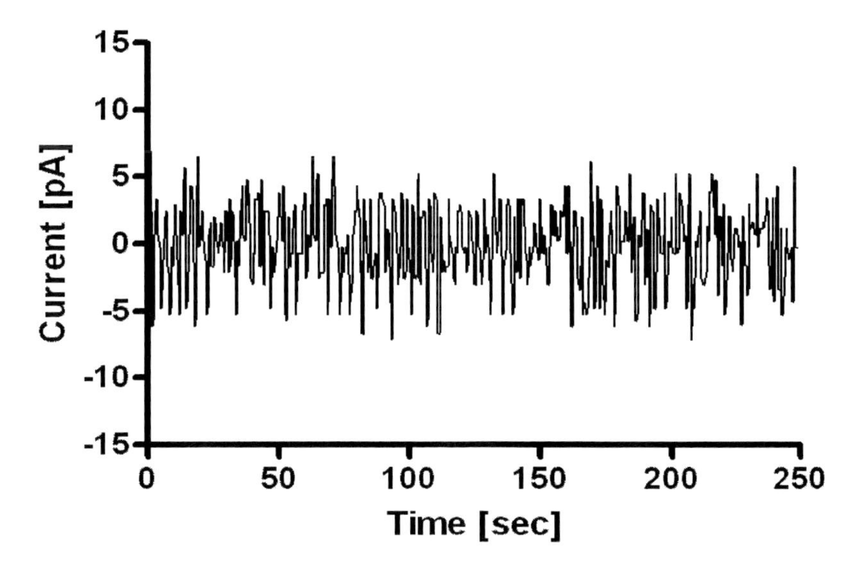

Figure 20. Background-subtracted graph of electronics baseline recordings obtained setting the dummy cell I/V resistor to 100 MΩ and applying +250mV to the WE. The noise was quantified around 10 pA.

AMPEROMETRIC MICROSENSORS AND BIOSENSORS CONSTRUCTION AND CALIBRATION

DA microsensors (∅=30μm; length≈500μm) were constructed similarly to those described by Yavich and Tiihonen [14], and were tested for selectivity against ascorbate then calibrated in PBS using standard DA solutions. Figure 21 shows a typical DA microsensor calibration performed under standard conditions.

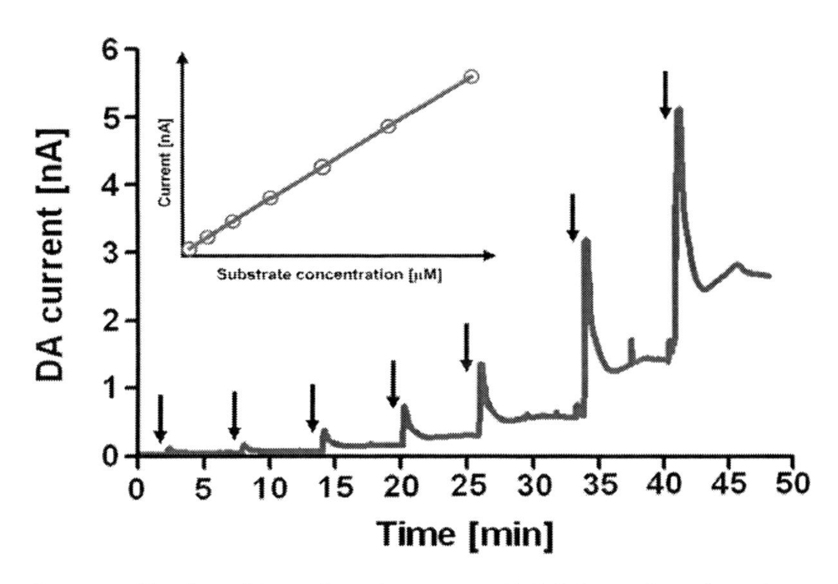

Figure 21. *In vitro* calibration of dopamine microsensors (0–2.5μM; n=4) performed in PBS at room temperature (25 ◦C). Seven consecutive injections of DA were made resulting in a slope of 0.968±0.008 nA·μM^{-1} demonstrating an excellent linearity (Inset, r^2 = 0.998).

Nitric Oxide microsensors were built as previous described [33]. A carbon fibre electrode (∅=30μm; length≈500μm) was modified with poly-o-phenylenediamine (p-OPD) and Nafion®. After p-OPD and Nafion® treatments, only electrodes with NO detection limits <75 nM and selective against ascorbic acid, dopamine and nitrite are used. The calibration of NO microelectrodes was performed by adding known volumes of the NO donor S-nitroso-N-acetyl-penicillamine (SNAP) in a saturated CuCl solution. Platinum cylinder (∅=125μm; length≈500 μm) glutamate biosensors were prepared as previously described [27-31]. As illustrated in Figure 22, glutamate calibrations were carried out in PBS using standard solutions of glutamate.

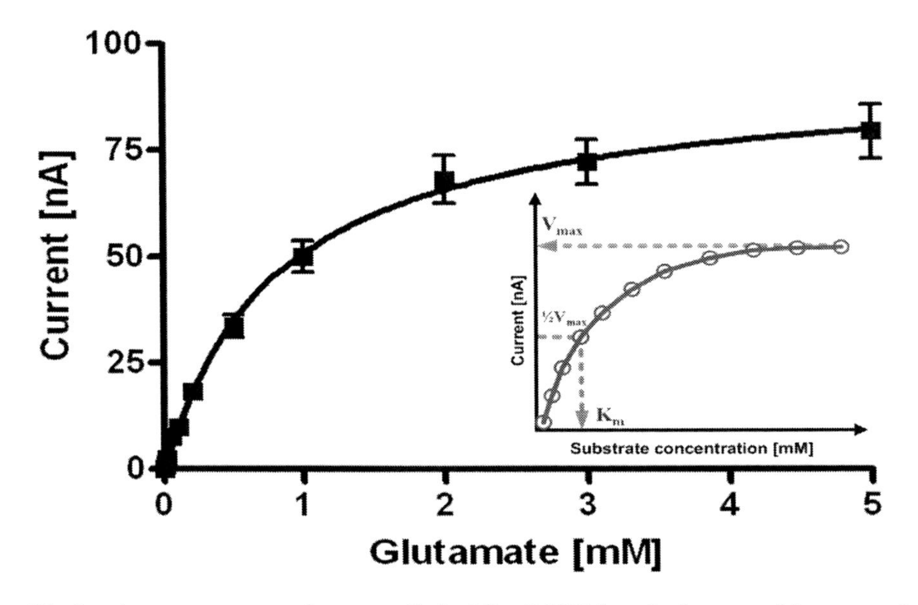

Figure 22. *In vitro* response to glutamate (0–5mM) of PPD-based glutamate biosensors (n = 4) connected to the biotelemetry unit. Ten consecutive injections of glutamate were made after a stabilization period in PBS at room temperature (25 °C). As illustrated in the inset, the calibration showed classical Michaelis-Menten kinetics ($r^2 = 0.997$) with a Vmax of 93.33±2.81 nA and a K_m of 0.84±0.038mM.

The electrosynthesis of the shielding polymer (p-OPD) was performed in nitrogenated PBS at +700mV. All the *in vitro* calibrations were preformed indoor, at room temperature, with a linear distance between the biotelemetry unit and the PC of ~25 m.

Conclusion

In this chapter we describe a distributed sensor NET, composed by biotelemetry units based on discrete components available at low cost and pre-assembled critical modules (XBee PRO™, USB-transceiver), so no complex calibrations of the transceiver are needed. The amperometric module has been optimized for low-voltage and single-supply operation by distribution of the gain between the I/V converter and the differentiator. At this stage of the development the sensor electronics exhibits low noise, high stability and good linear response. The choice of a multi-core 32-bit parallel processor makes the device suitable of further improvements such as the rapid implementation of different DSP algorithms. With a focused firmware development, several electrochemical techniques can be implemented to extend the analytical capabilities of the system. The CMOS technology permits a decrease in operating voltage and the power consumption. Studies are in progress to develop a low-power device using a single +3V lithium cell. The use of SM components and careful PCB design will make it possible to miniaturize the circuitry. The reduction of the weight and the size of the biotelemetry unit should permit its use as a lightweight portable/implantable device suitable for *in-vivo* applications.

REFERENCES

[1] FCC (Federal Communication Commission) (U.S.A.), Commission's Rules to Create a Wireless Medical Telemetry Service, Document FCC 00-211, Washington, DC, June 8, 2000.

[2] Leuher, D.C. (1983).Overview of biomedical telemetry techniques. *Eng. Med. Biol., 3,* 17–24.

[3] Shichiri M., Asasawa N., Yamasaki Y., Kawamori R., Abe H.. (1986). Telemetry glucose monitoring device with needle glucose sensor: a useful tool for blood glucose monitoring in diabetic individuals. *Diabetes Care, 9,* 298–301.

[4] Martin N. (1999). Telemetry monitoring in acute and critical care. *Crit. Care Nurs. Clin. North Am., 11,* 77–85.

[5] Matz P.G. (1997). Monitoring in traumatic brain injury, *Clin. Neurosurg., 44*, 267–294.

[6] Garris P.A., Ensman R., Poehlman J., Alexander A., Langley P.E., Sandberg S.G., Greco P.G., Wightman R.M., Rebec G.V. (2004). Wireless transmission of fast-scan cyclic voltammetry at a carbon-fiber microelectrode: proof of principle. *J. Neurosci. Methods, 140,* 103–115.

[7] Yun K.S., Gil J., Kim J., Kim H.J., Kim K., Park D., Kim M.S., Shin H., Lee K., Kwak J., Yoon E. (2004). A miniaturized low-power wireless remote environmental monitoring system based on electrochemical analysis. *Sens. Actuators B: Chem., 102,* 27–34.

[8] Bard A.J. and Faulkner L.R. (1980). Electrochemical Methods: Fundamentals and Applications, John Wiley and Sons, New York.

[9] O'Neill R.D. (2005). Long-term monitoring of brain dopamine metabolism in vivo with carbon paste electrodes. *Sensors, 5,* 317 – 342.

[10] Wightman R.M. and Robinson D.L. (2002). Transient changes in mesolimbic dopamine and their association with 'reward'. *J. Neurochem., 82,* 721-735.

[11] Rocchitta G., Migheli R., Mura M.P., Esposito G., Desole M.S., Miele E., Miele M., Serra P.A. (2004). Signalling pathways in the nitric oxide donor-induced dopamine release in the striatum of freely moving rats: evidence that exogenous nitric oxide promotes Ca2+ entry through store-operated channels. *Brain Res., 1023,* 243-252.

[12] Rocchitta G., Migheli R., Mura M.P., Esposito G., Marchetti B., Miele E., Desole M.S., Miele M., Serra P.A.. (2005). Signaling pathways in the nitric oxide and iron-induced dopamine release in the striatum of freely moving rats: role of extracellular Ca^{2+} and L-type Ca^{2+} channels. *Brain Res., 1047,* 18-29.

[13] Migheli R., Puggioni G., Dedola S., Rocchitta G., Calia G., Bazzu G., Esposito G., Lowry J.P., O'Neill R.D., Desole M.S., Miele E., Serra P.A. (2008). Novel integrated microdialysis-amperometric system for in vitro detection of dopamine secreted from PC12 cells: Design, construction, and validation. *Anal. Biochem.,* doi:10.1016/ j.ab. 2008.05.050.

[14] Yavich L., Tiihonen J. (2000). In vivo voltammetry with removable carbon fibre electrodes in freely-moving mice: dopamine release during intracranial self-stimulation. *J. Neurosci. Methods, 104,* 55–63.

[15] Song T., Hatano N., Kambe T., Miyamoto Y., Ihara H., Yamamoto H., Sugimoto K., Kume K., Yamaguchi F., Tokuda M., Watanabe Y. (2008). Nitric oxide-mediated modulation of calcium/calmodulin-dependent protein kinase II. *Biochem. J., 412,* 223-231.

[16] Kang J.S., Jeon Y.J., Kim H.M., Han S.H., Yang K.H.(2002). Inhibition of inducible nitric-oxide synthase expression by silymarin in lipopolysaccharide-stimulated macrophages. *J. Pharmacol. Exp. Ther. 2002, 302,* 138-144.

[17] [17] Zhang X., Cardosa L., Broderick M., Fein H., Davis I.R. (2000). A novel method to calibrate nitric oxide microsensors by stoichiometrical generation of nitric oxide from SNAP. *Electroanalysis, 12,* 425–428.

[18] Zhang X., Kislyak Y., Lin J., Dickson A., Cardosa L., Broderick M., Fein H.. (2002). Nanometer size electrode for nitric oxide and S-nitrosothiols measurement. *Electrochem. Commun., 4,* 11–16.

[19] Pantano P. and Kuhr W.G. (1995). Enzyme-modified microelectrodes for in vivo neurochemical measurements. *Electroanalysis, 7,* 405-412.

[20] Scheller F.W., Schubert F., Fedowitz J. (1997).Frontiers in Biosensorics I Fundamental Aspects, Birkhauser Verlag, Basel,

[21] Lowry J.P., Miele M., O'Neill R.D., Boutelle M.G., Fillenz M. (1998). An amperometric glucose oxidase/poly(o-phenylenediamine) biosensor for monitoring brain extracellular glucose: in vivo characterisation in the striatum striatum of freely-moving rats. *J. Neurosci. Methods, 79,* 65– 74.

[22] Magistretti P.J., Pellerin L., Rothman D.L., Shulman R.G. (1999). Energy on demand. *Science, 283*, 496-497.

[23] Fillenz M. (2005). The role of lactate in brain metabolism. *Neurochem. Int., 47,* 413-417.

[24] Pfeiffer D., Schubert F., Wollenberger U., Scheller F.W. (1996). Electrochemical sensors: enzyme electrodes and field effect transistors, in: R.F. Taylor, J.S. Schultz (Eds.), Handbook of Chemical and Biological Sensors, IOP, England.

[25] Wilson R., Turner A.P.F. (1992). Glucose oxidase: an ideal enzyme. *Biosens. Bioelectron., 7,* 165–185.

[26] O'Neill R.D., Lowry J.P., Rocchitta G., McMahon C.P. and Serra P.A. (2008). Designing sensitive and selective polymer / enzyme composite biosensors for brain monitoring in vivo. *Trends Anal. Chem.*, 1, 78-88.

[27] Kirwan S.M., Rocchitta G., McMahon C.P., Craig J.D., Killoran S.J., O'Brien K.B., Serra P., Lowry J.P., O'Neill R.D. (2007). Modifications of poly(o-phenylenediamine) permselective layer on Pt-Ir for biosensor application in neurochemical monitoring. *SENSORS, 7,* 1424-8220.

[28] McMahon C.P., Rocchitta G., Kirwan S.M., Killoran S.J., Serra P., Lowry J.P., O'Neill R.D. (2007). Oxygen tolerance of an implantable polymer/enzyme composite glutamate biosensor displaying polycation-enhanced substrate sensitivity. *Biosensors and Bioelectronics., 22,* 1466-1473.
[29] McMahon C.P., Rocchitta G., Serra P., Kirwan S.M., Lowry J.P., O'Neill R.D. (2006). Control of the oxygen dependence of an implantable polymer/enzyme composite biosensor for glutamate. *Anal.Chem., 78,* 2352-2359.
[30] Rocchitta G., McMahon C.P., Serra P., Kirwan S.M., Bolger F.B., Lowry J.P., O'Neill R.D. (2006). Significant Enhancement of Glutamate Biosensor Sensitivity, using a Polycation-Modified Polymer/Enzyme Composite Sensing Layer. *Monitoring Molecules in Neuroscience, 5,* 331-333.
[31] McMahon C.P., Rocchitta G., Serra P., Kirwan S.M., Lowry J.P., O'Neill R.D. (2006). The efficiency of immobilised glutamate oxidase decreases with surface enzyme loading: an electrostatic effect and reversal by a polycation significantly enhances biosensor sensitivity. *Analyst, 131,* 68-72.
[32] Bolger F.B., Serra P., Dalton M., O'Neill R.D., Fillenz M., Lowry J.P. (2006). Real-time monitoring of brain extracellular lactate. *Monitoring Molecules in Neuroscience, 5,* 286-288.
[33] Rocchitta G, Migheli R., Dedola S., Calia G., Desole M.S., Miele E., Lowry J.P., O'Neill R.D., Serra P.A. (2007). Development of a distributed, fully automated, bidirectional telemetry system for amperometric microsensor and biosensor applications. *Sens. Actuators B: Chem., 126,* 700-709.
[34] Serra P.A., Rocchitta G., Bazzu G., Manca A., Puggioni G.M., Lowry J.P., O'Neill R.D. (2007). Design and construction of a low cost single-supply embedded telemetry system for amperometric biosensor applications, *Sens. Actuators B: Chem., 1,* 118–126.
[35] Hebel M. and Serra P. (2007). Development of a Parallel-Computing embedded telemetry system for voltammetric microsensor and Biosensor applications. In: Baraton M.I. Sensors for Environment, Health and Security: Advanced Materials and Technologies. NATO-ASI Series. Dordrecht. Springer (Netherlands). *In press.*
[36] Bates R.G. and MacAskill J.B. Standard Potential of the Silver-Silver Chloride Electrode. *Pure and Applied Chem., 50,* 1701-1706.
[37] Backer B.C. (2000). Using Single Supply Operational Amplifiers in Embedded Systems, Application Note AN682, Document DS00682C, Microchip Technology Inc.
[38] Backer B.C. (2004). What Does "Rail-to-Rail" Operation Really Mean? Analog Design Note ADN009, Document DS21866A, Microchip Technology Inc.
[39] MCP6042 Datasheet. (2002). Document DS21669B, Microchip Technology Inc.
[40] Backer B.C., (2001) Optimizing Digital Potentiometer Circuits to Reduce Absolute and Temperature Variations, Application Note AN691, Document DS00691A, Microchip Technology Inc.
[41] King C.L., Haile E. (2001). Communicating with Daisy Chained MCP42XXX Digital Potentiometers, Application Note AN747, Document DS00747A, Microchip Technology Inc.
[42] Moghimi R., (2002). Ways to Optimize the Performance of a Differential Amplifier. Application Note AN589, Rev. 0, Analog Devices Inc.
[43] AD626 Datasheet. (1999). Rev. C, Analog Devices Inc.

[44] Crespi F., D'alessandro D., Annovazzi-Lodi V., Heidbreder C., Norgia M. (2004). In vivo voltammetry: from wire to wireless measurements. *J. Neurosci. Methods, 140,* 153–161.

[45] Black J., Wilkins M., Atanasov P., Wilkins E. (1996). Integrated sensor telemetry system for in vivo glucose monitoring. *Sens. Actuators B: Chem., 31,* 147–153.

Reviewed by Prof. Egidio Miele. University of Sassari.

In: Telemetry: Research, Technology and Applications
Editors: Diana Barculo and Julia Daniels
ISBN 978-1-60692-509-6

Chapter 6

BIOTELEMETRY RESEARCH ON UPSTREAM MIGRATION BEHAVIOR OF ADULT CHUM AND PINK SALMON IN A RE-MEANDERED SEGMENT OF THE SHIBETSU RIVER, JAPAN

***Yuya Makiguchi*[1], *Yoshifumi Konno*[1], *Hisaya Nii*[2], *Katsuya Nakao*[2], *and Hiroshi Ueda*[1, 3]**

[1]Division of Biosphere Science, Graduate School of Environmental Science, Hokkaido University, North 9 West 9, Kita-ku, Sapporo, Hokkaido 060-0809, Japan

[2]Hokkaido Aquaculture Promotion Corporation, North 3 West 7, Chuo-ku, Sapporo, Hokkaido 060-0003, Japan

[3]Laboratory of Aquatic Bioresources and Ecosystems, Field Science Center for Northern Biosphere, Hokkaido University, North 9 West 9, Kita-ku, Sapporo, Hokkaido 060-0809, Japan

ABSTRACT

The Shibetsu River located in eastern Hokkaido, Japan was channelized in the 1960s for the purpose of flood control and irrigation. However, a segment of the Shibetsu River was preliminary reconstructed to restore the natural meanders of the rivers and to improve the degraded stream segment to a more natural habitat in 2002. Chum (*Oncorhynchus keta*) and pink salmon (*O. gorbuscha*) are anadromous salmonid species migrating from the ocean to their spawning grounds in the Shibetsu River. To investigate the effects of river re-meandering on upstream migration behavior of chum and pink salmon, their behavior between channelized and reconstructed segments was compared using electromyogram (EMG) transmitters and depth/temperature (DT) loggers from 2001 to 2007. EMG transmitters allowed us to estimate swimming speeds of free-swimming salmon. Swimming behaviors (swimming speed, ground speed and swimming efficiency index) of chum and pink salmon did not differ between the two segments. Further, we developed energetic models using a swimming respirometer to estimate energy use swimming for pink salmon from EMG information. These models revealed that energy use during swimming for pink salmon did not differ between the two

segments. However, holding time of chum and pink salmon in the reconstructed segment increased in 2005 three years after re-meandering. Measurement of physical conditions in the two segments in 2004 and 2005 revealed that a greater diversity of current speed and water depth was observed in 2005. These results suggest that the physical river conditions developing in the reconstructed segment may be suitable for holding behavior of chum and pink salmon. In addition, chum salmon swimming speed exceeding critical swimming speed (U_{crit}) was observed prior to holding. Swimming depth of chum and pink salmon collected from retrieved DT loggers tended to be deeper during holding behavior than swimming. Since our results suggest that consecutive swimming of chum and pink salmon might be exhaustive during upstream migration, we propose that holding behavior in the reconstructed segment may contribute to muscle recovery during upstream migration behavior.

Keywords: upstream migration, chum salmon, pink salmon, reconstruction, biotelemetry, EMG transmitter, critical swimming speed, energetic cost

INTRODUCTION

Chum (*Oncorhynchus keta*) and pink salmon (*O. gorbuscha*) are anadromous fish with a wide geographic distribution, ranging from the east coast of Korea and Japan northward to the Arctic coasts of Russia and North America and then southward to Oregon [1, 2]. Pink salmon are the most abundant and chum salmon are the second most abundant in seven Pacific salmon species. The fry stay in their natal stream for a few weeks after emergence, and then migrate to the sea water [3]. After the ocean migratory period of one to five years for chum salmon and two years for pink salmon, they return to their natal river in fall. All chum and pink salmon are semelparous. Hokkaido is one of the most active regions for Pacific salmon propagation programmes. Most of the released fish are chum salmon, followed by pink and masu salmon (*O. masou*) [4]. Japan has one of the highest hatchery production rates of chum and pink salmon in the North Pacific region [5]. Some have claimed that the high number of Japanese hatchery-released salmon may have ecological impacts on other salmonid species. Furthermore, climate change has contributed to the recent increase in Japanese salmon catches [6]. Therefore, increasing the population size of wild chum and pink salmon in Japan is desirable for sustainable use as well as conservation of salmon resources.

It is well known that stream channel alterations have occurred in urban and rural rivers throughout Japan. The channelization stabilized the water levels and flow regimes, and drained a vast floodplain wetland, which was transformed into grassland. However, the channelization changed general fish communities and standing crops and resulted in low riparian and aquatic biodiversity [7]. The Shibetsu River, located in eastern Hokkaido, Japan, was channelized in the 1960s for the purpose of flood control and irrigation. Following the channel alteration, the project area consisted of cultivated farmland with fields for agriculture. However, interest in reconstructing rivers and streams is increasing as public opinion shifts toward support of less perturbed systems, and ecological experiments focus on hydrological reconstruction of river-floodplain systems to learn more about natural structure and function [8]. Planning for re-meandering construction projects of the Shibetsu River began to include ecological conservation in 1997. In 2002, preliminary reconstruction for river meandering was conducted in a short segment 9.0 km from the Shibetsu River mouth. The objectives of the reconstruction were to return the ecosystem and the natural hydrologic processes to a close approximation of its former mature condition. To reconstruct the river ecosystems, it is necessary to characterize the biotic elements in the stream after reconstruction, and chum and pink salmon are the main species that utilize the stream habitat in the Shibetsu River. However, there have been only a few studies documenting the effects of reconstruction in the Shibetsu River on upstream migration of chum and pink salmon assessed using EMG (electromyogram) transmitters and depth/temperature (DT) loggers simultaneously [9, 10].

Biotelemetry has been widely used for studies on migration patterns, behavior and physiology such as muscle activity and energetic status of fish in their natural environment [11]. Advances in technology have allowed transmitters to become smaller and lighter, which make monitoring smaller fish movements possible, and they can also provide information for individual identification [12]. Physiological telemetry has been used to examine activity levels and other correlates of oxygen consumption in adult fish [13]. EMG telemetry has proven to be especially effective for examining continuous swimming activity of sockeye salmon (*O. nerka*) [14], pink salmon [15], and chum salmon [10]. Muscle EMG is an indicator of the intensity of muscle activity in free-moving fish [16]. The swimming behaviors of migrating fish were classified into three major categories: sustained, prolonged and burst swimming [17]. Migrating fish use these swimming behaviors to control energy use during migration as the environmental situation demands [18]. EMG telemetry has also been used to describe energy uses associated with swimming behavior [19, 20] using bio-energetic models for adult sockeye salmon [21] and pink salmon [22], and found that the energy used during upstream migration of salmon is affected by swimming patterns of fish. Relationships between swimming performance, active metabolism and muscle EMG are used to determine various swimming behaviors of fish. The critical swimming speed (U_{crit}) is defined as the quantification of the sub maximum and largely aerobic swimming performance of fish, and is approximately the speed at which fish fatigue in an incremental velocity trial [17, 23, 24]. It is generally accepted that maximum oxygen uptake occurs at U_{crit} [25], and this allowed us to estimate the maximum aerobic capacity of fish [17].

In this study, the effects of the reconstruction conducted in a short segment on upstream migration of chum and pink salmon migrating in the Shibetsu River were investigated using EMG transmitters and DT loggers from 2001 to 2007. The main objectives of this study were to monitor the swimming behavior of chum and pink salmon during upstream migration over time, assess geomorphological and flow process in the reconstructed segment through the study periods, and to collect basic information that will help determine environmental factors, which affect upstream migration behavior and habitat use of chum and pink salmon in the Shibetsu River. Further, we attempt the energetic models estimating energy uses from EMG signals for adult pink salmon and evaluate effects of the reconstruction on their energetic costs during upstream migration.

Chart 1. Summary of mean fork lengths and body weights of tagged fish, tracking dates, number of individuals, type of radio transmitters, type of data, type of data loggers and releasing point (distance from the Shibetsu River mouth) from 2001 to 2007

Species	Year	Tracking dates	N (female)	Mean fork length ±S.E. (cm)	Mean body weight ±S.E. (kg)	Type of radio transmitters	Type of data	Type of data loggers	Releasing point (km)
Chum salmon (*Oncorhynchus keta*)	2001	11/07-11/08	5(0)	69.3±1.3	4.7±0.2	MBFT-2	ID-only	-	6.8
	2002	09/19-09/21	5(0)	78.7±1.0	5.8±0.4	MBFT-2	ID-only	M190-DT	6.8
		10/09-10/11	6(0)	70.3±0.4	4.0±0.1	MBFT-4	ID-only	M190-DT	6.8
	2003	10/20-10/25	4(0)	71.0±2.4	4.1±0.4	CEMG-R11	EMG	M190-DT	6.8
			2(0)	74.1±5.0	4.9±1.2	MBFT-4	ID-only	M190-DT	6.8
	2004	10/26-10/31	5(0)	72.5±2.8	3.4±0.4	CEMG-R11	EMG	M190-DT	7.25
	2005	11/07-11/20	3(9)	63.2±1.0	2.6±0.2	CEMG-R11	EMG	M190-DT	7.25
	2006	10/28-10/28	3(3)	66.7±2.9	3.2±0.3	CEMG-R11	EMG	M190-DT	7.25
			3(4)	64.4±2.0	2.9±0.28	CEMG-R11	EMG	-	8.5
Pink salmon (*Oncorhynchus gorbuscha*)	2003	08/25-09/05	10(0)	54.9±2.0	2.0±0.2	MBFT-4	ID-only	M190-DT	6.8
	2004	09/07-09/11	5(0)	58.4±1.4	2.0±0.1	CEMG-R11	EMG	M190-DT	7.25
	2007	08/29-09/01	5(0)	52.6±0.85	1.5±0.08	CEMG-R11	EMG	-	7.25
			5(0)	50.9±2.1	1.5±0.21	CEMG-R11	EMG	-	8.44

Chart 2. Number of chum and pink salmon selecting each segment, and flow volume and mean water velocity of each segment in the confluence point

Species	Year	Total number of fish (female)	Number of fish upstream migrating (female)	Number of individual selecting each segment			Flow volume (m^3/s)	
				Channelized segment	Reconstructed segment	Selectivity rate of the reconstructed segment (%)	Channelized segment	Reconstructed segment
Chum salmon (*Oncorhynchus keta*)	2002	11(0)	3(0)	0	3	100	4.31	10.26
	2003	6(0)	4(0)	0	4	100	5.46	14.33
	2004	5(0)	4(0)	0	4	100	0.36	11.07
	2005	12(9)	10(8)	0	10	100	5.46	14.33
	2006	13(7)	12(7)	6	0	0	9.17	6.68
Pink salmon (*Oncorhynchus gorbuscha*)	2003	9(0)	5(0)	1	4	80	3.47	10.73
	2004	5(0)	3(0)	0	3	100	0.60	12.26
	2007	10(0)	3(0)	3	0	0	5.28	2.74

MATERIALS AND METHODS

Study Area

The Shibetsu River, located in eastern Hokkaido, Japan, drains 671km^2 from Mount Shibetsu (1061 m), running through the cities of Shibetsu and Naka-Shibetsu and flowing into the Nemuro Straight with a total length of approximately 77.9 km. The study area covered a reach between 7.25 km and 9.0 km from the river mouth and was 1.2 km long and composed of two consecutive segments: the 1.15 km channelized segment and a 0.5 km reconstructed segment (Figure 1). The Shibetsu River width is range from 16.3 to 49.0 m in the channelized segment and from 12.4 to 31.3 m in the reconstructed segment respectively. In June 2004, four artificial fallen trees (approximately 10 m in length) were installed separately in the reconstructed segment to promote fish habitats areas. Locations of the installed fallen trees were as follows: right bank 80 m, left bank 40 m, right bank 280 m and left bank 420 m from the confluence point with the channelized segment.

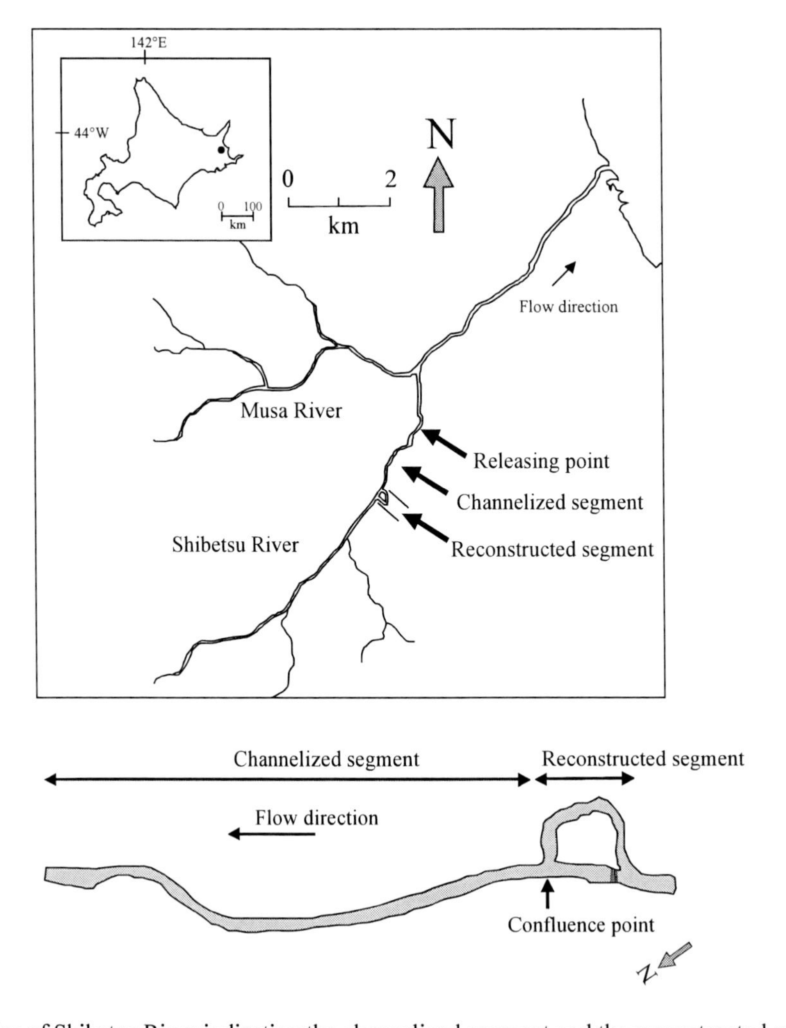

Figure 1. Map of Shibetsu River indicating the channelized segment and the reconstructed segment.

Experimental Animals

Chum salmon migrate through the study area from September to November. All chum salmon used in this study were captured in the Shibetsu River estuary and then transferred to outdoor tanks at the Shibetsu Salmon Museum until the experiment. Thirty six chum salmon were tracked through the channelized and reconstructed segment from September to November for five years (2001-2006; Chart 1). Pink salmon migrate through the study area from August to October. All pink salmon used in this study were captured in the Shibetsu River estuary and then transferred to outdoor tanks at the Shibetsu Salmon Museum until the experiment. Twenty five pink salmon were tracked through the two segments from August to September for three years (2003, 2004 and 2007; Chart 1). Individual fish were tracked continuously through the channelized and reconstructed segment along the river bank on foot and their horizontal positions recorded on maps.

Telemetry Equipment

EMGs were recorded using EMG radio transmitters (Lotek Engineering Inc., Newmarket, Ontario). The EMG transmitters consisted of a cylindrical, epoxy-coated transmitter package with a pair of Teflon-coated electrodes with brass muscle-anchoring tips (dimension 5×1 mm). The CEMG model that we used is equipped with a differential muscle probe, a signal conditioning circuit, a digitizer, a micro-controller and a radio transmitter. The voltage corresponding to muscle activity is rectified and sampled from the beginning to the end of every 2 s time interval. Individual samples are summed and temporarily stored. At the end of the time interval, the average value is calculated and assigned an activity level (EMG signal) ranging from 0 to 50 (no units). This calculated result is then transmitted to a radio receiver (model SRX_400; Lotek Engineering Inc.). Each fish was equipped with a cylindrical, epoxy-encased EMG transmitter and a micro data logger which simultaneously records depth and temperature (Chart 1). Sampling rate of the micro data logger was 1 second in all channels. Chum salmon tracked in 2001 and 2002 were attached with a radio transmitter, which allows us an individual identification. Fish position was monitored using received signals from EMG transmitters or radio transmitters by a radio receiver, and swimming depth was monitored using the micro data logger. These radio transmitters and DT loggers allowed us to demonstrate horizontal and vertical position of a fish. Both an EMG transmitter and a DT logger were externally attached to the body surface positioned anterior to the dorsal fin.

Transmitter Attachment Procedure

EMG transmitter and radio transmitter attachment procedure were similar to Makiguchi et al.[10]. Experimental fish were anaesthetized using FA100 (eugenol; Tanabe Seiyaku Co. Ltd, Osaka, Japan) at a concentration of 0.5 ml l^{-1} in the Shibetsu River water, and measured their fork length and weight (Chart 1). External attachment is suitable for short-term research, which allows us to reduce the effects of handling of the fish [26] and this has also proved useful for research on gravid females [27]. Once attachment location was determined, experimental fish were placed upright on a surgical table. Their gills were irrigated with water containing diluted FA100 to maintain sedation during the attachment procedure. Two stainless needles, large enough hold to restraining nylon ties, were pushed through the dorsal muscular to secure the EMG transmitter and DT logger which were sutured together using nylon ties and epoxy resin with silicon pads. The silicon pads were attached to minimize abrasion. The nylon ties were passed through the needles and tied on the opposite side.

The EMG electrodes were inserted subcutaneously using a hypodermic needle at about a 0.7 ratio of the body length on the left side of the fish. The electrodes detect electro-potentials within the axial dark muscle composed by red muscle tissue, with the amplitude and frequency of these pulses being directly correlated to the level of muscle activity. Paired electrode tips were positioned approximately 10 mm apart, secured in the lateral red muscle toward the rear of the fish, which is primarily used in steady, non-bursting aerobic swimming activity [28]. The electrodes also appear to be sensitive to contractions of the relatively larger white muscle [14, 29], which is recruited primarily during burst, anaerobic swimming. EMG signals can therefore generally be related to swimming speed [30]. The electrodes were sutured to avoid entangling vegetation and/or structure in the environment. An aminoglycoside antibiotic (Akiyama Seisakujyo Co. Ltd, Tokyo, Japan) was applied to the skin around the punctures and stitches. The attachment procedures usually required approximately five minutes to complete. After recovering from anesthesia, the behavior of the chum and pink salmon with EMG transmitters and DT loggers were observed in a hatchery pond and, were found no differences in behavior manner between the experimental fish and the intact fish.

Calibration of EMG Signals to Swimming Speeds

Following the recovery period, an individual calibration curve was analyzed to convert EMG signals from the fish into swimming speeds. To quantify the relationship between swimming speeds and EMG signals, a swim chamber (PT-110R, West Japan Fluid Engineering Laboratory Co. Ltd, Nagasaki, Japan: 110 cm length, 30 cm width, and 30 cm depth) was used. The Shibetsu River water entered the chamber before each trial. Water volume was 400 l and water velocity was generated by a centrifugal pump whose motor frequency was controlled by a variable speed drive unit. Water velocities (up to 1.2 m s^{-1}) at selected motor frequencies were verified using an impeller connected to a pre-calibrated frequency counter. Experimental fish were placed individually into the swimming section of the swim chamber and measured 10 EMG signals at 0 m s^{-1} were measured when the fish maintained a holding position using the EMG radio receiver. After this period, the trial was started at 0.3 m s^{-1}, and water velocity was incrementally increased by 0.3 m s^{-1} to measure 10 EMG signals at each increment. Fish readily swam against the current and rarely came in contact with the grid at the back of the swim chamber. The trial was finished at 1.2 m s^{-1}. The relationship between swimming speeds and EMG signals output from tagged fish was plotted and a standard curve was used to estimate swimming speeds of free swimming fish using the collected EMG signals [31]. Once the trial was completed, fish were transferred to a recovery live-box and allowed to recover at the release point of the Shibetsu River at least for 24 h to recover from the trial.

Field Study

Each chum and pink salmon used in the EMG calibration were individually released 6.8 or 7.25 km upstream from the river mouth and tracked upstream to the end of the reconstructed segment on foot using a hand-held directional Yagi antenna. EMG radio receivers recorded EMG signals at 2-5 s intervals. Migration time through both segments was measured for each fish by subtracting time of reach exit from time of reach entry. Ground speed was then calculated by dividing the segment length by the migration time. Cessation of swimming of chum and pink salmon during their upstream migration was often observed. In this study, more than three minutes of swimming cessation during their upstream migration were defined as holding behavior. To clarify upstream migration behavior of chum salmon, calculation of swimming and ground speed did not include holding time. A swimming efficiency index (SEI) was also calculated by dividing each individual's ground speed by its swimming speed. These data are useful for estimating an individual's energy utilization strategy. For instance, values below 1 indicate that individuals are swimming relatively hard to achieve forward progress, values equal to 1 imply relatively easy for forward progress (analogous to moving through still water), and values greater than 1 indicate that fish are receiving some type of forward assistance. EMG signals were converted to swimming speed using equations established from the EMG calibration individually. During tracking, characteristics of habitat cover, for example a fallen tree, used by a migrating salmon was also recorded.

After tracking, water depth and water velocity were also measured at 20-100 m intervals along the two segments in 2004 and 2005. Water velocity was measured at 0.2 m intervals in the water column and at 1 m intervals along the cross-section of the river using an electromagnetic current meter (ES7603; Yokogawa Navi-tech Co. Ltd, Tokyo, Japan.). Water depth was measured at 1 m interval along the cross-section of the river.

Critical Swimming Speed Trials

Critical swimming speed (U_{crit}) trials were conducted on individual fish using the swim chamber at the same time as the EMG calibration trial. Thirteen adult chum salmon (mean fork length ± S.D. 64.8±4.10 cm; mean body weight ± S.D. 3.0±0.54 kg) were used to estimate U_{crit}. Of thirteen fish used for the swimming performance tests, eight were males and five were females. The swim camber was filled with Shibetsu River water at 10.1 °C on an average. Initial and final temperatures were monitored for each trial and temperature did not increase more than 2 °C during the course of any individual trial. Mean temperature values ranged from 9.5 to 10.7 °C for the trials. Fish were placed individually into the swimming section of the swim chamber. In all cases, fish were acclimated for an hour at 0.3 m s^{-1} before use in all U_{crit} trials to minimize handling effects. After this period, water velocity was increased to 0.75 m s^{-1} and fish was swum for 30 min. At the completion of each 30 min period, the water velocity was increased by an additional 0.25 m s^{-1} and was maintained at the new speed for 30 min or until the fish became fatigued and was unable to swim against the current. The water velocity and point in time within the 30 min period were used in the calculation of U_{crit}. After each trial was completed, body mass, fork length, width and depth were determined. Water velocity in the swim chamber increases in the presence of a fish that occupies part of the cross-sectional area of the working section (solid blocking effect) [32].

Water velocities were corrected for the solid blocking effects of the fish on water flow in the chamber using these measurements [33]. Blocking effects were calculated according to the equation by Bell and Terhune [34]:

$$V_c = V_m(1 + E_s)$$

where V_c is the corrected water velocity, V_m is the measured water velocity determined in the absence of a fish in the swim chamber and E_s in the fractional error due to the solid blocking effect of the fish. Assuming the cross-sectional area and length of a fish to be best represented by an elliptical cylinder, E_s, as previously described by Bell and Terhune [34] is:

$$E_s = \mathrm{T}L(A_0\, A_t^{-1})^{1.5}$$

where T is a dimensionless factor depending on flume cross-sectional shape, L is the shape factor for the fish (described by 0.5×(fork length/fish depth)), A_0 is the maximum cross-sectional area of the fish (described by 0.25×fish depth ×fish width) and A_t is the cross-sectional area of the cylindrical swim section. For any sectional shape, T=0.8 to one decimal place [34].

U_{crit} was calculated, after correction for blocking effects, in relative (normalized for body length [BL]) units, using the formula described by Beamish [35] as:

$$U_{\mathrm{crit}} = U_{\mathrm{p}} + (T_{\mathrm{p}}\, T_{\mathrm{i}}^{-1}) \times U_{\mathrm{i}}$$

where U_{p} is the velocity at which the fish last swam for the full period , U_{i} is the velocity increment (0.3 m s^{-1}), T_{p} is the length of time in min that the fish was able to swim against the water velocity which produced fatigue, and T_{i} = the time between velocity increments (30 min).

New Approaches Estimating Energy Costs of Swimming from EMG Values

Swimming Respirometry Trials Associated with EMG Signals and Swimming Speeds

To estimate energetic cost associated with swimming behavior of adult male pink salmon from EMG values, we quantify the relationships between rate of oxygen uptake, swimming speed and EMG values. Twenty male pink salmon (mean fork length ± S.D. 53.1±4.67 cm; mean body weight ± S.D. 1.6±0.54 kg) were used for the experiment, and equipped with EMG transmitter (CEMG-R11, Lotek Engineering Inc., Newmarket, Ontario: 18.3 g in air, 16.2 mm in diameter, 53.0 mm in length) using the same procedure previously described. The swimming respirometer used was the swim chamber at the same time as the EMG calibration trial sealed with an acrylic board so that no gas exchanged occurred. Dissolved oxygen concentration in the swim chamber was measured using a multi water quality sensor probe (U-21XD; HORIBA Ltd, Kyoto, Japan.) housed outside the swim chamber in a flow-through. The oxygen probe had automatic temperature compensation.

Prior to fish introduction, the swimming respirometer was run to remove any bubbles. The water used was 15.0°C Shibetsu River water on an average. Background oxygen consumption without a fish in the swim chamber was measured repeatedly but found to be negligible (<1%). Individual fish equipped with EMG transmitters placed in the respirometer. Protocol was same in the U_{crit} trials. Fish were then acclimated 1h in the respirometer at a velocity 0.3 m s^{-1}. After recovery, the water velocity was increased to 0.75 m s^{-1} and the fish was swum for 30 minutes. Water velocity was then increased 0.25 m s^{-1} every 30 minutes until the fish fatigued. Fatigue was defined as when the fish rested or became impinged on the downstream screen and would not leave for more than 10 seconds. During the trials dissolved oxygen (mg l^{-1}) were recorded manually every 10 minutes. At the end of any speed increment, if the air-saturation was more than 70 % and water temperature had not increased more than 2 °C from the starting condition, the water velocity was increased to the next speed increment and data collection resumed. If these conditions were not met, the water velocity was reduced to 0.3 m s^{-1} and the respirometer was flushed with fresh water until air-saturation was more than 90% and water temperature was reduced to starting conditions. The rate of oxygen consumption (MO_2 in mgO_2 h^{-1} kg^{-1}) for an individual fish during a velocity increment was calculated as:

$$MO_2 = [O_2]v/m$$

where the rate oxygen concentration $[O_2]$ is measured in $mg \cdot O_2 \cdot l^{-1}$ h^{-1}, v is water volume of the swim chamber (l), and m is body mass of the fish (kg).

The relationship between individual values of MO_2 and swimming speed (U) was described using an exponential function as:

$$MO_2 = ae^{bU}$$

where a is the estimate of the standard metabolic rate (SMR), which is the minimum maintenance oxygen consumption of a resting, i.e. MO_2 at zero speed, U is the swimming speed and b is a constant. To establish the relation between COT and U an exponential function describing MO_2 as a function of U (km h^{-1}) were divided by U in kilometers per hour. For the exponential function was calculated as:

$$\mathrm{COT}(U) = ae^{bU} U^{1}$$

The expression for COT(U) was differentiated, and the value where this differential equation equaled zero was taken as the optimal swimming speed (U_{opt}) where the cost of transport is at a minimum (COT_{min}). U_{opt} was then determined as the U where the first derivative of this equation equals zero

$$U_{opt} = 1/b$$

and COT_{min} given by inserting U_{opt} into COT(U).

During swimming respirometry trials, EMG radio transmitter signals were detected using a radio receiver (model SRX 600; Lotek Engineering Inc.) to document the relationship between EMG activity and swimming speed. EMG values ranging from 0 to 50 were recorded at 2 second intervals in the receiver memory. Fish often did not swim at a constant pace in swimming respirometer (e.g., they accelerate and decelerate), particularly at the higher water velocities. Fluctuation of EMG values should therefore be larger at the higher water velocities and be associated with the swimming speed. Thus, we developed a model predicting the swimming speed from EMG information (mean values and standard deviation of EMG values) using generalized liner modeling (GLM) and the influence of mean values and standard deviation of EMG values was evaluated. GLM is a predictor that is based on combinations of measurements that are free to vary in response to other variables [36]. GLM is based on the formula

$$y = B_0 + B_1X_1 + B_2X_2..B_nX_n,$$

where y is the predicted (or probable) outcome (dependent variable), B_0 is the y-intercept, and B_x is the slope of the regression, which indicates the change in the mean of the probability distribution of Y per unit increase in X. X_1 represents the independent variables that were evaluated. The resulting models were then compared using Akaike Information Criterion (AIC) of the model to decide the best fit model. Using these energetic models, energy use of adult male pink salmon migrating through the channelized and reconstructed segments was estimated by inputting averaged swimming speeds.

All statistical analyses were performed using R statistical software (http://www.r-project.org). Regression analysis was performed by simple regression of EMG signals on swimming speed. Correlation coefficients were obtained using simple regression analysis. Statistical significance was determined using Student's t test, and achieved when $P<0.05$. Values are presented as means ± standard deviation (S.D.) or standard error of mean (S.E.).

RESULTS

Calibration of EMG Signals to Swimming Speeds

From EMG data, muscle activity levels were strongly correlated with swimming speed using the simple regression analysis for all year fish (average of 2 chum salmon in 2003 r^2=0.78, average of 4 chum salmon in 2004 r^2=0.77, average of 15 chum salmon in 2005 r^2=0.87, average of 6 chum salmon in 2006 r^2=0.987, average of 5 pink salmon in 2004 r^2=0.78, average of 10 pink salmon in 2007 r^2=0.96; $P<0.05$). The equations from the simple regression analysis were individually used to convert EMG signals of the field study to each fish's swimming speed (For example, U=0.053EMG– 0.44 r^2=0.95 for # 8 chum salmon in 2005 and U=0.078EMG– 0.723 ^{2}=0.90 for # 3 pink salmon in 2007 where U is swimming speed (BL s^{-1}) and EMG is EMG values, Figure 2).

Selectivity of the Reconstruction Segment

Before the reconstruction in 2001, 2 of 4 chum salmon moved upstream from the release point in the Shibetsu River and were tracked. One of fish reached the confluence point with the channelized and reconstructed segment, while the other fish moved upstream only approximately 500 m from the release point. Consequently, data on swimming behavior of chum salmon before the reconstruction in 2001 could not be enough and excluded it from data analysis. Before 2005 except 2001, all of chum and pink salmon migrating up the channelized segment, reached the confluence point with the reconstructed segment, and then entered the reconstructed segment (Chart 2). After 2006, all of chum and pink salmon migrating up the channelized segment and reached the confluence point with the reconstructed segment, but they migrated into the channelized segment. To collected swimming behavior data, additional chum salmon (3 males and 4 females) in 2006 were released in the reconstructed segment 100 m upstream from the confluence point with the channelized segment, and additional 5 male pink salmon in 2007 were released in the reconstructed segment 40 m upstream from the confluence point with the channelized segment. Chart 2 also shows the annual flow volume in the channelized and reconstructed at the confluence point from 2001 to 2007. Before 2005, flow volume in the reconstructed segment were greater than in the channelized segment, however, after 2005 flow volume in the reconstructed segment were lower than that in the channelized segment.

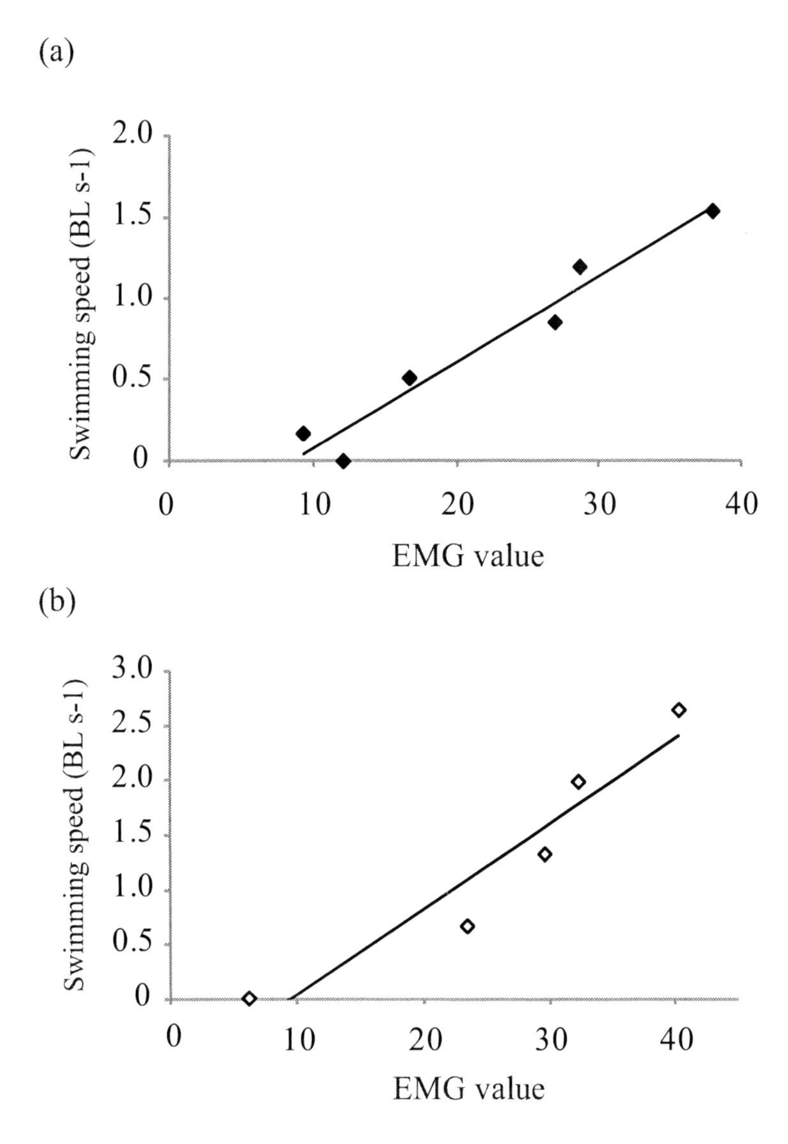

Figure 2. Relationships between electromyogram (EMG) signals and swimming speed (BL s^{-1}) for #8 chum salmon in 2005 (a) and #3 pink salmon in 2007 (b).

The Swimming Behavior of Chum and Pink Salmon in the Channelized and Reconstructed Segment

In 2004 and 2005, EMG data for chum salmon were collected in both segments, but in 2006 these data could not be collected in the reconstructed segment. Swimming speed calculated from EMG signals, ground speed and SEI calculated from swimming and ground speed as swimming behavior of chum and pink salmon were compared between the channelized and reconstructed segment (Chart 3). Although mean values of ground speed for chum salmon in the channelized segment were higher than that in the reconstructed segment in 2006, no differences for swimming speed, ground speed and SEI were observed between the channelized and reconstructed segment. Duration of holding behavior of chum and pink salmon between the channelized and reconstructed segment from 2002 to 2007 were compared for chum salmon (Figure 3) and pink salmon (Figure 4). Before 2005, duration of holding behavior of chum and pink salmon in the reconstructed segment was little observed. However, after 2005 duration of holding behavior of chum and pink salmon in the reconstructed segment tended to be longer than that in the channelized segment.

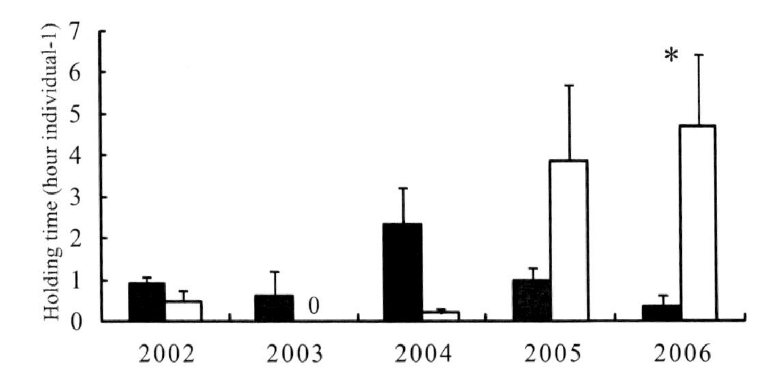

Figure 3. Holding time (hour) per individual of chum salmon from 2002 to 2006 for the channelized segment (solid bars) and the reconstructed segment (open bar). 0 indicates that no holding behavior was observed.

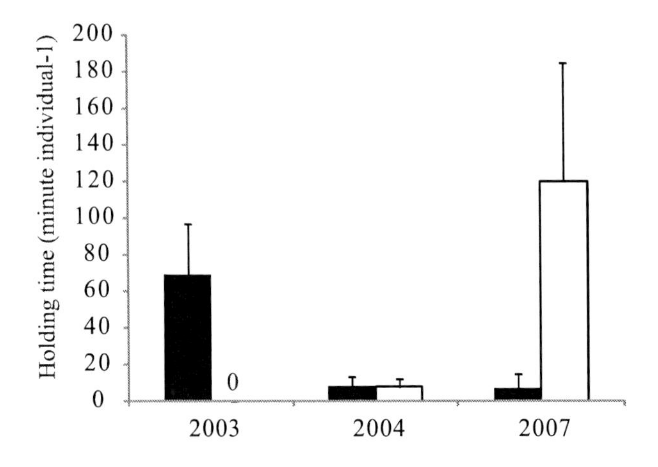

Figure 4. Holding time (minute) per individual of pink salmon in 2003, 2004 and 2007 for the channelized segment (solid bars) and the reconstructed segment (open bar). 0 indicates that no holding behavior was observed.

Chart 3. Mean swimming speed, ground speed and swimming efficiency index (SEI) of chum and pink salmon from 2002 to 2007 for the channelized segment and the reconstructed segment

Year	Swimming speed (BL/s)			Ground speed (BL/s)			Swimming efficiency index (SEI)		
	Channelized segment	Reconstructed segment	P-value	Channelized segment	Reconstructed segment	P-value	Channelized segment	Reconstructed segment	*P*-value
2002	-	-	-	0.07±0.03 (3)	0.12±0.04 (3)	0.42	-	-	-
2003	0.68±0.23 (2)	0.75±0.34 (2)	-	0.10±0.01 (3)	0.14±0.04 (3)	0.44	0.18±0.060 (2)	0.19±01.001 (2)	-
2004	0.80±0.21 (4)	0.61±0.09 (4)	0.44	0.28±0.05 (4)	0.15±0.02 (4)	0.35	0.45±0.11 (4)	0.36±0.07 (4)	0.50
2005	0.92±0.16 (9)	0.87±0.14 (10)	0.82	0.15±0.02 (9)	0.16±0.03 (9)	0.76	0.19±0.02 (9)	0.22±0.05 (9)	0.51
2006	0.51±0.04	-	-	0.43±0.07 (5)	0.12±0.02 (5)	0.012*	0.84±0.12 (5)	-	-
2003	-	-	-	0.20±0.04 (5)	0.16±0.03 (5)	0.42	-	-	-
2004	1.39±0.09 (3)	1.56±0.27 (3)	0.55	0.34±0.05 (3)	0.21±0.03 (3)	0.12	0.25±0.05 (3)	0.15±0.04 (3)	0.19
2007	1.10±0.57 (3)	0.30±0.08 (3)	0.28	0.38±0.01 (3)	0.33±0.05 (3)	0.43	0.42±0.09 (3)	0.58±0.08 (3)	0.33

Data are means ± S.E.

* indicates significant difference of means between the canalized and reconstructed segment by student's t-test.

Chart 4. Characteristics and numbers of cover used for holding behavior of chum (a) and pink salmon (b) in each segment

	Year										
	2002		2003		2004		2005		2006		
Cover	Channelized segment	Reconstructed segment	Channelized segment	Reconstructed segment	Channelized segment	Reconstructed segment	Channelized segment	Reconstructed segment	Channelized segment	Reconstructed segment	Total number (%)
Fallen tree	13	8	6	1	3	3	11	2	1	8	56 (44.8)
Artificial fallen tree	-	-	-	-	-	-	-	11	-	-	11 (8.8)
deep pool	1	-	-	-	-	-	-	4	-	5	10 (8.0)
vegetation	11	-	1	-	1	-	1	1	-	2	17 (13.6)
Block along the river bank	3	-	4	-	3	-	15	-	2	-	27 (21.6)
Large stone	-	-	-	-	-	2	1	1	-	-	4 (3.2)

(a)

	Year						
	2003		2004		2007		
Cover	Channelized segment	Reconstructed segment	Channelized segment	Reconstructed segment	Channelized segment	Reconstructed segment	Total number (%)
Fallen tree	6	3	1	4	-	9	23 (42.6)
Artificial fallen tree	-	-	-	1	-	4	5 (9.3)
deep pool	-	-	-	3	-	13	16 (29.6)
vegetation	3	-	-	-	-	3	6 (11.1)
Block along the river bank	1	-	1	-	2	-	4 (7.4)
Large stone	-	-	-	-	-	-	0

(b)

Geomorphological and Flow Process in the Reconstructed Segment from 2004 to 2005

To reveal the factor affecting the behavioral changes in the reconstructed segment, the water velocity distribution and water depths in the reconstructed segments were compared between 2004 and 2005. Figs. 5 and 6 show the calculated cross-sectional water velocity distribution and water depths in the reconstructed segments in 2004 and 2005 using measurements of flow velocities (1721 points in 2004 and 1653 points in 2005) and water depths data (561 points in 2004 and 587 points in 2005) respectively. Cross-sectional water velocity distribution was described by an average value for the water column. Intermediate values were obtained by linear interpolation from the neighboring points. More diverse flow conditions in the reconstructed segment were found in 2005 compared with 2004 (Figure 5). More diverse water depth conditions in the reconstructed segment were also found in 2005 compared with 2004 and deep pools were especially formed at the bending section (80, 180, 280, 420 m from the confluence point with the channelized segment (Figure 6). Chart 3 shows the resulting habitat utilization for holding behavior of chum and pink salmon in the channelized and reconstructed segment respectively from 2002 to 2007. Blocks along the river bank were frequently used as covers for holding behavior of chum and pink salmon in the channelized segment. In the reconstructed segment, natural fallen trees, installed artificial fallen trees and deep pools were relatively used more as covers for holding behavior.

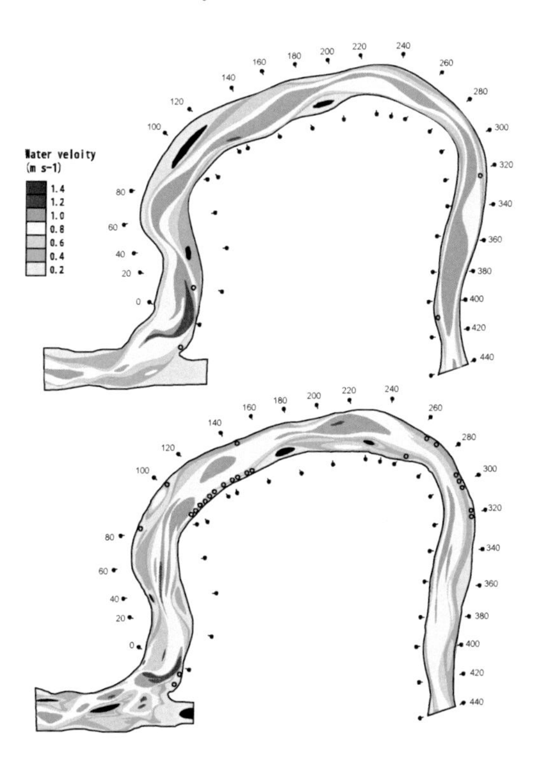

Figure 5. Two-dimensional water velocity distributions for base flow conditions in the reconstructed segment in 2004 (above) and 2005 (below). Numbers along the river indicate the a distance from the confluence point with the channelized segment. Open dots indicate holding positions of chum salmon.

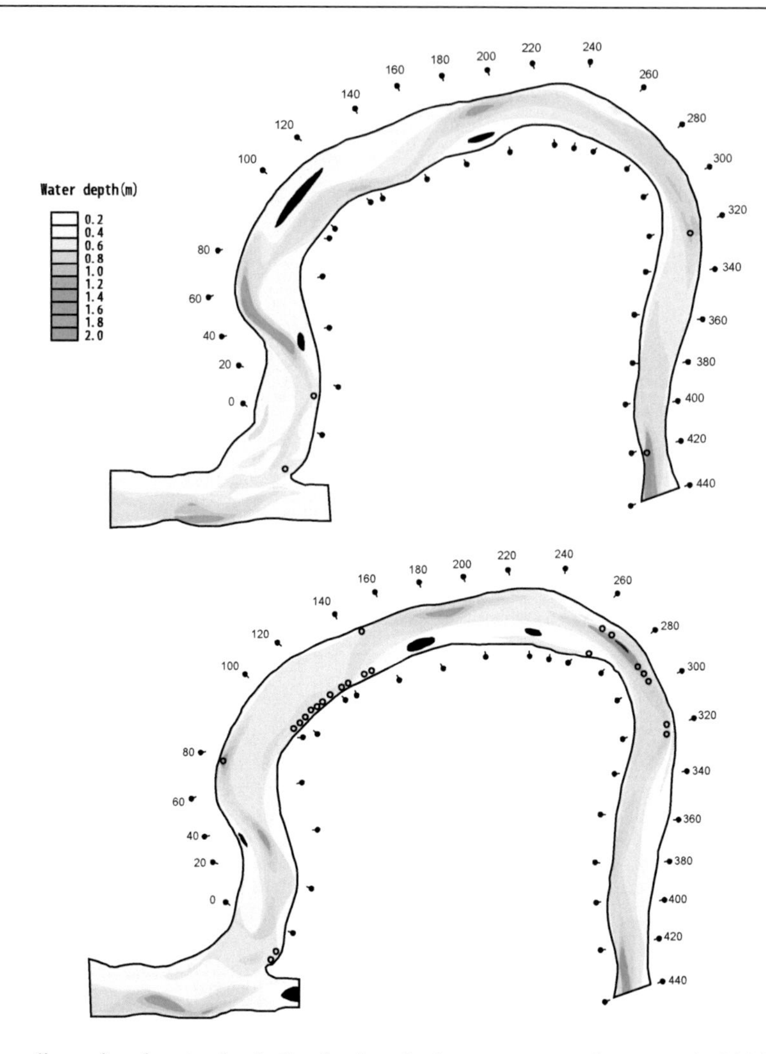

Figure 6. Two-dimensional water depth distributions in the reconstructed segment in 2004 (above) and 2005 (below). Numbers along the river indicate the distance from the confluence point with the channelized segment. Open dots indicate holding positions of chum salmon.

Mean water velocity in the channelized and reconstructed segments in 2005 was 0.65 ± 0.30 m s^{-1} and 0.69 ± 0.23 m s^{-1} (mean ± S.D.) respectively. The average water depth in the channelized and reconstructed segments in 2005 was 0.49 ± 0.29 m and 0.49 ± 0.31 m (mean ± S.D.) respectively. Figure 7 shows mean values of water velocity and depth and CV of water velocity and depth calculated from measurement of water velocity and depth every 20-100 m in the channelized and reconstructed segment. Mean values of water velocity and depth in the reconstructed segment were similar to those in the channelized segment. However, there was a tendency that CVs of water velocity and depth in the reconstructed segment were more variable both longitudinally and laterally than those in the channelized segment (Figure 7).

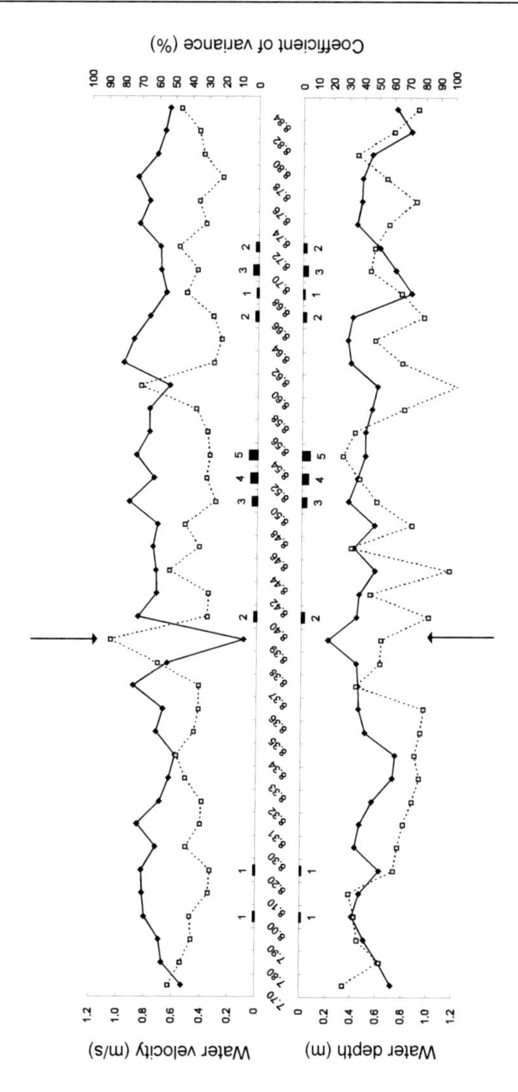

Figure 7. Mean water velocity (solid line) and coefficient of variance (CV) of water velocity (dotted line) measured along the channelized and reconstructed segments every 20-100 m (above). Mean water depth (solid line) and coefficient of variance (CV) of water depth (dotted line) measured by along the channelized and reconstructed segment every 20-100 m (below). Vertical arrows indicate the confluence point with the channelized and reconstructed segment. Numbers on black bars indicate the number of observed holding behavior.

Temporal Description of Swimming Speed and Migration Paths

Chum and pink salmon tended to migrate along the river banks, and sometimes crossed from one bank to another. Figure 8 shows an example of a detailed migration path and temporal description of swimming speed calculated from EMG signals of chum salmon in the reconstructed segment. Holding behavior was often observed following high swimming speed exceeding U_{crit}. Swimming speeds tended to be stable below U_{crit} during holding behavior. Further, swimming speed calculated from EMG signals was compared between moving and holding behavior of chum and pink salmon. Data of swimming speed in each individual was pooled in all years studied. Swimming speed during holding behavior of chum (0.70±0.92 BL/s; N=19) and pink salmon (0.91±0.15 BL/s; N=19) tended to be lower than that during moving of chum (0.53±0.05 BL/s; N=18) and pink salmon (0.56±0.11 BL/s; N=7) (p=0.013, 0.136, respectively).

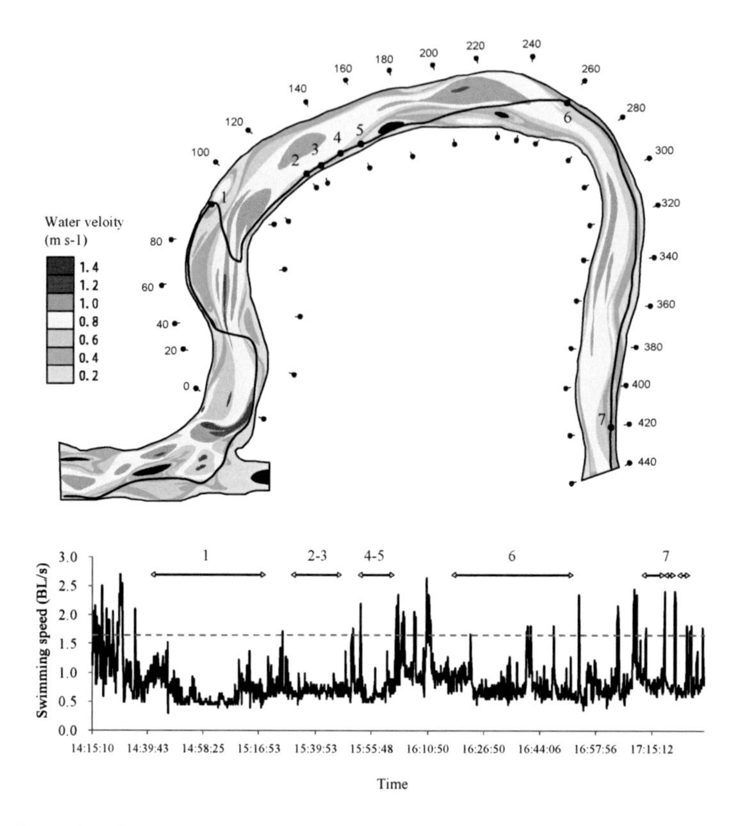

Figure 8. Two-dimensional water velocity distributions for base flow conditions in the confluence point with the channelized and reconstructed segment in 2005. The solid line indicates of path way the #2 chum salmon during upstream migration. Lower figure shows time-series plots of swimming speeds for #2 chum salmon, and dotted line represents the mean critical swimming speed of chum salmon calculated from the experiment using swimming chamber. Black dots and numbers on black dots indicate holding position and an order of holding behavior for #2 chum salmon.

Comparison of Swimming Depth during Moving and Holding Behavior

Swimming depth of chum and pink salmon were compared between moving and holding behavior from from retrieved DT loggers. Thirteen DT loggers for chum salmon and four DT loggers for pink salmon could be retrieved in this study from 2001 to 2007. Swimming depth during moving and holding were compared using pooled data for all years. A Student's t test revealed that swimming depth (mean ± S.E.) of chum salmon was not significantly different between moving for chum salmon (0.50 ± 0.05 m) and pink salmon (0.64 ± 0.07 m), and holding for chum salmon (0.63 ± 0.13 m; P=0.13) and for pink salmon (0.72 ± 0.20 m; P=0.72). Swimming depth during holding of chum salmon relatively tended to be deeper than during moving.

Critical Swimming Speeds

The U_{crit} (mean ± S.D.) of chum salmon ranged from 1.34 to 1.95 BL s^{-1} (1.07± 0.13 m s^{-1}, 1.65 ± 0.19 BL s^{-1}). Mean value of male U_{crit} was 1.63 ± 0.14 BL s^{-1} and mean value of female U_{crit} was 1.67 ± 0.27 BL s^{-1}. A Student's *t* test revealed that the U_{crit} values did not differ by sex ($P > 0.05$).

The U_{crit} (mean ± S.D.) of pink salmon ranged from 1.89 to 2.19 BL s^{-1} (1.16± 0.09 m s^{-1}, 2.19 ± 0.23 BL s^{-1}).

Respirometry Metabolism of Pink Salmon

MO_2 (mg O_2 kg^{-1} h^{-1}) significantly increased with increasing swimming speed ($p<0.001$; Figure 9). The relationship between MO_2 and swimming speed was explained almost equally well by the exponential equation $MO_2 = 174.2e^{1.26U}$ where the constant a=174.2 is an estimate of the weight-specific SMR (mg O_2 kg^{-1} h^{-1}).

COT was calculated as MO_2 U^{-1} at each measured swimming speed (Figure 10). The relationship between COT and swimming speed could described by $COT(U) = (174.2e^{1.26U})U^{-1}$ where U_{opt} equals 0.79 m s^{-1} and COT_{min} 596.6 mg O_2 kg^{-1} h^{-1}.

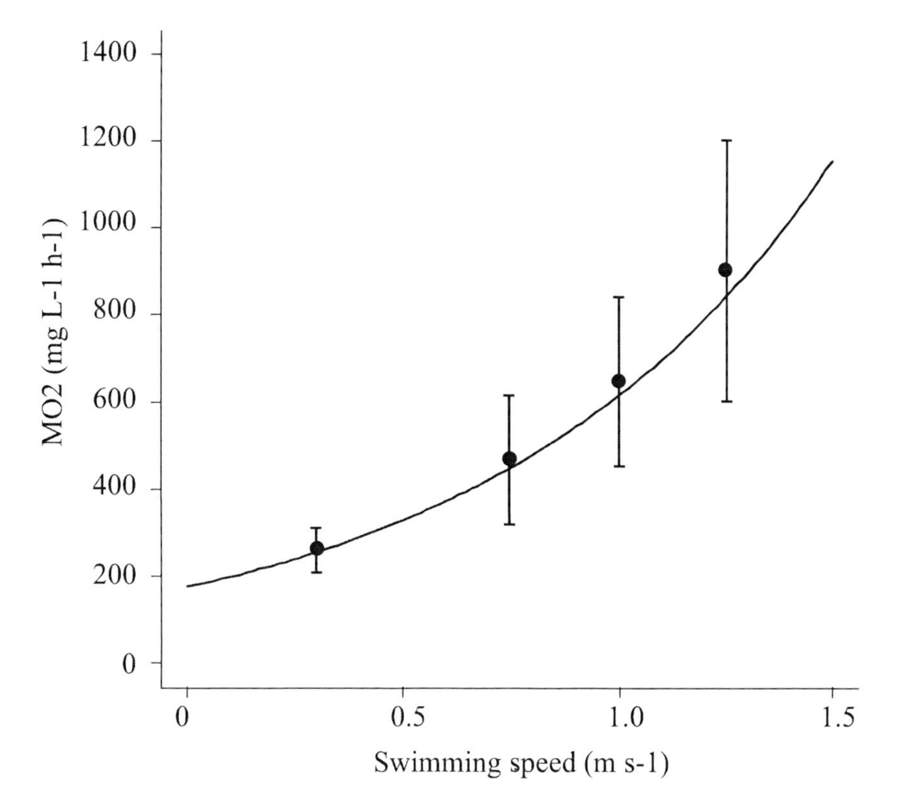

Figure 9. MO_2 (mean ± S.E.) of adult pink salmon as a function of swimming speed (m s^{-1}) at 15 °C.

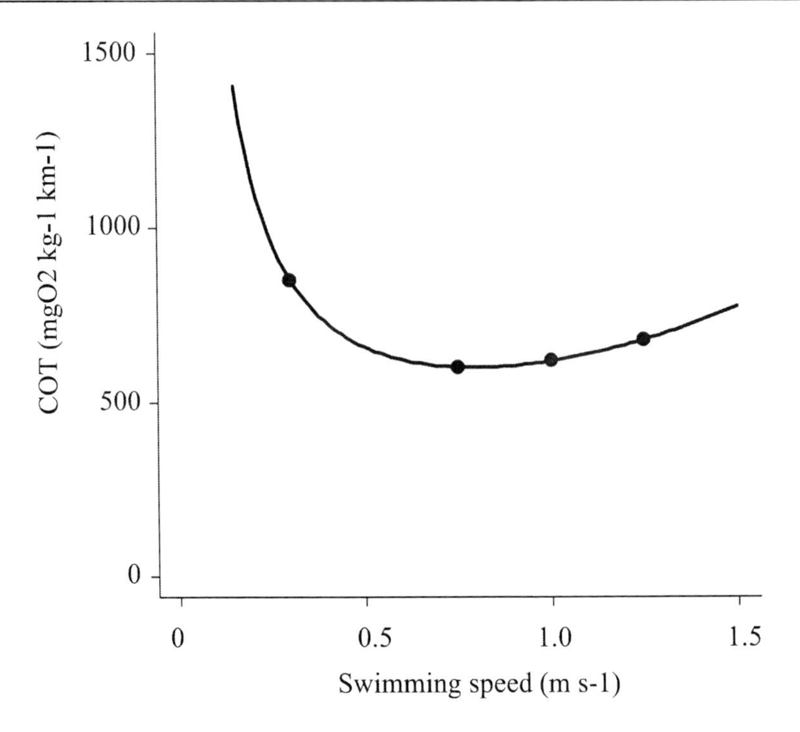

Figure 10. Costs of transport (COT), the amount of oxygen used per unit mass per unit distance, as a function of swimming speed (m s^{-1}).

EMG-swimming Speed Relationships

Figure 11 shows an example of temporal EMG values of pink salmon during respirometry trials. As increasing swimming speeds, fluctuation of EMG values tended to increase. Figure 12 shows the mean values and standard deviation of EMG values of pink salmon in relation to swimming speed. GLM revealed that mean values and standard deviation of EMG values significantly increased with increasing swimming speed ($p<0.001$). The model containing mean values and standard deviation of EMG values (AIC:12.94) was a better model to describe the swimming speed of pink salmon rather than the model containing only mean values of EMG values (AIC:22.78). The relationship between EMG values and swimming speed could described by

$$U = 0.126EMG_M+0.078EMG_{SD}-0.013$$

where EMG_M is mean values of EMG values and EMG_{SD} is standard deviation of EMG values. Using these models, energy uses of adult male pink salmon migrating in the channelized and reconstructed segments in 2004 and 2007 were estimated from EMG values, and compared (Figure 13). Although energy use of pink salmon migrating in the reconstructed segment tended to be lower than that in the channelized segment, mean values of energy use of pink salmon between the channelized and reconstructed segment were not significantly different in 2004 ($p=0.42$) and 2007 ($p=0.16$) respectively.

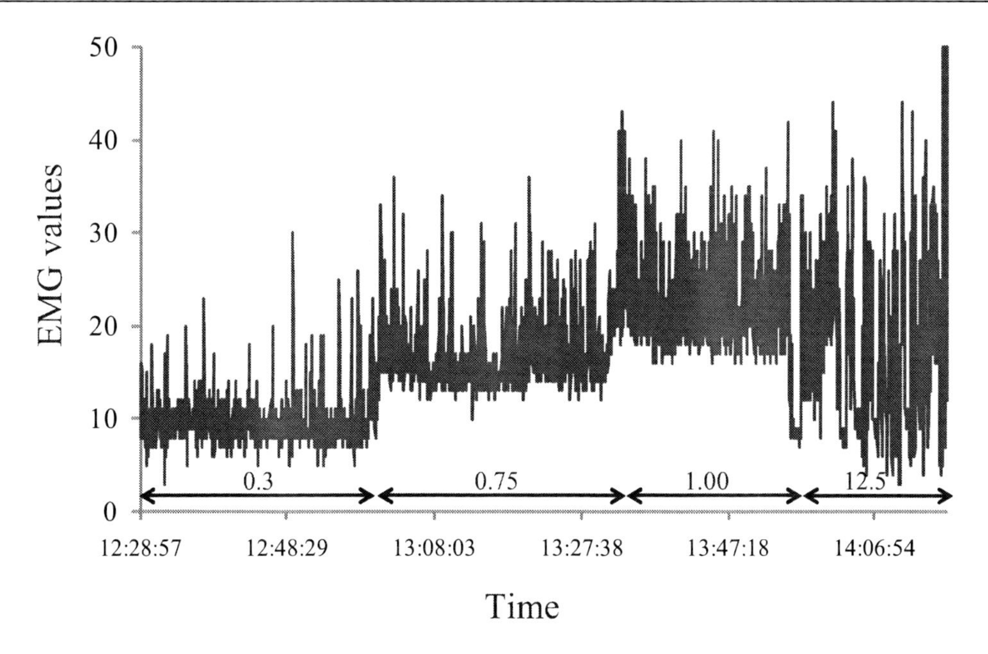

Figure 11. Time-series plots of swimming speeds for #11 pink salmon during the respirometry trial. Numbers on arrows indicate swimming speed (m s^{-1}) the fish were swimming.

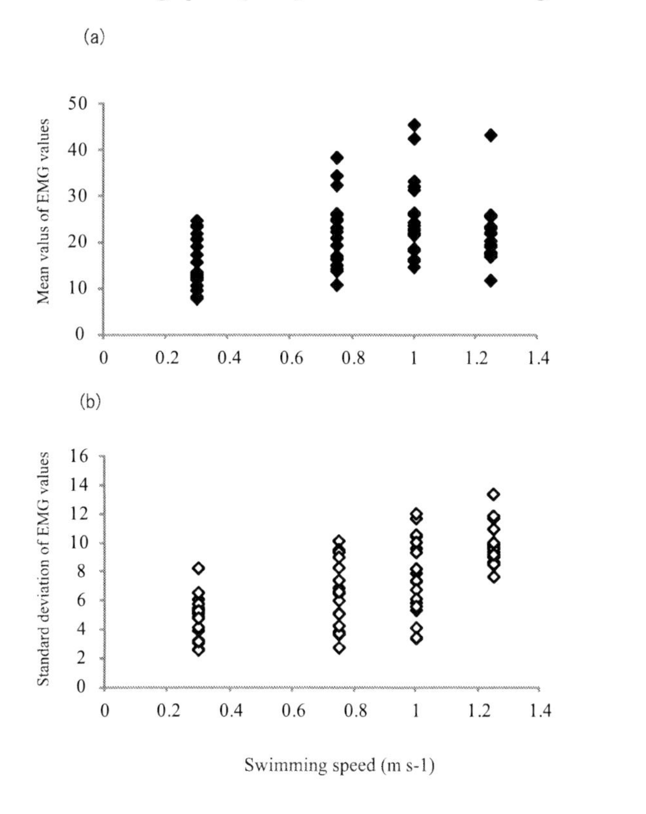

Figure 12. The observed mean values (a) and standard deviations (b) of EMG values for 20 adult pink salmon in relation to swimming speed (m s^{-1}).

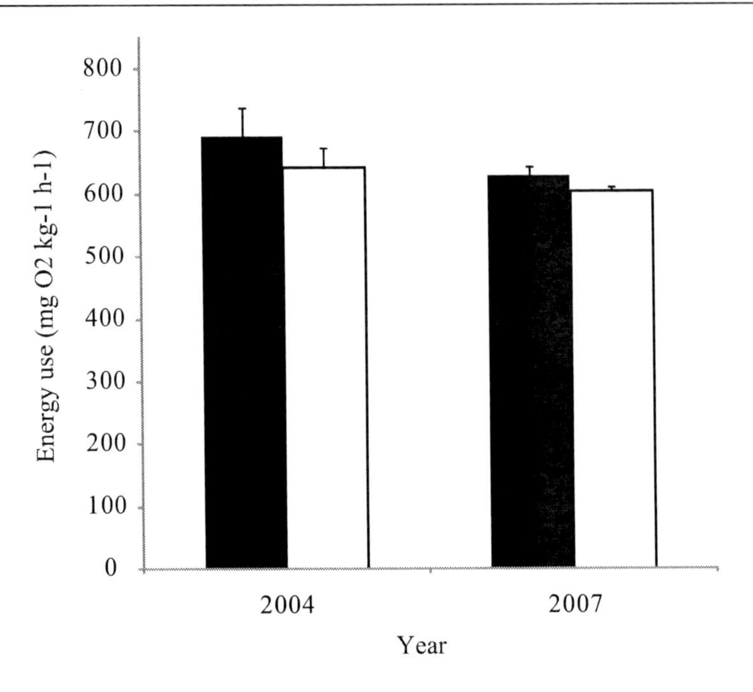

Figure 13. Mean energy uses estimated from the energetic models of adult pink salmon in the channelized segment (solid bars) and the reconstructed segment (open bar) in 2004 and 2007 respectively.

CONCLUSION

In this study, the effects of re-meandering construction of the Shibetsu River on upstream migration behavior of chum and pink salmon was examined using telemetry techniques from 2001 to 2007.

Influence of Hydrological Conditions on Upstream Migration

This study found that chum and pink salmon migrating in the Shibetsu River selected the segment with the highest water volume. This result indicates that upstream migration of chum and pink salmon may be affected by water volume when selecting migration pathways. Increasing water volume (water velocity) of the river is a factor affecting upstream migration of adult salmonids [37]. Weaver [38] reported that steelhead trout (*O. mykiss*), chinook salmon (*O. tshawytscha*) and coho salmon (*O. kisutch*) have a preference selecting a channel with the higher water velocity. It is generally accepted that increasing water volume (water velocity) induces positive rheotaxis of upstream migration of adult salmonids. A short channelized segment was retained for the purpose of preventing flooding along the reconstructed segment in the Shibetsu River. These results imply that adult salmonids migrating in the Shibetsu River may not select the reconstructed segment and locate holding sites during their upstream migration if water volume of the reconstructed segment is lower than that of the channelized segment.

Effects of the Reconstructed Segment on Swimming Behavior of Chum and Pink Salmon

Mean values of swimming speed calculated from EMG signals, ground speed and SEI of chum and pink salmon were not different between the channelized and reconstructed segment through the study periods. Time of holding behaviors for chum and pink salmon were rarely observed in the reconstructed segment before 2005; however, after 2005 a longer time of holding behavior tended to be observed in the reconstructed segment than in the channelized segment. These results indicate that there were few suitable holding sites for migrating chum and pink salmon in the reconstructed segment before 2005, and suitable sites for their holding behavior in the reconstructed segment has been developing by 2005. Upstream migration of anadromous adult salmon to the spawning sites is energetically expensive and must rely on stored energy reserves to carry out their upstream migration, to complete their gonad development, and to spawn because they stop feeding prior to upstream migration [39]. Therefore, efficient swimming during upstream migration is crucial for anadromous adult fish to minimize energetic costs of swimming wherever possible. This study found that swimming speeds during holding behavior of chum and pink salmon tended to be lower than during moving. This result indicates that chum and pink salmon decrease muscle activity at holding to reduce energetic costs of swimming. It is generally assumed that water velocity decreases as the riverbank and river bed approaches. Our study showed that temporal swimming speed calculated from EMG signals and migration paths, in which swimming speed of chum salmon exceeded U_{crit}, decreased rapidly during holding behavior. These results suggest that chum and pink salmon might become fatigued and unable to swim against temporarily imposed currents during upstream migration. It is generally accepted that maximum oxygen uptake occurs at the U_{crit} [25]. Thus U_{crit} is presumed to be a relatively close measurement of the maximum aerobic capacity of the fish. Hammer [17] also suggested that aerobic metabolism predominates in salmonids (i.e., red muscle activity) at swim speeds up to 70–80% of the U_{crit}, but anaerobic metabolism is initiated during swimming speeds equal to or exceeding 80% of the U_{crit} [40, 41]. Webb [40] also indicated that maximum efficiency of swimming will occur at about 70% of the U_{crit}. Adult salmonids swim below and above U_{crit} with only short periods during upstream migration, using burst swimming, which is performed anaerobically [15, 18, 42]. Therefore, swimming behavior of salmon prior to holding behavior might be exhausting, and observed holding behavior may be important for muscle recovery of chum and pink salmon during upstream migration.

Geomorphological Evolution of the Reconstructed Segment

More diversity of water velocity and depth conditions was observed in 2005 rather than in 2004 in the reconstructed segment of the Shibetsu River (Figs. 5 and 6). In this study, there was a tendency for the CVs of water velocity and depth in the reconstructed segment to vary more longitudinally and laterally than those in the channelized segment in 2005 (Figure 7). These results indicate that physical diversity had been created by the effects of the re-meandering construction, and habitat structures for holding chum and pink salmon may have developed in the reconstructed segment of the Shibetsu River over the three year period. This agrees with previous studies that suggest a physically stable environment can be created rapidly (less than three years) after the re-meandering of a segment such as that in the River Gelså in Denmark [43]. During the initial phase, biotic elements and morphology exhibit large spatial and temporal variation in contrast to the reconstructed ecosystems [43-45]. Meandering water flow and modifications promote morphological diversity to create and maintain the physical habitat [46]. The creation of morphological features and meander bends may encourage development of a physical environment which is enriched by longitudinal and lateral variations in velocity and water depth. These aspects promote the generation of secondary currents, turbulence and variations in both the local water velocity and water depth: essential habitat conditions for the life-cycle of fish [47]. There are some studies suggesting that fish prefer to swim under unsteady flows [20, 48]. Wake zones and vortices created by bends and woody debris provided pool habitats and cover essential for larger fishes [49]. Liao et al. [50] demonstrated the interactions between fish swimming and vortices, which are generated in unsteady flow, in a laboratory study, and suggested that during vortex exploitation muscle activity of fish is reduced to decrease the cost of locomotion. Fishes in natural habitats are faced with vortices that arise from a variety of obstacles in the flowing water. Hinch et al. [18] indicated that small reverse-flow vortices may even reduce swimming drag of adult salmonids, making it easier for upstream passage. Our results indicate that diversity of water velocity and depth created by the reconstruction of the Shibetsu River may be useful and suitable for holding sites of chum and pink salmon during upstream migration.

Energy Costs Associated with Swimming of Pink Salmon in the Channelized and Reconstructed Segment

Oxygen consumption rates of adult pink salmon migrating in the Shibetsu River increased with swimming speed, and ranged from a minimum of 174.4 mg O_2 kg-^{1} at 0.3 m s^{-1} to a maximum of 946.3 mg O_2 kg-^{1} s^{-1} at 1.25 m s^{-1}. Our results suggest that oxygen consumption rates of adult pink salmon are consistent with other study results at similar temperature and swimming speed of pink salmon. Oxygen consumption for adult pink salmon (1.6 kg) at 15°C ranged from 150 to 830 mg O_2 kg-^{1}, implying that adult pink salmon have a wide scope of activity [51].

Energy use of pink salmon migrating in the reconstructed segment tended to be lower than that in the channelized segment; however, there were not significant differences in energy use of pink salmon between the channelized and reconstructed segment. Migrating salmon achieve U_{opt}, which is the swimming speed where the energy use per unit distance is minimized, though only in environments with slow currents less than 0.25 m s^{-1} [18, 52, 53]. Hinch et al. [18] also indicated that adult salmon can locate and exploit very small reverse-flow vortices, created by rough substrates or banks. Mean values of water velocity were similar between the channelized and reconstructed segment; however, more diversity of water velocity and depth conditions were observed in the reconstructed segment. Therefore, we speculated that very small water velocities and vortices on a small scale may have been created in the reconstructed segment and salmon could locate and exploit the physical characteristics to reduce the energy uses during upstream migration. Although we could not fully understand energy uses of pink salmon migrating in the two segments because the sample sizes were small, we speculated that physical characteristics such as more diversity of water velocity and depth conditions created in the reconstructed segment may affect energy use of salmon during upstream migration, and further research studies are needed.

This study, conducted for seven years, provides essential information on the success of the reconstruction scheme of biotic and abiotic elements, but extrapolations and conclusions from these studies have to be made with caution until the reconstructed ecosystems reach equilibrium. The reconstructed segment in the Shibetsu River has been gradually developing suitable holding sites for chum and pink salmon during upstream migration. The habitat quality in the reconstructed segment for migrating chum and pink salmon will likely further improve and grow close to the natural meandering of the river, and the results of this study will help the success of the river reconstruction scheme for migrating adult salmon.

ACKNOWLEDGEMENTS

We thank the following people for their help and co-operation: M. Mizoguchi and A. Nitta (Japan NUS Co., LTD) for technical support, Dr. J. B. K. Leonard (Northern Michigan University) for assistance with the study design, H. Kakizaki (Nemuro Salmon Propagative Association) for providing experimental animals and Dr. M. Nagata and K. Kasugai for providing the swimming chamber and facilities of Hokkaido Fish Hatchery Doto Research Branch. We are also grateful to Dr. N. G. Miles (Hokkaido University) for valuable comments on the manuscript. This study was supported in part by a Research Fellowship for Young Scientists to Y. M. (195295) from the Japanese Society for the Promotion of Science (JSPS) as well as Grant In-Aid for Scientific Research (A) (18208017) from JSPS, and from the Foundation of Riverfront Improvement and Restoration to H. U.

REFERENCES

[1] Salo EO. Life history of chum salmon (*Oncorhynchus keta*). In: Groot C, Margolis L, eds. Pacific salmon life history,vol Vancouver, British Columbia, University of British Columbia Press, 231-309; 1991.

[2] Groot C, Margolis L. Pacific Salmon Life Histories In Vancouver, BC, University of British Columbia Press, 1991.

[3] Miller RJ, Brannon EL. The origin and development of life history patterns in Pacific salmonids. In Brannon EL, Salo EO, eds. Proceedings of the Salmon and Trout Migratory Behaivor Symposium,vol School of Fisheries, University of Washington, Seattle, 296-309, 1982.

[4] Morita K, Saito T, Miyakoshi Y, Fukuwaka MA, Nagasawa T, Kaeriyama M. A review of Pacific salmon hatchery programmes on Hokkaido Island, Japan. ICES *Journal of Marine Science* 63:1353-1363, 2006.

[5] Zaporozhets OM, Zaporozhets GV. Interaction between hatchery and wild Pacific salmon in the Far East of Russia: A review. *Reviews in Fish Biology and Fisheries* 14(3): 305-319, 2004.

[6] Morita K, Morita SH, Fukuwaka MA. Population dynamics of Japanese pink salmon (*Oncorhynchus gorbuscha*): are recent increases explained by hatchery programs or

climatic variations? *Canadian Journal of Fisheries and Aquatic Sciences* 63(1): 55-62, 2006.

[7] Mizuno N, Gose K. Ecology of streams. Tokyo, Tsukiji-shokan, 1981.

[8] Bayley PB. Understanding Large River Floodplain Ecosystems. *Bioscience* 45(3): 153-158, 1995.

[9] Akita M, Makiguchi Y, Nii H, Nakao K, Sandahl JF, Ueda H. Upstream migration of chum salmon through a restored segment of the Shibetsu River. *Ecology of Freshwater Fish* 15(2): 125-130, 2006.

[10] Makiguchi Y, Nii H, Nakao K, Ueda H. Upstream migration of adult chum and pink salmon in the Shibetsu River. *Hydrobiologia* 582: 43-54, 2007.

[11] Cooke SJ, Hinch SG, Wikelski M, Andrews RD, Kuchel LJ, Wolcott TG, Butler PJ. Biotelemetry: a mechanistic approach to ecology. *Trends in Ecology and Evolution* 19(6): 334-343, 2004.

[12] Moore A, Russell IC, Potter ECE. The effects of intraperitoneally implanted dummy acoustic transmitters on the behavior and physiology of juvenile Atlantic salmon, *Salmo Salar L. Journal of Fish Biology* 37(5): 713-721, 1990.

[13] Lucas MC, Johnstone ADF, Priede AG. Use of physiological telemetry as a method of estimating metabolism of fish in the natural environment. *Transactions of the American Fisheries Society* 122: 22-833, 1993.

[14] Hinch SG, Diewert RE, Lissimore TJ, Prince AMJ, Healey MC, Henderson MA. Use of electromyogram telemetry to assess difficult passage areas for river-migrating adult sockeye salmon. *Transactions of the American Fisheries Society* 125(2): 253-260, 1996.

[15] Hinch SG, Standen EM, Healey MC, Farrell AP. Swimming patterns and behaviour of upriver-migrating adult pink (*Oncorhynchus gorbuscha*) and sockeye (*O. nerka*) salmon as assessed by EMG telemetry in the Fraser River, British Columbia, Canada. *Hydrobiologia* 483(1-3): 147-160, 2002.

[16] Weatherley AH, Rogers SC, Pincock DG, Patch JR. Oxygen consumption of active rainbow trout, *Salmo gairdneri* Richardson, derived from electromyograms obtained by radiotelemetry. *Journal of Fish Biology* 20(4): 479-489, 1982.

[17] Hammer C. Fatigue and exercise tests with fish. *Comparative Biochemistry and Physiology a-Physiology* 112(1): 1-20, 1995.

[18] Hinch SG, Rand PS. Optimal swimming speeds and forward-assisted propulsion: energy-conserving behaviours of upriver-migrating adult salmon. *Canadian Journal of Fisheries and Aquatic Sciences* 57(12): 2470-2478, 2000.

[19] Standen EM, Hinch SG, Healey MC, Farrell AP. Energetic costs of migration through the Fraser River Canyon, British Columbia, in adult pink (*Oncorhynchus gorbuscha*) and sockeye (*Oncorhynchus nerka*) salmon as assessed by EMG telemetry. *Canadian Journal of Fisheries and Aquatic Sciences* 59(11): 1809-1818, 2002.

[20] Hinch SG, Rand PS. Swim speeds and energy use of upriver-migrating sockeye salmon (*Oncorhynchus nerka*): role of local environment and fish characteristics. *Canadian Journal of Fisheries and Aquatic Sciences* 55(8): 1821-1831, 1998.

[21] Beauchamp DA, Stewart DJ, Thomas GL. Corroboration of a bioenergetics model for sockeye salmon. *Transactions of the American Fisheries Society* 118: 597-607, 1989.

[22] Brett JR. Energetics. In: Groot C, Margolis L, Clarke WC, eds. Physiological ecology of Pacific salmon,vol Vancouver, B.C., University of B.C. Press, 1-68; 1995.

[23] Brett JR. The respiratory metabolism and swimming performance of young sockeye salmon. *Journal of the Fisheries Research Board of Canada* 21: 1183-1226, 1964.
[24] Brett JR. Swimming performance of sockeye salmon (*Oncorhynchus nerka*) in relation to fatigue time and temperature. *Journal of Fisheries Research Board of Canada* 24: 1731-1741, 1967.
[25] Farrell AP, Steffensen JF. An analysis of the energetic cost of the branchial and cardiac pumps during sustained swimming in trout. *Fish Physiology and Biochemistry* 4(2): 73-79, 1987.
[26] Bridger CJ, Booth RK. The effects of biotelemetry transmitter presence and attachment procedures on fish physiology and behavior. *Reviews in Fisheries Science* 11(1): 13-34, 2003.
[27] Winter JD. Advances in underwater biotelemetry. In: Murphy BR, Willis DW, eds. Fisheries Techniques, 2nd Edition,vol Bethesda, Maryland, American Fisheries Society, 555-590; 1996.
[28] Beddow TA, McKinley RS. Importance of electrode positioning in biotelemetry studies estimating muscle activity in fish. *Journal of Fish Biology* 54(4): 819-831, 1999.
[29] McKinley RS, Power G. Measurement of activity and oxygen consumption for adult lake sturgeon (*Acipenser fulvescens*) in the wild using radio-transmitted EMG signals. In: Priede AG, Swift SM, eds. Wildlife Telemetry: Remote Monitoring and Tracking Animals,vol 307. West Sussex, U.K., Ellis Horwood, 1992.
[30] Økland F, Finstad B, McKinley RS, Thorstad EB, Booth RK. Radio-transmitted electromyogram signals as indicators of physical activity in Atlantic salmon. *Journal of Fish Biology* 51(3): 476-488, 1997.
[31] McFarlane WJ, Cubitt KF, Williams H, Rowsell D, Moccia R, Gosine R, McKinley RS. Can feeding status and stress level be assessed by analyzing patterns of muscle activity in free swimming rainbow trout (*Oncorhynchus mykiss Walbaum)? Aquaculture* 239(1-4): 467-484, 2004.
[32] Smit H, Amelink-Koutstaal JM, Vijverberg J, Von Vaupel-Klein JC. Oxygen consumption and efficiency of swimming goldfish. *Comparative Biochemistry and Physiology* 39A:1-28, 1971.
[33] Gehrke PC, Fidler LE, Mense DC, Randall DJ. A respirometer with controlled water-quality and computerized data acquisition for experiments with swimming fish. *Fish Physiology and Biochemistry* 8(1): 61-67, 1990.
[34] Bell WM, Terhune LDB. Water tunnel design for fisheries research. *Technical Report Fisheries Research Board of Canada* 195: 1-69, 1970.
[35] Beamish FWH. Swimming capacity. In: Hoar WH, Randall DJ, eds. *Fish Physiology*,vol 7. New York, Academic Press, 101-187; 1978.
[36] Dobson AJ. An introduction to Generalized Linear Models, 2nd ed. Boca Raton, CRC Press, 2002.
[37] Banks JW. A review of the literature on the upstream migration of adult salmonids. *Journal of Fish Biology* 1:85-136, 1969.
[38] Weaver CR. Influence of water velocity upon orientation and performance of adult migrating salmonids. *Fishery Bulletin* 63(1):97-121, 1963.
[39] Hasler AD, Scholz AT, Ross MH. Olfactory imprinting and homing in salmon. *American Scientist* 66:347-355, 1978.

[40] Webb PW. The swimming energetics of trout: II. Oxygen consumption and swimming efficiency. *Journal of Experimental Biology* 55:521-540, 1971.

[41] Burgetz IJ, Rojas-Vargas A, Hinch SG, Randall DJ. Initial recruitment of anaerobic metabolism during sub-maximal swimming in rainbow trout (*Oncorhynchus mykiss*). *Journal of Experimental Biology* 201(19): 2711-2721, 1998.

[42] Hinch SG, Bratty J. Effects of swim speed and activity pattern on success of adult sockeye salmon migration through an area of difficult passage. *Transactions of the American Fisheries Society* 129(2): 598-606, 2000.

[43] Friberg N, Kronvang B, Hansen HO, Svendsen LM. Long-term, habitat-specific response of a macroinvertebrate community to river restoration. *Aquatic Conservation-Marine and Freshwater Ecosystems* 8(1): 87-99, 1998.

[44] Hughes FMR, Colston A, Mountford JO. Restoring riparian ecosystems: The challenge of accommodating variability and designing restoration trajectories. *Ecology and Society* 10(1), 2005.

[45] Petranka JW, Murray SS, Kennedy CA. Responses of amphibians to restoration of a southern Appalachian wetland: Perturbations confound post-restoration assessment. Wetlands 23(2): 278-290, 2003.

[46] Downs PW, Thorne CR. Design principles and suitability testing for rehabilitation in a flood defense channel: The River Idle, Nottinghamshire, UK. *Aquatic Conservation-Marine and Freshwater Ecosystems* 8(1): 17-38, 1998.

[47] Bockelmann BN, Fenrich EK, Lin B, Falconer RA. Development of an ecohydraulics model for stream and river restoration. *Ecological Engineering* 22(4-5): 227-235, 2004.

[48] Fausch KD. Experimental analysis of microhabitat selection by juvenile steelhead (*Oncorhynchus mykiss*) and coho salmon (*Oncorhynchus kisutch*) in a British-Columbia stream. *Canadian Journal of Fisheries and Aquatic Sciences* 50(6): 1198-1207, 1993.

[49] Schlosser IJ. Fish community structure and function along 2 habitat gradients in a headwater stream. *Ecological Monographs* 52(4): 395-414, 1982.

[50] Liao JC, Beal DN, Lauder GV, Triantafyllou MS. Fish exploiting vortices decrease muscle activity. *Science* 302(5650): 1566-1569, 2003.

[51] Williams IV, Brett JR. Critical swimming speed of Fraser and Thompson River pink salmon (*Oncorhynchus gorbuscha*). *Canadian Journal of Fisheries and Aquatic Sciences* 44(2): 348-356, 1987.

[52] Quinn TP. Estimated swimming speeds of migrating adult sockeye salmon. *Canadian Journal of Zoology-Revue Canadienne De Zoologie* 66(10): 2160-2163, 1988.

[53] Madison DM, Horrall RM, Stasko AB, Hasler AD. Migratory movements of adult sockeye salmon (*Oncorhynchus nerka*) coastal British Columbia as revealed by ultrasonic tracking. *Journal of Fisheries Research Board of Canada* 29: 1025-1033, 1972.

In: Telemetry: Research, Technology and Applications
Editors: Diana Barculo and Julia Daniels
ISBN 978-1-60692-509-6

Short Communication

BIO-TELEMETRY OF INSHORE FISH IN POLAR REGIONS

Hamish Campbell[1] and Stuart Egginton[2]
1. Department of Biological Sciences, University of Queensland, Brisbane, QLD, Australia
2. Department of Physiology, Birmingham University, Birmingham, UK

ABSTRACT

Bio-telemetry studies have provided valuable insight into the biology of marine and freshwater fish. The majority of this research has focused on commercially important temperate and tropical species, and the movements and ecology of polar fish remains poorly understood. This lack of bio-telemetry research on polar species in general, and Antarctic fish in particular, is undoubtedly due to logistical constraints imposed by harsh and unpredictable environmental conditions, isolated locations, and a short field season. Moreover, due to the limited capacity of polar science logistic operations, researchers often have only a single opportunity to obtain field data. This essay aims to encourage both fisheries and polar biologists to engage in acoustic telemetry studies on polar fish by sharing our experience of the issues surrounding environmental constraints, equipment limitations, tracking protocols, choice of species, and safety in the field.

INTRODUCTION

Sonic telemetry is a valuable tool in understanding the ecology of marine fish. Behavioural parameters such as size and dimensions of the home range, foraging patterns, and the influence of environmental conditions are all amenable to such tracking techniques. Furthermore, the remote measurement and data logging of physiological variables such as heart-rate, muscle activity and feeding, can provide information that may illustrate the energetic requirements of free-ranging fish. Current off-the-shelf telemetry equipment, and most of the published techniques, has been developed in temperate and tropical fish species, and little information exists for researchers planning to track marine fishes inhabiting sub-zero inshore waters. Both the study subjects and the polar environment provide unique problems, which need to be overcome to enable a successful field study. However, due to the high cost of polar research and the rotational booking of individuals within a research program, scientists will often not have the opportunity to undergo pilot studies, and the isolated location makes the replacement of faulty or inappropriate equipment impractical.

This essay aims to facilitate the preparation for polar fish telemetry projects by presenting knowledge gained from researchers that have worked extensively on a number of different inshore fish species around the Antarctic continent. The results are presented within the context of a guide for telemetry application, and we therefore emphasise the methodology and logistics of tracking fish in the Southern Ocean rather than present an extensive discussion of Antarctic fish behaviour. It is hoped that by elucidating the pitfalls, limitations and frustrations from our own experience we may encourage others to productively conduct bio-telemetry research on fish inhabiting the polar regions. .

Environmental Context

The study site conditions in inshore polar seas will range from open water, through which a small vessel can easily navigate, to the other end of the spectrum, where it is covered by a thick layer of sea-ice traversable by ski-doo. Realistically, for much of the year polar inshore waters will be covered by shifting brash and berg ice that will inhibit boat movement and be too thin for sea-ice travel. Thus, the ability to move through the inshore to track fish movements may be prohibited on a number of days, depending on wind and ice conditions.

Around the majority of the Antarctic continent sea-ice begins to form in May, but there exists huge variability in inshore conditions between location, season and years. For example, along the Antarctic Peninsula recent years have seen no winter sea–ice thick enough for long-distance sea-ice travel. Yet, at the time of writing this essay, the inshore sea-ice that usually breaks up mid-summer in McMurdo Sound, had not broken out for 6 years. This has hampered a number of fish research studies within the sound, yet facilitated others. For example, inter-annual sea ice provides safe sea-ice travel, but makes drilling ice holes for catching or tracking fish very hard work.

The Antarctic inshore region is subjected to katabatic winds that blow off the ice shelf. These winds can change sea and ice conditions rapidly over a small geographical range, causing brash ice to blow quickly off-shore, and even thick sea-ice to break out along tidal cracks. This has resulted in the stranding of research personnel, and on some occasions the outcome has been fatal..The unpredictable nature of the waters around the Antarctic continent has been recognised and a strict limit is set by all Antarctic research agencies regarding small boat and sea-ice travel. For example, the British Antarctic Survey (BAS) set a 25 km wind speed limit on both boat and ski-doo travel, and a 25 cm minimum thickness before research personnel can venture on to the sea-ice.

Essentially, the unpredictable nature of the environment may result in your chosen research platform being out of action for a significant proportion of the field season, and hence you should plan for alternative tracking methods. We studied the movements and heart-rate of free-ranging Antarctic cod *(Notothenia coriiceps)* in the inshore waters along the

Antarctic Peninsula over a full annual cycle, where tracking required a number of different strategies to enable continual monitoring (Campbell et al. 2008).

Figure 1. A chainsaw is used to cut a hole through the sea-ice to enable placement of an acoustic hydrophone. Operator wears full protective clothing over an insulated and waterproof boating suit.

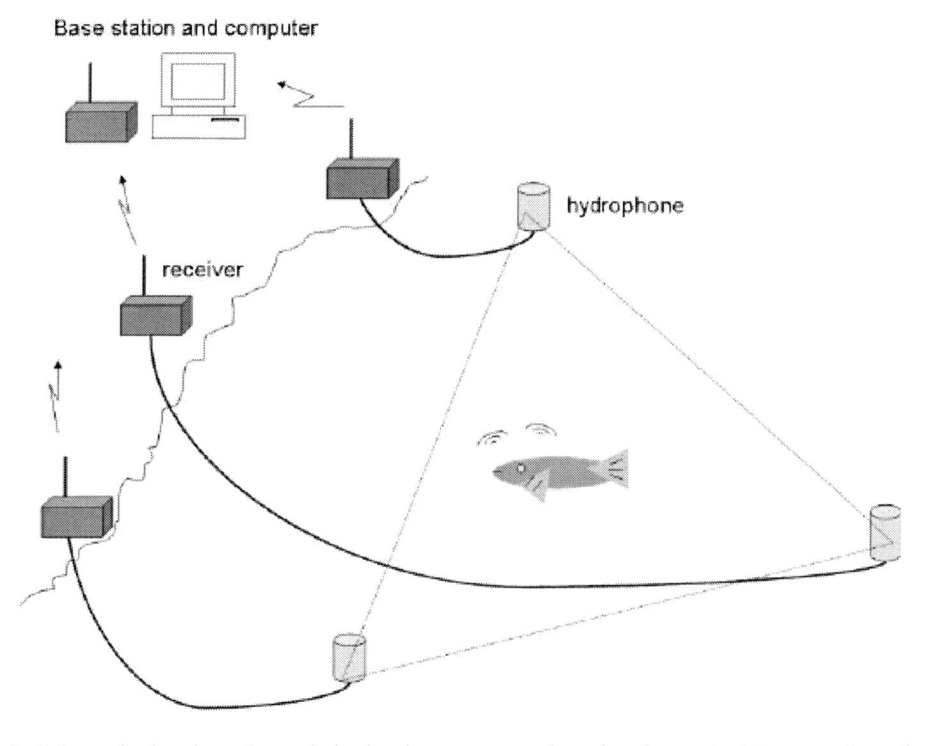

Figure 2. Schematic drawing of a static hydrophone array anchored to the sea-bed in an equilateral triangle formation. The data is transmitted to a shore based receiver by underwater cables and then from these receivers the signal is transmitted by radio to the base station. The acoustic pulse is decoded here, and from the difference in arrival time of the signal between the hydrophone the fish location is calculated.

Equipment and Tracking Protocol

As well as weather conditions, tracking methodology will depend largely on the level of logistical support available, often supplied by the research agency. Summer research for inshore telemetry work, especially along the Peninsula and sub-Antarctic islands, is best suited to a small vessel with an outboard motor. At Rothera Research Station (BAS), a rigid inflatable of 3.5 m, with two 40 HP 4 stroke engines has proved very efficient. This size of vessel is sufficiently large to push through surface brash ice, but still light and small enough to be winched free from the water after use. This is important to avoid crushing by drifting pack ice, and because wind driven surface ice will quickly break a boat free from its moorings.

A pole mounted omni-directional hydrophone is ideal for tracking from a small vessel and should be fixed to the side of the boat in a bracket that allows quick placement and retrieval from the water. The VEMCO (Nova Scotia, Canada) VR60 or VR100 receivers, housed in a tough waterproof case are ideal for polar marine conditions. However, do note then on very cold days (< -10 ^{0}C) the battery life of the receiver will be significantly reduced. The tracking technique is very similar to that employed in temperate of tropical waters. That is, an omni-directional hydrophone will provide 360^0 coverage around your vessel and provide information as to the presence or absence of your tagged fish. A sweep survey along the coastline with a uni-directional hydrophone will provide more precise information on specific fish location. However, shore access will not always be possible due to ice-cliffs and grounded ice-bergs, and it is recommended to keep the vessel at least 50 m distance from ice structures, as falling ice or turning bergs can generate a large swell that would swamp a small boat.

The sea-ice begins to form between May and June, and during this time the ability to get a boat on the water becomes ever more restricted. An alternative option for tracking fish is to travel over the sea-ice and put the hydrophone though pre-drilled holes set out along the inshore site in a grid formation. The potential for sea-ice travel depends very much on available logistics and local safety restrictions. Each member of the team should be fitted with a waterproof insulated boating suit and carry an ice axe, ice screws, and a length of rope. Firstly, the thickness of the sea-ice must be determined by a drilling a hole with an adequately sized drill bit. Once through the sea-ice, insert a pole with graduations, and a spring-loaded lever on the end. The lever will release when the end of the pole exits under the ice, and the thickness of the sea-ice can be determined. An attached cable is needed to reset the lever and allow the pole to be retrieved. A post drill or chainsaw is required to enable you to catch fish and place your hydrophone through the sea-ice. Alternatively, if you are based at the American (National Science Foundation) McMurdo station, a 2 meter diameter hole can be drilled through the sea-ice by mobile heavy plant, and a warmed cabin with a removable section in the floor, situated over the hole for your convenience.

For our study, ice holes were cut 150 m apart using a chainsaw with a 1 m blade. The operator wore chainsaw protective clothing over an insulated waterproof suit (Figure 1). Prior to cutting the ice hole, ice-screws were inserted towards the center, and later used as an anchor to fasten a rope to remove the (0.5m X 0.5 m) ice-blocks from the hole. After the ice-hole was cut, it was then covered with a wooden board to reduce further ice formation. Breaking and removing surface ice daily kept the ice hole open for a number of weeks. Using this technique, a grid pattern of hydrophone access points could be created throughout the inshore area. The distance between each hole was set by the range of the transmitter implanted in the fish.

Lack of access to the inshore study sites due to unpredictable sea-ice and weather conditions can seriously jeopardise a long-term telemetry study in polar waters. An alternative option to manual tracking of the fish is to use a static hydrophone array and automated recording system. This will continuously monitor acoustic signals, and can provide precise location data over a range of a few kilometers (Figure 2). VEMCO provide a system that consists of a central base station with PC interface, and three VR60 receivers. The latter are embedded within floatation aids, and these are anchored to the sea bed in an equilateral triangle formation. The precise longitude and latitude of each hydrophone is determined by either a satellite navigation device (GPS, GARMIN) or a theodilite, and the geographic coordinates entered into a projected image of the study, which can be either be drawn to scale or superimposed onto an orthorectified aerial photograph (Figure 3). To fix position of fish carrying an acoustic transmitter within the study area the system uses triangulation based on the position of each hydrophone relative to each other, the speed of sound through sea-water at sub-zero temperature, and the difference in the timing of the acoustic pulse emitted from the transmitter arriving at each hydrophone. The time and pulse information is then radioed to a central base station on-shore. The resolution within the triangle formation of the hydrophones is less than 1 meter, and the resolution decreases the further outside the triangle formation the acoustic pulse is transmitted. The system that we used could continuously track up to 12 fish simultaneously.

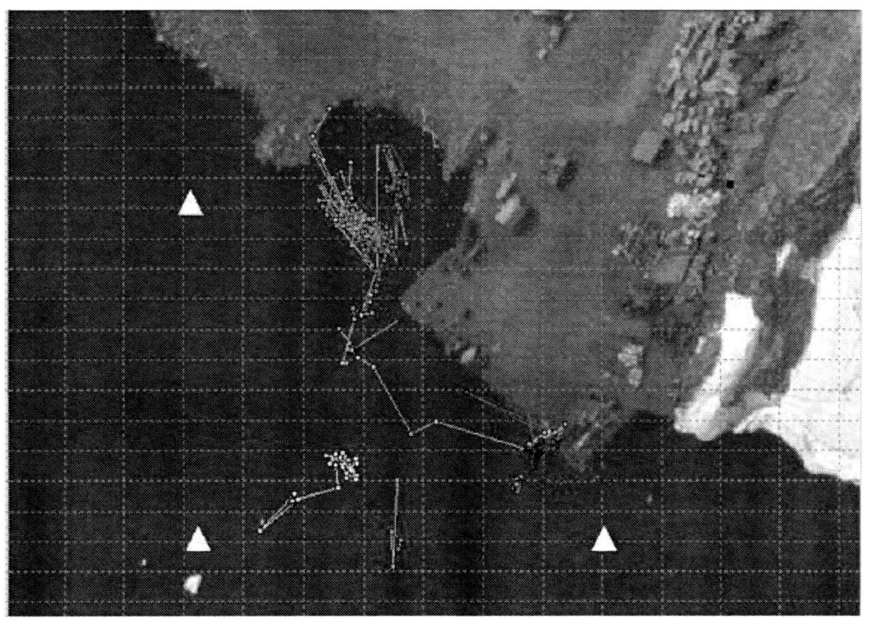

Photo courtesy of MAGIC, British Antarctic Survey.

Figure 3. Orthorectified aerial photograph of the inshore study area, Adelaide Island, Antarctic Peninsula. The dark area indicates water and the light grey area the rocky shore line and wharf. Grid cells are 10 m long and 5 m high. The position of each hydrophone is represented by a white triangle. The movements of 11 individual Antarctic cod *(N. coriiceps)* are shown, squares show actual fixed positions and the shortest distance between each positional fix is also shown.

This type of static system is ideal for inshore Antarctic fishes, many of which show high site fidelity. However, there needs to be some modification in design before using a radio acoustic positioning system in polar waters, as any hardware at or near the sea-surface will be quickly damaged or removed by sea-ice. We deployed a static hydrophone array in Antarctic waters by using SCUBA divers. The hydrophones were fixed to the sea-bed by hammering in rock pitons at about 20 m depth. The cable was then run along the sea-bed and returned to shore through hollow steel poles (3 m long x 0.1 m diameter). It should also be noted that during periods of high ice build up in the inshore due to long periods of on shore winds, or when a large ice berg traverses the study area, acoustic transmissions can be ricocheted back to the hydrophones by the ice and generate erroneous data in regards to fish position. This type of error should be deleted during the post-processing of the data.

Sonic telemetry tags can be purchased from a number of sources, and those used for tropical and temperate fish work just as well for polar fish, although we have found that tags with silver oxide batteries have a reduced power output and charge in sub-zero waters. The life of the implanted tag will depend on a combination of battery size, range of transmission and frequency of the acoustic pulse. In our experience the fish which inhabits sub-zero waters move at relatively slow rates compared to temperate and tropical species, therefore, a longer time interval between emitted pulses is a possible alternative to extend tag battery life. In our study, the pulse rate from tags was set between 30 – 60 seconds and the battery lasted up to 13 months (Campbell et al. 2008).

Subject Species and Tag Implantation

The group Notothenidae comprises about 80 % of all fish species in the Southern Ocean and nearly, if not, all of the inshore fish (Clarke and Johnstone, 1996). Therefore, this group of fish will be the most likely study species when engaging in telemetry studies around the Antarctic continent. They are characterised by a demersal body form, often with a relatively large head and reduced trunk, and have a low metabolic rate compared to that of temperate and tropical fish (Clarke and Johnstone, 1999). Their body size can vary greatly from the 5 cm long *Harpagifer* species up to the 1.5 meter long *Dissostichus mawsonii*. The mode of propulsion of Notothenidae is primarily by lateral motion of the pectoral fins, andsub-carangiform swimming is only used for short periods during prey capture or predator evasion (MacDonald et al. 1987). Inshore fishes of the genus *Notothenia* or *Trematomus*, tend to have a restricted home range and mark/recapture studies have shown high site fidelity (Campbell et al. 2007, 2008). This makes the use of implanted data loggers a very viable option, because retrieval rates are potentially very high. The ice fishes, or channichthyids, are less frequent inshore visitors and present greater logistical challenges due to their tendency to live on the sloping shelf at depths >100m. Although their crepuscular vertical migrations mean they can be caught, their thin integument is less resilient to sutures carrying any great strain, such as that imposed by external devices. The pelagic species have other problems, e.g. the large tooth fish (*Dissostichus*) appears to have little site fidelity and hence retrieval of tagged fish has so far proved very difficult (De Vries, personal communication).

To capture inshore Antarctic fish a rod and baited hook is very effective. The various species inhabit different parts of the water column and independent species can be captured by hook placement within the water column. For example, the active *Pagothenia borchgrevinki* inhabits the platelets immediately below the sea-ice, whereas the *Trematomus* species *such as T. bernacchii, T. hansoni* and *T. pennellii* live on the sea bed, and baits must be positions here for captured. Another method that enables the capture of a large number of demersal fishes is a fyke net. This net requires SCUBA divers for deployment, and consists of a weighed skirt that is set along the sea bed with a netted enclosure at either end. The bottom dwelling nature of the fish results in them being corralled into the end enclosures, keeping them alive and in good physical condition until collection. During winter months hand capture of fish by SCUBA divers is also possible (Figure 4).

After capture it is probably required to transport the fish back to the laboratory for tag implantation or attachment. During this time it important to ensure that the fish do not come into contact with any ice crystals, and air temperatures may be lower than sea water temperatures. Therefore, it is important to transport the fish in an insulated container. Also avoid handling the fish without gloves as this can also cause thermal stress damage to the fish epidermis. We also recommend that all surgical procedures and tag attachments be undertaken within a refrigerated room below 5 ^{0}C.

Anesthesia is required prior to implantation or attachment of electronic tags. MS 222 is suitable but from our experience notothenioids require a lesser dose than temperate or tropical species (~ 0.1 mg L^{-1}). The low metabolic rate of fish from sub-zero waters also results in a very long recovery time after exposure to anesthetic.

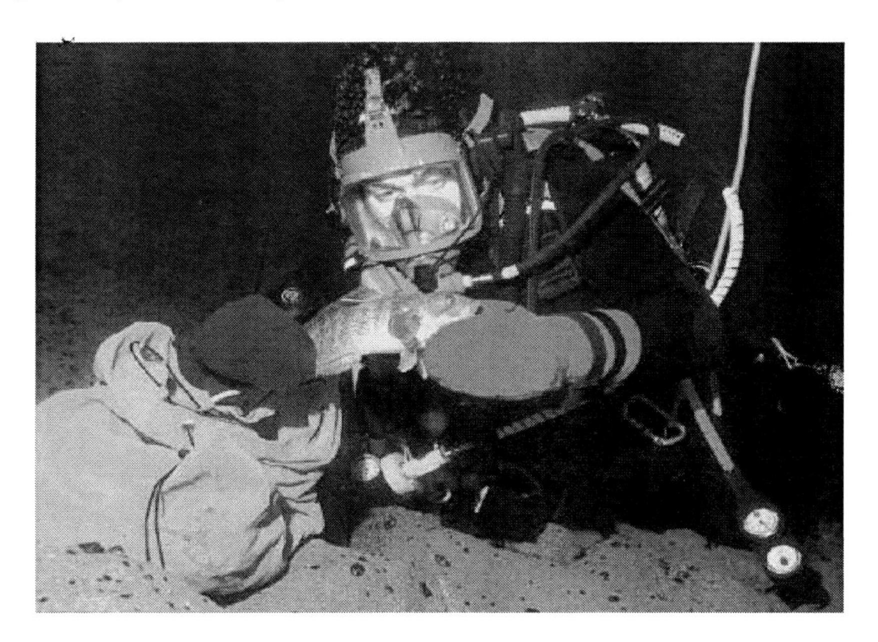

Figure 4. SCUBA diver hand collects quiescent *Notothenia coriiceps* from nesting burrows during winter on the Antarctic Peninsula.

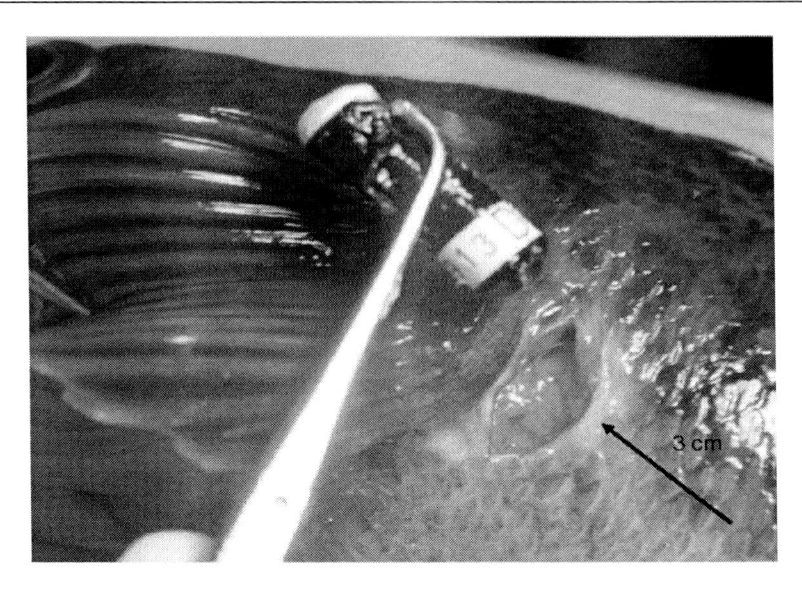

Figure 5. Antarctic Cod (*N. coriiceps*) showing incision behind the pectoral fin and a coded sonic telemetry tag (VEMCO, Canada) being inserted into the peritoneal cavity. The tag emitted a pulse every 30 seconds, the range was 300 m, and the battery life was approximately 13 months.

Most of the telemetry studies on temperate or tropical fish use an incision on the ventral surface to implant a tag into the peritoneal cavity. The bottom-dwelling habits of notothenioids results in the fish rubbing the wound on the sea bed. This can cause inflammation and unraveling of sutures (Campbell et al. 2005). However, the body shape of notothenioids allows a large semi-circular incision to be made in the area behind their large pectoral fin (Figure 5). Access to the peritoneal cavity can be made from this incision, and a telemetry device inserted. We have found that in incision made in this area heals very quickly. For suture material, catgut is very effective and allows for a tight wound closure, and the sub-zero water results in catgut suture holding cohesion for a much longer period than may be expected in warmer environments.

The expanded head region of notothenioids means that the cavity occupies a relatively short proportion of the body total length, so the use of internal telemetry devices may be limited in certain species or age classes. A viable alternative is to attach external tags (Campbell et al. 2005, 2007). This is possible because the Notothenioid fish swim at low speeds, primarily sculling through the water using the pectoral fins. This results in the body being fairly straight whilst swimming. Consequently, externally mounted tags are less likely to become displaced compared to a salmoniform or sub-carangiform fish species, where the extra drag may also induce stress and raise basal oxygen consumption. Opting for an external tag allows the use of housing large enough to incorporate a sufficient air space to make the tag neutrally buoyant in sea water. In a pilot study, we found that *Notothenia angustata* exhibited a lower stress response and more normal swimming activity with a neutrally buoyant, externally mounted tag than an internal tag that was only 1.7 % of the body mass (Campbell et al. 2005). In contrast, internal tags as heavy as 4 % of the body mass has been considered suitable for temperate marine fish that posses a swim bladder.

a.

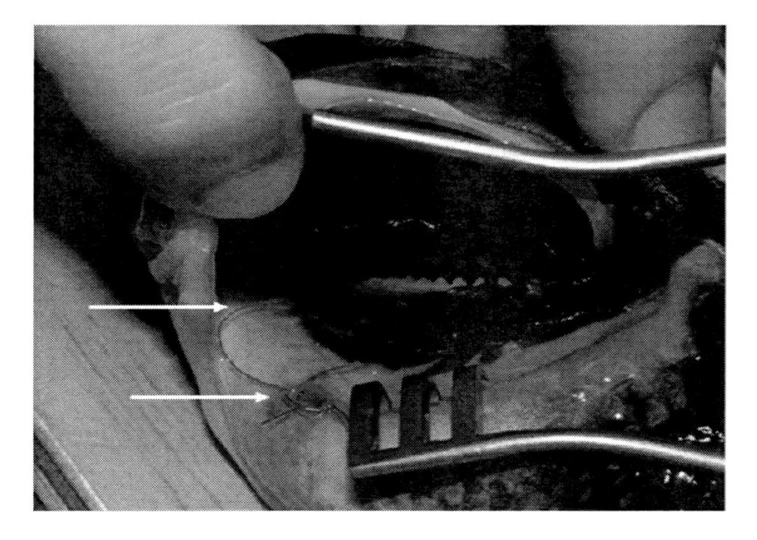

b.

Figure 6. a. The Antarctic cod *(N. coriiceps)* with neutrally buoyant data logger attached to the dorsal surface of the body. b. Lower picture shows the operculum behind the 4th gill arch. White arrows indicate the location of subcutaneous electrodes inserted to record the instantaneous electrocardiogram.

In Figure 6*a* *N. coriiceps* is fitted with a dorsally mounted external tag that comprises a miniature electronic microprocessor controlled data logger capable of making high resolution recordings of the electrocardiogram (ECG) during free-ranging activity. It was attached with cable ties to a rubber saddle that was secured through the large dorsal rays with nylon T-bar tags (FD-64, Floy tag Seattle, U.S.A). The ECG electrodes (0.2 mm; A-M systems, Carlsberg, U.S.A) were located subcutaneously through the septum behind the 4th left gill arch (Figure 6*b*). The saddle and electrodes were permanently attached to the fish, and for data download and battery replacement the DL could be changed quickly requiring only light anesthesia of the animal (Campbell et al. 2007). The device enabled the continual monitoring of water temperature and the instantaneous heart rate of the free-ranging fish (Figure 7).

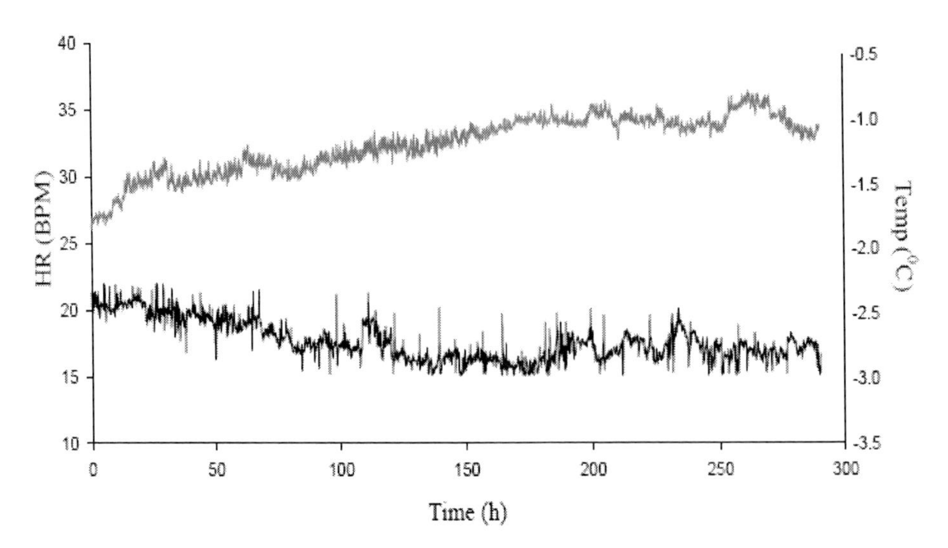

Figure 7. A 280 hour section of instantaneous heart rate (black) and water temperature (grey) recorded by a dorsally attached data logger from a free-ranging Antarctic cod *(N. coriiceps)*. Heart rate was recorded on a 20 min duty cycle each hour.

Cardiac output is a robust indicator of metabolic rate and the notothenioid fish modulate cardiac output primarily through changes in heart rate (Axelsson et al. 1994). Therefore, the remote recording measuring of heart rate is a viable strategy for estimating metabolic rate in free-ranging fish (Campbell et al. 2007, 2008). Such information is important in illustrating fish ecology and life history strategies.

Safety

Throughout this essay we have emphasised the danger of undertaking fisheries research in the inshore environment of polar seas. Research organisations are all too familiar with this issue and usually provide protective clothing, safety protocols and emergency action plans. They may even provide logistics personnel, such as, a general assistant, boat handler, dive officer and ski-doo driver, all of which should be familiar with local environmental conditions. Additionally, all personnel movement within the Antarctic inshore environment should be provided with accurate meteorological forecasts, this should be particular to the exact area because inshore conditions can vary rapidly in space and time.

Conclusions

Our knowledge of the ecology and life-histories of Polar fish remains sparse. This is unfortunate as the presumed limited adaptive response of these fish to changing water temperatures is of concern due to predicted shifts in climatic temperatures, and the cold-water specialisation and stenothermal physiology of polar fish make them particularly vulnerable (Somero and De Vries, 1967). Moreover, recent data indicate considerable increases in shallow sea temperatures in polar waters (Meredith and King, 2005). Inshore polar fish are a significant link in the marine food chain, making up a large portion of the diet for summer migratory birds and marine mammals. It is imperative that sound knowledge about movement patterns and ecology exists to enable informed decisions to be taken in regards to ecological management and predicting future climatic effects. It is precisely this type of data that could be acquired through sonic tracking techniques.

At the time of writing at least 32 countries had a polar research station. Most stations engage in some sort of fish research, and the physiology and molecular biology of polar fish has received a great deal of attention due to their unique characteristics, and endemic speciation fostered by isolation. However, a broad spectrum exists in the facilities and logistical capabilities between stations and the capacity to support a fish telemetry study may only be applicable to a few. Furthermore, given the labour-intensive nature of such studies, under-resourced projects are unlikely to be successful. This has resulted in the ecology of these fish only just beginning to be discovered. This essay has focused on our experiences from Antarctic fish research, but the principles are also applicable to Arctic fish research. It is hoped it will encourage the use of bio-telemetry and expand our knowledge of the biology of inshore polar fish.

References

Axelsson M., Davison W., Forster M.E., and Farrell A.P., (1992) Cardiovascular responses of the red-blooded Antarctic fishes *Pagothenia bernacchii* and *P. borchgrevinki*. *Journal of Experimental Biology* 167: 179-201.

Campbell, H.A., Fraser, K.P.P, Bishop, C. M., Peck L.S., and Egginton, S. (2008). Hibernation in an Antarctic fish: On ice for winter. *Public library of Science ONE* 3(3): e1743.doi:10.1371/journal.pone.0001743.

Campbell, H.A., Fraser, K.P.P., Peck L.S, Bishop, C.M. and Egginton, S. (2007). Life in the fast lane: The free-ranging activity, heart rate and metabolism of an Antarctic fish tracked in temperate waters. *Journal of Experimental Marine Biology and Ecology* 349: 142-151.

Campbell, H.A., Bishop, C.M., and Egginton, S. (2005). Recording long-term heart rate in the Black cod *(P. angustata)* using an electronic data logging device. *Journal of Fish Biology* 67: (4) *1150-1156.*

Clarke, A. and Johnston, N.M. (1999). Scaling of metabolic rate with body mass and temperature in teleost fish. *Journal of Animal Ecology* 68: 893-905.

Clarke, A. and Johnston, I.A. Evolution and adaptive radiation of Antarctic fishes (1996). Trends in Ecology and Evolution 11: (5): 212-2128.

Macdonald, J. A., Montgomery, J.C. and Wells, R.M.G. (1987). Comparative physiology of Antarctic fishes. *Advances in Marine Biology* 24: 321-388.

Meredith, M.N. and King, J.C. (2005). Rapid climate change in the ocean west of the Antarctic Peninsula during the second half of the 20th century. *Geophysical Research Letters*. 32; (19), L19604.

Somero, G.N. and De Vries A.L. (1967). Temperature tolerance of some Antarctic fishes. *Science* 156: 257-258

In: Telemetry: Research, Technology and Applications
Editors: Diana Barculo and Julia Daniels
ISBN 978-1-60692-509-6

Short Communication

REPRODUCTIVE STATE-SPECIFIC HABITAT USE BY ADULT JAPANESE FLUVIAL SCULPIN COTTUS POLLUX (PISCES: COTTIDAE), IN RELATION TO BOTTOM SUBSTRATE CONDITION

Takaharu Natsumeda[1] and Yoshikazu Nagata
Department of Biology, Osaka Kyoiku University,
4-698-1 Asahigaoka, Osaka, 582-8582, Japan

ABSTRACT

We examined the patterns of habitat use by 24 adult Japanese fluvial sculpin, Cottus pollux (large-egg type) by direct observation on a single night attaching a set of luminous diode and small lithium battery on the skin just beyond the 1st dorsal fin of fish. Home-range size, which was calculated by the minimum convex polygon method (range: 0.3-79.9 m2; mean: 9.8 m2), was positively correlated with the number of focal points (range: 3-22 points, mean: 10.5 points). Home-range size was not different among three reproductive states (gravid and normal females, non-nesting males). Of four categories of substrate condition (boulder, gravel, sand, and unspecified), boulder was the most common substrate category used by sculpin. Mature females spent most of their time boulder-associated substrate regardless of their reproductive condition. Non-nesting males, on the other hand, exhibited lower dependence of boulder-associated substrate than mature females. Since nesting males spent most of the time in and around their nests (i.e., boulder-associated substrate), lower dependence of boulder-associated substrate of non-nesting males may reflect reproductive state-specific life-history tactic related to foraging during the breeding season.

1 Present address: Inland Division, National Research Institute of Fisheries Science, 1088 Komaki, Nagano 386-0031, Japan. e-mail: natsumed@affrc.go.jp.

Introduction

Many animal species often exhibit unique requirements for habitat characteristics during the breeding season. In nest-holding animals such as birds and fishes, acquisition of suitable habitat for reproduction is prerequisite for ensuring successful reproduction (Sargent and Gebler 1980; Searcy and Yasukawa 1994). Despite growing evidence for the importance of substrate characteristics for nest-site selection by territorial nest owners (Mori 1994; Östlund 2000), little attention has paid for substrate use by non-nest owners (e.g., non-nesting males and females).

The Japanese fluvial sculpin, *Cottus pollux*, is a small bottom-dwelling freshwater fish endemic to the mountain streams of Honshu, Kyushu, and Shikoku and it spawns a small number of large eggs (i.e., large-egg type; Goto 1989). They are characterized as rock nesters (Balon 1975), with mature males maintaining breeding space under cobbles or boulders as nests and they mate with females (Natsumeda 1998a). From the viewpoint of feeding ecology, they are characterized as nocturnal foragers remaining on the bottom as they wait for suitable prey organisms to cross the boundary of its strike space (Natsumeda 1998a; 1998b). Although previous studies have reported the preference for substrate condition of the species at seasonal scale during non-breeding season (Natsumeda 1998a; 1998b), little attention has paid for their substrate use during the breeding season. Artificial placement of cobbles on the substrate before the breeding season may substantially enhance the quality of spawning habitat for freshwater sculpin (Knaepkens et al. 2002), implying that clarification of substrate use by the sculpin during the breeding season would provide useful information for establishment of management implication for their habitat enhancement during that season.

The present study aims to clarify the patterns of habitat use by non-nest owners (i.e., non-nesting males and females) of Japanese fluvial sculpin related to bottom substrate condition at a finer temporal scale (i.e., a single night) during the breeding season.

Materials and Methods

We conducted the study at two sites (Inabe and Kochidani) along the upper reaches of the Inabe River, central Japan (35°10'N, 136°31'E) during the spring of 1993. Inabe (77 m long, 7.5-10.0 m width, 578.9 m^2) located about 1 km downstream of Kochidani (73 m long, 6.2-11.0 m width, 572.9 m^2). Kochidani had steeper course gradient (0.4%) than Inabe (0.2%), but both study sites included raceway channel habitat which is used for spawning ground for the species (Natsumeda 1999; 2001). Prior to the censuses, we mapped all rocks on the streambed larger than 30 cm in maximum diameter in both study sites. Japanese fluvial sculpin was one of the common species of fishes at both study sites (Natsumeda et al. 1997; Natsumeda 1998a).

Sculpins were located with a glass-based viewing box and a flashlight at night. They were captured by a hand net and anaesthetized with a 0.01% ethylene-glycol-monophenyl-ether solution. Prior to marking procedures, we measured standard length (SL) and wet weight (W) to the nearest 0.5 mm and 0.1 g, respectively, and determined sex from external sexual characteristics with males having a different genital papilla than females (Natsumeda et al. 1997). We determined maturity of the males by examination of the gonad condition during the breeding seasons, and whether or not the fish oozed milt when squeezed at the abdomen (Goto 1987). Since males held nests (i.e., nesting males) spent most of the time in their nests even at night (Natsumeda 2001), mature males that did not occupy the beneath nests and exposed entire their bodies on the substrate were considered as non-nesting males. To examine the effect of reproductive condition of females on home-range size, we classified females as follows: (1) gravid females (= abdomen considerably swollen because of the presence of ripe eggs); or (2) normal females (= abdomen not swollen, ripe eggs not present).

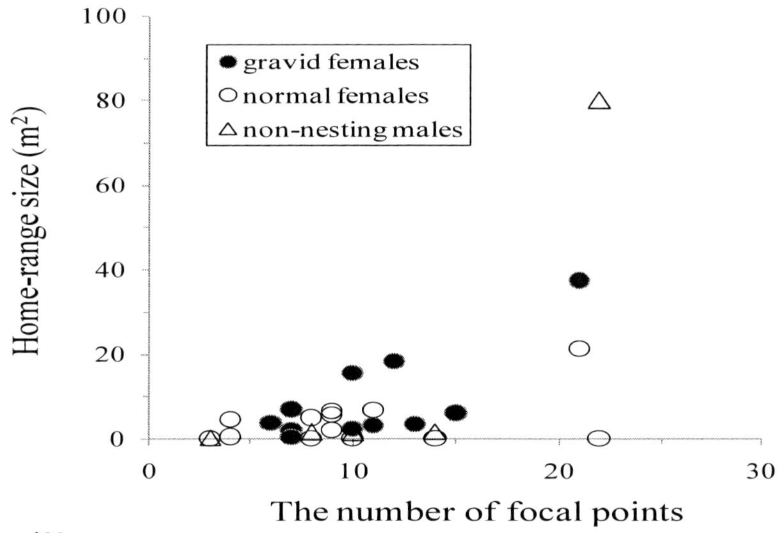

Natsumeda and Nagata.

Figure 1. Relationship between the number of focal points and nocturnal home-range size of Japanese fluvial sculpin.

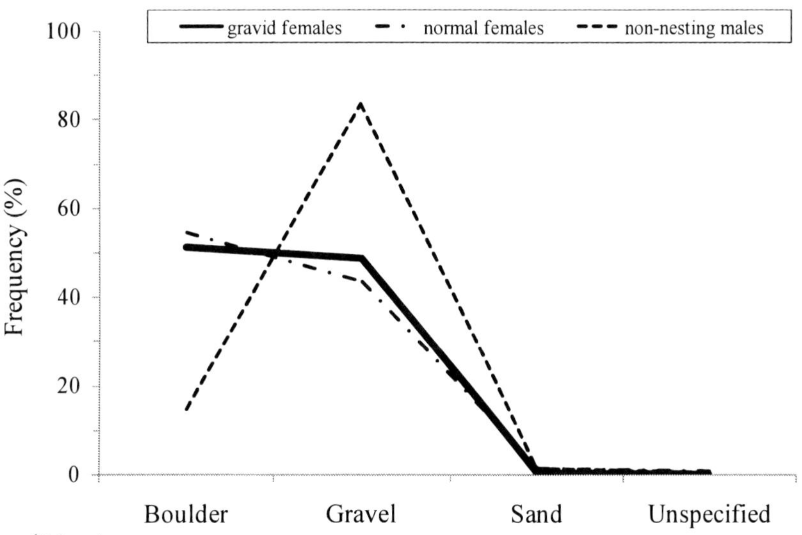

Natsumeda and Nagata.

Figure 2. Comparison of the relative frequency of four types of substrate use by Japanese fluvial sculpin among three reproductive states.

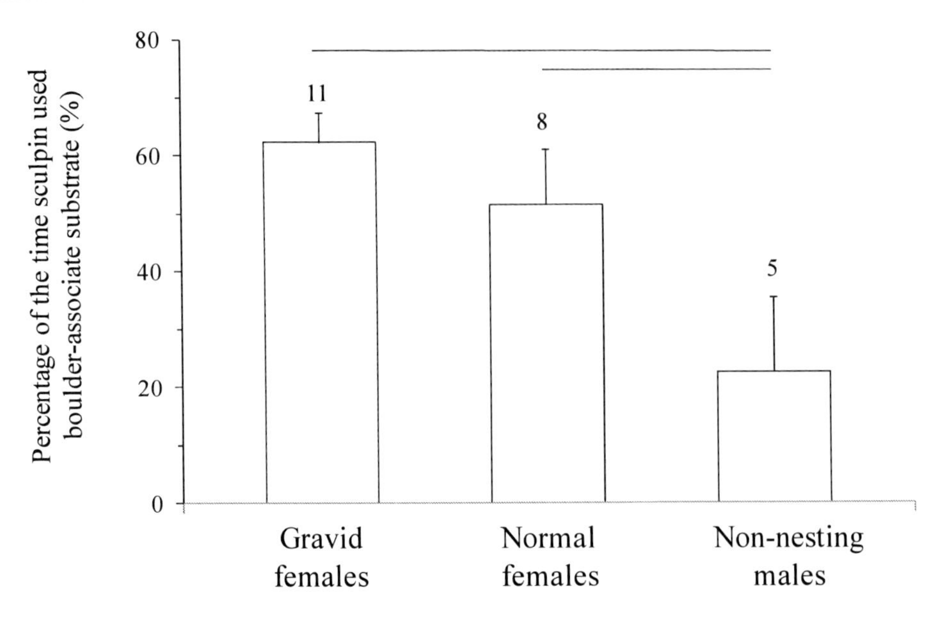

Figure 3. Comparison of the percentage of the time used by 24 adult Japanese fluvial sculpin for boulder- associate substrate among three reproductive states. Vertical bars indicate mean + SE. Numbers in the figure indicate sample size. Horizontal bars represent significant differences between groups by pairwise t-test ($P < 0.05$).

Marking procedures involved removing the first and second dorsal fin spines and rays (Goto 1985), and (2) attaching a set of luminous diode (φ2.0 mm) and small lithium battery (National BR435, 3V, φ4.2 mm × 35.9 mm long) on the skin just beyond the 1st dorsal fin using transparent monofilament sutures (total weight: 1.6 g in air). The former method ensures that each individual receives a permanent mark; the latter tag enables us to determine the location of individual fish throughout the night without a flashlight (Natsumeda 2007a). Thereafter, we kept the fish in a small mesh cage (15 × 15 cm, 5 mm mesh) which was set on a streambed at the location of capture for at least 12 hours until it had recovered from anesthetic and physical stress.

Preliminary direct observations showed that the sculpin moved actively from after dusk to before dawn (from 18:00 to 06:00 h); they hardly moved in daytime (from 08:00 to 16:00 h; Natsumeda 2007a). Based on this finding, we located one or two individuals per each survey by direct observation from 18:00 to 06:00 h.

Twenty-four sculpin representing 18 different adult individuals (mean 87.4 mm SL; range 72.5-101.5 mm; mean 15.4 g, range 9.4-27.0 g) were located during 15 surveys from 15 February to 28 May 1993. This period includes the breeding season for the sculpin. Each individual was freed at 18:00 by opening the cage, and we traced the swimming paths and subsequent location of each sculpin on the map until dawn (at least 06:00 h). If the individual stayed at the point more than one minute, we counted the point as the 'focal point'. We recorded the dominant substrate, occupying > 50% of bottom surface area within a radius of 50 cm of each focal point used by sculpin as follows: (1) boulder-associated: boulder and rubbles (2) gravel-associated: gravel and pebbles, (3) sand-associated: sand and silt, and (4) unspecified.

Although diode tag weight were in excess of 2 % (mean 7.3 %, range 5.9-9.2 %) of body weight commonly recommended for external tags (Winter 1996), we detected minor the effects of tags on home-range behavior of the species by comparing the proportion of home-range overlap of each individual before and after tag attachment (Natsumeda 2007a).

We assessed the home-range size of each individual by the minimum convex polygon method (Schoener 1981) using HomeRange Program v. 2.0 (Huber and Bradbury 1996). Location data of each individual into the x- and y-coordinates (see White and Garrott 1990). As the home-range size of the sculpins did not differ between Inabe (10.8 m^2 ± 18.8 SD, n = 17) and Kochidani (7.4 m^2 ± 13.3 SD, n = 7; t-test, t = 0.43, $P > 0.6$), we pooled these data and then we compared the home-range size by using one-way ANOVA (factor = gravid females, normal females, non-nesting males). Data were transformed ($\log_{10}$) to satisfy concerns about normality and variance heteroscedasticity.

RESULTS

Nocturnal home-range size of the Japanese fluvial sculpin varied from 0.3 to 79.9 m^2 among 24 adult sculpin (9.8 $m^2 \pm$ 17.2 SD) and they used an average of 10.5 (± 5.2 SD; range 3-22) focal points at night and remained for an average of 93.1 min (± 52.7 SD; range 2-637) at each location. Overall, home-range size was positively correlated with the number of focal points (r = 0.729, $P < 0.001$, n = 24; Figure 1). Home-range size was not different among gravid females, normal females, and non-nesting males (One-way ANOVA, $F_{2,21} = 0.57$, P = 0.57).

At group level, substrate use by gravid and normal females were similar: they spent over half of the time on boulder-associated habitat (Figure 2). Non-nesting males, on the other hand, spent over 80% of their time on gravel-associate substrate and exhibited considerably lower dependence for boulder-associate substrate than females.

At individual level, the percentage of boulder-associate substrate used by the Japanese fluvial sculpin was significantly different among there reproductive states (One-way ANOVA, $F_{2,21}$=50.4, P = 0.016), and pairwise student t-test showed that non-nesting males exhibited significantly lower dependency of boulder-associated substrate than gravid and normal females (Figure 3).

DISCUSSION

Despite no significant differences in nocturnal home-range size of Japanese fluvial sculpin among three reproductive states (i.e., non-nesting males, gravid and normal females), our results showed apparent differences in the duration of substrate use among them. Both gravid and normal females spent longer duration at boulder-associated substrate than non-nesting males. Although several empirical studies have noted the differences in substratum use by freshwater sculpins with regard to interspecific comparison (Daniels 1987; Brown 1991), ontogenetic shift (Brandt 1986; Freeman and Stouder 1989), and comparison of day and night during non-breeding season (Natsumeda 1998b), no studies have noted for either sexual or state differences in substrate use by freshwater sculpin during the breeding season. Therefore, this is the first report refer to the presence of state-specific substrate use by freshwater sculpin during the breeding season.

Low dependence of boulder-associated substrate by non-nesting males of Japanese fluvial sculpin may reflect their life-history tactics during the breeding season (Natsumeda 2007b). Since larger males of the species successfully hold suitable nests through male-male competition and spent most of the time in and around their nests after breeding (Natsumeda 1998a, 2001), nesting males consequently exhibit higher dependence for boulder-associated substrate during the breeding season. Non-nesting males, on the other hand, are often smaller and competitively inferior than larger nesting males (Natsumeda, 1998c). Such smaller, non-nesting males actively forage even in the breeding season and attain larger body size by the next breeding season (i.e., delayed reproduction; Natsumeda 2007b).

CONCLUSION

We found state-specific habitat use by non-nest owners of Japanese fluvial sculpin at a fine scale during the breeding season. Lower dependence of boulder-associated substrate of non-nesting males may reflect life-history tactic related to foraging (i.e., delayed reproduction) during the breeding season.

ACKNOWLEDGMENTS

We are grateful to M. Yuma, S. Mori, M. Hori, T. Kondo, K. Watanabe, Y. Koya, D. Tahara, R. Fujii, Y. Hayakawa, H. Yanbe, H. Matsubara, T. Yabe, A. Goto, and T. Narita, for their kind advice; T. Ito and Y. Shimizu for valuable information during the field survey; and anonymous reviewers for helpful comments on the text.

REFERENCES

Balon, E.K. (1975). Reproductive guilds of fishes: a proposal and definition. *J. Fish Res. Bd. Can.* 32:821-864.

Brandt, S.B. (1986). Ontogenetic shifts in habitat, diet and diel-feeding periodicity of slimy sculpin in Lake Ontario. *Trans Amer. Fish Soc.* 115:711-715.

Brown, L.R. (1991). Differences in habitat choice and behavior among three species of sculpin (*Cottus*) in artificial stream channels. *Copeia* 1991:810-819.

Daniels, R.A. (1987). Comparative life histories and microhabitat use in three sympatric sculpins (Cottidae: *Cottus*) in northeastern California. *Environ. Biol. Fish* 19:93-110.

Freeman, M.C. and Stouder, D.J. (1989). Interspecific interactions influence size specific depth distribution in Cottus bairdi. *Environ. Biol. Fish* 24:231-236.

Goto, A. (1985). Individual identification by spine and ray clipping for freshwater sculpins. *Jpn. J. Ichthyol.* 31:161-166.

Goto, A. (1987). Polygyny in the river sculpin, *Cottus hangiongensis* (Pisces: Cottidae), with special reference to male mating success. *Copeia* 1987:32-40.

Goto ,A .(1989). Genus *Cottus*. In: Kawanabe, H. and Mizuno, N (eds) *Freshwater fishes of Japan* (in Japanese), Yamakei Publisher, Tokyo, pp 648-668 .

Huber, R. and Bradbury, J. (1996). HomeRange Program. Bowling Green State University.

Knaepkens, G., Bruyndoncx, L., Bervoets, L. and Eens, M. (2002). The presence of artificial stones predicts the occurrence of the European bullhead (*Cottus gobio*) in a regulated lowland river in Flanders (Belgium). *Ecol. Freshw Fish* 11: 203-206.

Mori, S. (1994). Nest site choice by the three-spined stickleback, *Gasterosteus aculeatus* (form leiurus), in spring-fed waters. *J. Fish Biol.* 45: 279-289.

Natsumeda, T. (1998a). Life history and reproductive ecology of the Japanese fluvial sculpin, *Cottus pollux*, with special reference to spatial-temporal availability of nest resources. Ph. D. Thesis, Kyoto University.

Natsumeda, T. (1998b). Home range of the Japanese fluvial sculpin, *Cottus pollux*, in relation to nocturnal activity patterns. *Environ. Biol. Fish* 53:295-301.

Natsumeda, T. (1998c). Size-assortative nest choice by the Japanese fluvial sculpin in the presence of male-male competition. *J. Fish Biol.* 53:33-38.

Natsumeda, T. (1999). Year-round local movements of the Japanese fluvial sculpin, *Cottus pollux* (large egg type), with special reference to the distribution of spawning nests. *Ichthyol. Res.* 46:43-488.

Natsumeda, T. (2001). Space use by the Japanese fluvial sculpin, *Cottus pollux*, related to spatio-temporal limitations in nest resources. *Environ. Biol. Fish* 62:393-400.

Natsumeda, T. (2007a). Estimates of nocturnal home-range size of the adult Japanese fluvial sculpin, *Cottus pollux* (Pisces: Cottidae) in relation to bottom topography and sampling intervals. *J. Ethol.* 25:87-93.

Natsumeda, T. (2007b). Variation in age at first reproduction of male Japanese fluvial sculpin induced by the timing of parental reproduction. *J. Fish Biol.* 70:1378-1391.

Natsumeda, T., Kimura, S. and Nagata, Y. (1997). Sexual size dimorphism, growth and maturity of the Japanese fluvial sculpin, *Cottus pollux* (large egg types) in the Inabe River, Mie Prefecture, central Japan. *Ichthyol. Res.* 44:43-50.

Östlund, S.(2000). Are nest characters of importance when choosing a male in the fifteen-spined stickleback (*Spinachia spinachia*)? *Behav. Ecol. Sociobiol.* 48: 229-235.

Sargent, R. C. and Gebler, J. B. (1980). Effects of nest site concealment on hatching success, reproductive success, and parental behaviour of the three-spined stickleback, *Gasterosteus aculeatus*. *Bahav. Ecol. Sociobiol.* 7: 137-142.

Searcy, W. A. and Yasukawa, K. (1994). Polygyny and sexual selection in red-winged blackbirds. Princeton Univ. Press.

Schoener, T.W. (1981). An empirically based estimate of home range. *Theor. Popl. Biol.* 20:281-325.

White, G.C., Garrott, R.A. (1990). Analysis of wildlife radio-tracking data. Academic Press, London.

Winter, J. (1996). Advances in underwater biotelemetry. In: Murphy, B.R. and Willis, D.W. (eds) Fisheries techniques second edition, American Fisheries Society, Bethesda, pp 555-590.

In: Telemetry: Research, Technology and Applications
Editors: Diana Barculo and Julia Daniels
ISBN 978-1-60692-509-6

Short Communication

A NOVEL TELEMETRIC SYSTEM FOR RECORDING BRAIN ACTIVITY IN SMALL ANIMALS

Damien Lapray, Jürgen Bergeler, Erwan Dupont, Oliver Thews, and Heiko J. Luhmann[1]
Institute of Physiology and Pathophysiology,
University of Mainz, Germany

ABSTRACT

Wireless technology recording systems are as comfortable as possible to the awake experimental animal and are in many aspects advantageous and more valuable when compared to conventional recording devices. We describe here the different steps taken in the realization of a telemetric system to record an electroencephalogram (EEG) from adult freely-moving rats. Our system consists of an implantable transmitter that communicates bidirectionally with a receiver via radio transmission over a distance of up to three meters. The impact of the system on the animal movements was tested and has shown that the device did not restrict limb movements during locomotion. The sleeping posture was also maintained, proving the lack of discomfort produced by our system. The current system is optimized for recording electrical activity from the animal's brain, but can be easily modified to record other physiological parameters. Using our expertise, we also discussed here the points that have to be clarified before starting the production of such a system. Building a wireless device requires the consideration of a number of options concerning the experimental design and the recording environment to integrate the right parameters.

[1] Correspondence to: Heiko J. Luhmann, Institute of Physiology and Pathophysiology, University of Mainz, Duesbergweg 6, D-55128 Mainz, Germany. Phone: +49 6131 39 26070; fax: +49 6131 39 26071; e-mail: luhmann@uni-mainz.de.

INTRODUCTION

One desirable goal of experimental research in neurophysiology is to perform electrophysiological recordings from awake, freely-moving animals without the restrictions of a large recording set-up. Electroencephalogram (EEG) and field potential (FP) recordings in freely-moving mammals is currently an important issue in neuroscience, because a number of different brain rhythms have been described in the last two decades (for review, see [1-4]), but the physiological function and the behavioural context of the different oscillatory activity patterns are not completely understood (for comprehensive review, see [5]). Fully implantable telemetric systems offer the advantage to monitor physiological parameters in an unrestricted manner in freely-moving animals performing their natural explorative behaviour. Wireless technology recording systems are as comfortable as possible to the awake experimental animal and are in many aspects advantageous and more valuable when compared to conventional recording devices.

In this chapter we describe the technology of the telemetric system, which we developed in Mainz, the general concepts behind its technology and its advantages compared to traditional recording methods. We also discuss which requirements telemetric systems should fulfill in the near future to be valuable for electrophysiological monitoring and simultaneous behavioural analyses.

1. THEORETICAL CONSIDERATIONS OF A TELEMETRIC SYSTEM

A telemetric recording system should fulfill a number of requirements in order to guarantee a long-term and powerful analysis of physiological parameters in freely-moving animals under stress-free conditions: (a) All devices, which have to be implanted in the animal, should be as small and light as possible in order to minimize or better prevent stress and pain to the animal. The system should be also fully bio-stable and bio-compatible in order to allow telemetric recordings over long periods, such as weeks or even months. (b) The energy consumption of the implanted components should be as low as possible and it should be possible to control the energy source (usually a battery) from a distance by switching the system on and off or by recharging the battery. (c) The system should transmit the recorded signals over at least a few meters in order to allow free exploratory behaviour of the animals in a defined environment, e.g., a conventional open field for rodents or in their natural environment. (d) Internal components should be re-usable and the external recording system should be as simple as possible to minimize the costs. (e) All external components of the telemetric system (e.g., the receiver) should be as small as possible and portable to allow experiments outside the conventional laboratory settings. (f) In order to allow maximal flexibility in the experimental design and in the recording protocols, the properties of the signals sent out from the transplanted transmitter should be adjustable from outside.

2. Why Is Wireless Technology Preferable?

The advantage of the fully implantable telemetric systems in comparison to the anaesthetized or restrained recording methods will not be discussed further here, since the latter methods are limited in the collection of physiological parameters under unstressed, normal conditions [6]. The traditional in vivo recording method in unrestrained laboratory animals consists of the implantation of an electrode or transducer connected with wires to the recording set-up. These tethering systems have been used for many years now, and despite the lack of direct measurement of stress hormone levels to compare it to telemetric implantations, this technique has been documented to be a stressor [7]. Many studies have documented an increase in heart rate in rodents [8] or non-human primates [9] during conventional in vivo recordings. Implantable telemetry has also the advantage of minimizing the risk of infections and the recording of movement artefacts produced by wires and connectors. Another advantage is the possibility of recording animals continuously or during certain periods of the day without the involvement of any personnel since the animals don't need to be unplugged for any maintenance of the system and telemetric devises can be turned on and shut down by external PC control. In B6 mice it has been shown [10] that sleep diurnal ratio with telemetry was reduced when compared with conventional recordings with external cables. The authors have concluded that this result was due to either greater or relatively more sleep in the dark period for telemetric-implanted animals. As stressed in this chapter, tethered recordings can be even more problematic for small animals like rodents compared to larger ones since rodents are fast-moving with more vertical movements such as climbing. Cable systems can limit such movement and can be difficult to adequately counterweight.

3. Current Technical Limitations of such Recording Systems

A telemetry system for continuous measurement of physiological signals consists of a battery, an analogue processing circuit and the transmitter. Depending on the experimental design, power consumption, complexity of the electronic circuit and the size and weight of the system may differ. On the market are systems that are fully implantable and some have externally fixed units, e.g., "backpack"-style units. The external units have disadvantages and limitations:

- The animal is always mechanically affected by the unit and therefore the behaviour will not be natural.
- The connection between the electrodes and the unit penetrate the skin and the affected area is very sensitive to infections and injuries.

Although the implantation of the whole unit has many advantages, this telemetric systems also possess some unavoidable restrictions:

- the connection between the wire from the unit to the electrode or transducer,
- the electrical connections, which may react with blood and other body fluids,

- the problem of encapsulation of the unit,
- the difficulty in exchanging the battery for multiple usage,
- the high expenses when the implanted unit is used only once.

A telemetric system is able to send signals and possibly also to receive external data. Depending on the demands of the data exchange rate and capacity, the technical realization, especially to handle EEG-data, may be rather restricted. If the data exchange rate does not need to be fast, e.g. to monitor body temperature, passive transmitter systems can be used similar as in RFID (Radio Frequency IDentification) technology. EEG-data contain frequencies up to 80 Hz, so an upper limit of 100 Hz is commonly necessary to be sure not lo loose relevant information. Only some very special approaches require frequencies up to 200 Hz. In all cases the only way for data communication with acceptable results is radio transmission.

The technical design and final implementation of the telemetric system mainly takes account of three aspects:

3.1. Frequency

The choice of the transmitter frequency depends on the available frequency bands that are legally usable, the supported and available components on the market and at least on physiological parameters given by attenuation of the tissue. Commonly used in Europe are the so called ISM-bands (**I**ndustrial, **S**cientific, and **M**edical Band) with frequencies between 433.05 and 434.79 MHz or between 868 and 870 MHz. The higher frequency range has the main advantage of better transmission quality, higher data rate and a smaller antenna size. The limitation of the signal attenuation by the tissue in the 868 MHz frequency is acceptable and the data rate is limited to about 1 Mbit/s. Higher data rates are possible with the highest ISM-band between 2.4 and 2.483 GHz. Commercial systems for video transmission of signals from small camera units passing the intestinal part of the body operate in this frequency range. The main disadvantage of this high frequency is its use by many other applications (WLAN, microwave oven or standard TV-video transmitters) and therefore more sophisticated technical solutions have to be developed to suppress signal interferences.

3.2. Data Decoding

The EEG signal can be transmitted analogue or digital and the RF-modulation may be amplitude or frequency based. In general, the analogue signal transmitting design requires simpler circuits with a smaller number of electronic parts resulting in smaller systems with less power consumption. Every approach based on digital technology needs an analogue to digital data conversion unit, a microcomputer and a suitable transmitter unit. Therefore the power consumption of digital systems is usually much higher as the analogue version.

The main limitations of the analogue systems are the sensitivity for EMI (**E**lectro**M**agnetic **I**nterference), the difficult implementation of multi-channel capability, the rather fixed system parameters, the poor data reliability and the lack of error control and repair methods. In contrast, digital systems are able to use the complete range of digital data transmission technology, like data compression and error repair algorithms. The number of channels is only limited by the data bandwidth and can be easily changed by programming as all other parameters too.

3.3. Uni- or Bidirectional Communication

With a unidirectional communication the receiver has no control on the transmitter. Even switching the unit on or off has to be performed separately and there is no way to change any feature of the implanted unit. Bidirectional data communication adds the possibility to configure the unit and allows a high level of flexibility in the operation mode of the system. Switching on or off is easy to realize with a minimum of power consumption. Any further handling of the animal can be avoided, since direct contact is not necessary.

Whereas the transmission of analogue signals with amplitude or frequency modulation is very easy, receiving analogue signals is not. The main reason for that is the poor quality of the analogue signals and the requirement of complex receiver units with special features. Therefore, no bidirectional analogue systems are available on the market and this important property is reserved for digital microcomputer based systems.

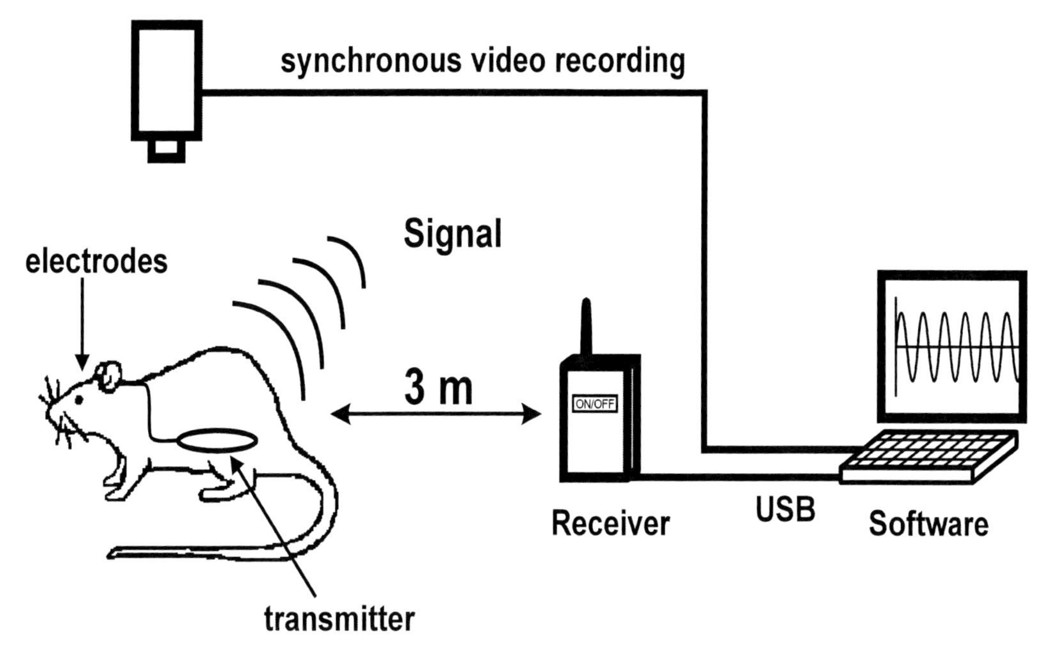

Modified from Lapray et al., 2008.

Figure 1. Schematic illustration of the main components of the telemetric system. The recording electrodes are implanted on the cortical surface and connected to a transmitter placed in the abdominal cavity of the rat. The transmitter is switched ON/OFF by the receiver. The recorded signal is directly amplified and digitized by the transmitter, emitted in all directions and detected by the receiver at a distance of up to 3 meters. The data are transferred from the receiver to a computer via an USB connection and can be visualised online with our software. The animal behaviour is synchronously recorded with a video camera and movies are stored on-line on the PC.

Property	
Weight	4 g (with batteries and wires)
Dimensions	40mm x 8mm x 5 mm
Power Supply	2 x 1.55 V (silveroxide)
Carrier Frequency	868.35 MHz ISM (Europe) 916.5 MHz ISM (USA)
Run Mode Current	1.6 mA
Standby Mode Current	2 µA
Life Time Battery	25h at 500Hz sampling rate

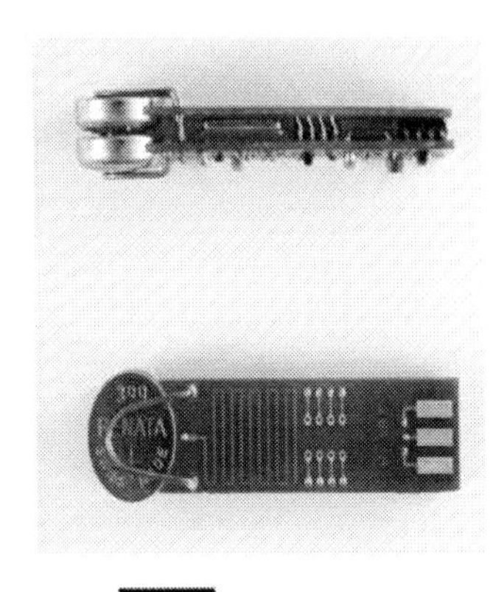

Modified from Lapray et al, 2008.

Figure 2. Properties of the transmitter unit. The system operates at sampling rates sufficient for physiological recordings of EEG activity (250 to 2000 Hz). The light transmitter contains a battery that on average allows 25 hours of recording.

The power supply for external or implanted units originates from commercially available standard batteries. The use of rechargeable batteries is restricted by the smaller energy density and the loading technology. Implantable units can only be recharged with an electromagnetic field. The main problem is the distance and finally the coupling of the transmitting and the receiving coil. In a static situation a battery recharging via an electromagnetic field may be practical, but with an unfixed animal this technique is rather complicated. Currently there are no such systems on the market, only a commercial system for loading portable devices positioned on a special mat ("splashpower ®").

4. Our System

Although telemetric systems are suited to record a variety of different physiological parameters, subsequently we will focus on a system suitable for EEG, electrocorticogram (ECoG) and field potential (FP) recordings in freely moving rats (Figure 1). This system is fully implantable and does not affect the animal´s natural behavior [11]. Most important, the EEG signal is transmitted at a distance of up to 3 meters, thus cortical electrical activity and behaviour can be synchronously recorded and subsequently directly correlated on- and off-line.

4.1. Weight and Size

In order to minimize the size of the system most of the parts are Surface Mount Devices with a size of case type "0402" resulting in a 4 g system (with batteries and wires) with reduced dimensions (4 x 0.8 x 0.5 cm) (Figure 2). The transmitter is covered with two layers of a biocompatible silicone (Elastosil N2010, Wacker Chemie AG, Munich, Germany) to avoid any internal tissue disturbances.

4.2. Power Consumption

This is a critical parameter and the choice of the transmitted signal has a major impact on power consumption. Analogue or digital signals, distance of transmission, sampling rate are all energy consuming parameters of the system that have to be considered in accordance with the planned experiments. Our interest was mainly focused on the recording of freely moving animals in a large environment like a standard open field for rats. In this regard the implanted part of the telemetric system we developed consists of a microcomputer (PIC12F675, Microchip®), an analogue input amplifier and a transceiver. We avoided the use of an external analogue-digital converter and amplifier allowing flexibility in the use of the system in a wide variety of environments like the animal facility. At present our system allows 25 h of continuous recording or weeks of short daily sessions.

4.4. The Surgical Implantation

Surgery was performed following deep anaesthesia of the animals [11]. The area from the part between the eyes, back between the ears and across the neck as well as the abdominal skin were shaved and cleaned with iodine. The animal was then placed on a heating pad and an incision was made in the left part of the abdomen's mid-line, 1 cm caudal to the xyphoid cartilage, and the transmitter was placed in the peritoneal cavity. This procedure was preferred to the back implantation between skin and muscles in order to minimize the impact of the system on the animal´s movements. The unit could be sutured to the body wall with non-absorbable nylon to limit the impact of the system on the gut. After secured in the stereotaxic apparatus a 4 cm mid-sagittal incision was made on the scalp and the skin reflected with hemostats to expose the entire skull. A way for the leads was made between the skin and muscles by moving aside connective tissue and transmitter leads were slipped subcutaneously from the abdomen to the incision made on the head. The skull was then cleaned and dried. The wires, after removing tips insulation, can be directly used as electrodes. The assembling was anchored in place with grip cement (Dentsply Caulk International, Milford, USA). Both incision sites were closed using 4-0 Resolon (Resorba, Nürnberg, Germany). Surgery lasted a maximum of 3 h from induction of anaesthesia and the success rate was 100% for recovery from anaesthesia and surgery.

4.5 Impact on the Animal´s Behavior

It has been shown that telemetric systems don't restrict the animals´ movements and this was also the case for our system [6;11]. The implantations were followed by a lost of weight with a maximum peak 2 days after the surgery (94.6±0.4% of pre-surgery body weight). Five days after the implantation the animals gained weight again and reached the pre-surgery's one after 10 days. No infections or mortality due to the implantation were observed on the 6 rats used for this study. To test the impact of the system on the animal's movements, rats were placed for 10 min in an empty open field (60x70 cm^2) and behavioural recordings were performed with EthoVision (Noldus Information Technology, Berlin, Germany). The total running distance and the running velocity were then compared between implanted and non-implanted rats in the same experimental condition. No significant differences were observed in these two parameters (30.3±1.7 m and 5.1±0.3 cm/s for the control animals versus 29.2±1.7 m and 4.9±0.3 cm/s for the implanted animals, each group n=6). The device did not restrict limb movements during locomotion. The telemetry device did not cause any discomfort to the animals, which were still able to curl up to sleep, a prominent sleeping posture in rodents [12;13].

4.6 Collecting the Data

The receiving and control system is composed of a personal computer and an interface box connected via standard USB. The PC software consists of three independent modules for (1) capturing data, (2) replaying/analyzing data, and (3) software servicing of the control system and the implanted devices. The replay software module reads the EEG data together with the stored video stream and displays them synchronously on the screen. With a time mark the user can indicate time intervals of interest. The accompanying EEG data of these time frames can either be stored separately on the hard disk or be exported in a format compatible for use with other analysis software (Spike2, MatLab, etc.). In this module the incorporation of individual specific analysis routines can be easily implemented.

5. Future Developments of This Technology

Desirable developments include a larger number of recording channels with high sampling rates, a further miniaturization to allow telemetric recordings in small mammals such as mice, the implementation of smaller batteries with higher capacity and longer lifetime, the reduction of power consumption and the implementation of a bidirectional communication between receiver and transmitter to allow maximal experimental flexibility. The possibility to record several animals at the same time and the stimulation of specific brain areas are also important issues for animal research. All of these technological improvements are currently in the process of development, and the steady progress in electronics will lead to powerful telemetric systems in the near future.

Acknowledgements

DL is member of the neuroscience graduate school at the University of Mainz (DFG GRK 1044). This work was supported by the Stiftung Rheinland-Pfalz für Innovation and by the EC (LSH-CT-2006-037315, EPICURE).

REFERENCES

[1] Buzsaki G, Draguhn A. Neuronal oscillations in cortical networks. *Science* 2004; 304(5679):1926-1929.

[2] Singer W. Neuronal synchrony: a versatile code for the definition of relations? *Neuron* 1999; 24(1):49-25.

[3] Steriade M. Sleep, epilepsy and thalamic reticular inhibitory neurons. *Trends Neurosci.* 2005; 28(6):317-324.

[4] Uhlhaas PJ, Singer W. Neural synchrony in brain disorders: relevance for cognitive dysfunctions and pathophysiology. *Neuron* 2006; 52(1):155-168.

[5] Buzsáki G. Rhythms of the Brain. Oxford University Press, 2006.

[6] Kramer K, Kinter LB. Evaluation and applications of radiotelemetry in small laboratory animals. *Physiol. Genomics* 2003; 13(3):197-205.

[7] Brockway BP, Hassler CR. Application of Radiotelemetry to Cardiovascular Measurements in Pharmacology and Toxicology. New Technologies and Concepts for Reducing Drug Toxicities. *Informa Healthcare*, 1992: 109-132.

[8] Bohus B. Telemetered heart rate responses of the rat during free and learned behavior. *Biotelemetry* 1974; 1(4):193-201.

[9] Adams MR, Kaplan JR, Manuck SB, Uberseder B, Larkin KT. Persistent sympathetic nervous system arousal associated with tethering in cynomolgus macaques. *Lab. Anim. Sci.* 1988; 38(3):279-281.

[10] Tang X, Sanford LD. Telemetric recording of sleep and home cage activity in mice. *Sleep* 2002; 25(6):691-699.

[11] Lapray D, Bergeler J, Dupont E, Thews O, Luhmann HJ. A novel miniature telemetric system for recording EEG activity in freely moving rats. *J. Neurosci. Methods* 2008; 168(1):119-126.

[12] Morton DB, Hawkins P, Bevan R, Heath K, Kirkwood J, Pearce P et al. Refinements in telemetry procedures. Seventh report of the BVAAWF/FRAME/RSPCA/UFAW Joint Working Group on Refinement, Part A. *Lab. Anim.* 2003; 37(4):261-299.

[13] Tang X, Yang L, Sanford LD. Sleep and EEG spectra in rats recorded via telemetry during surgical recovery. *Sleep* 2007; 30(8):1057-1061.

In: Telemetry: Research, Technology and Applications
Editors: Diana Barculo and Julia Daniels
ISBN 978-1-60692-509-6

Short Communication

AUDIBLE-WAVE TELEMETRY WITH PC SOUND CARD FOR REMOTE ANALYSIS APPLICATIONS

***Natchanon Amornthammarong*[1] §,†**
***Duangjai Nacapricha*†, *Kamonthip Sereenonchai*†,**
***Peerapat Anujarawat*†, *and Prapin Wilairat*†**
§ Ocean Chemistry Division, Atlantic Oceanographic and Meteorological Laboratory, National Oceanic and Atmospheric Administration (NOAA), Miami, Florida and Cooperative Institutes of Marine and Atmospheric Studies, Rosenstiel School of Marine and Atmospheric Science, University of Miami, Miami, Florida
† Flow Innovation-Research for Science and Technology Laboratory (FIRST Laboratory), Department of Chemistry, Faculty of Science, Mahidol University, Bangkok 10400, Thailand

ABSTRACT

This chapter describes the development of a cost-effective telemetric system through a combination of a wireless microphone for signal transmission and a computer sound card for recording of signals in the audible range. Three common communication systems, which are normally used for voice transmission, were compared for data transmission. The final developed telemeter provides a high potential for remote monitoring up to a distance of 30 m with a sampling rate of 10 Hz and 100% accuracy with low noise. The working signal range was from 0 to 2 volts, with resolution of more than a 10 bit A/D. A satisfactorily good precision of 0.1% RSD was achieved. The system works well for wireless monitoring of output from a spectrophotometer and pH meter. This work also demonstrated successful applications of the telemetric system with various chemical analyses in our laboratory.

[1] Corresponding author e-mail address: natchanon.amornthammarong@noaa.gov, natchanon9@gmail.com; Tel: +1-305-361-4537; Fax: +1-305-361-4447.

Keywords: *telemetry, audible-wave, remote measurement, sound card*

1. Introduction

Remote sensing is increasingly used in many disciplines, mostly in scientific monitoring, such as in rocket, aircraft, nuclear power stations as well as in plants for nuclear waste treatment. Apparently, remote sensing is useful for inaccessible places or for in situ analysis. In addition, in life science some experiments must entail monitoring of live animals. Field scientists are often required to track animals for research purposes. In order to do so, wireless transmitters are attached to wildlife and re-released into the native environment. For instance, in analysis of brain stimulants, in which sensing probes with cables are still widely used to connect to the subjects, the deployment of a remote monitoring device can minimize distraction and emotional distress to the animal [1-3]. Because the goal is to observe animals with as little disturbance to them as possible, the transmitters of the remote monitoring system need to be optimized for size and weight depending on the animal. A small-size telemetric system is often more attractive and more useful to users, and it should be capable of signal transmission at a target distance with high resolution and low noise.

Wireless measurement is widely known as telemetry. A telemetric system consists of a transmitter for sending the detected signal with a carrier wave (mostly radio frequency) to a receiver. Then the data is decoded from the carrier wave at the receiver before being transferred as digital signals to a computer for data recording and/or processing. Typically, a serial port is used for the digital signal interfacing from the receiver to the computer. Recently, we have reported the utilization of sound cards that are commonly attached to desktop computers and laptops for signal interfacing from an absorption spectrometer [4,5]. The output voltage signals of the spectrometer are transformed to audio or audible frequency (AF) signals by a voltage-controlled oscillator (VCO) before sending the transformed signals to a microphone channel of the sound card. An audio or audible frequency is characterized as a vibration whose frequency is audible to the average human; the generally accepted standard range is 20 to 20,000 hertz. The sound card that is connected to the VCO converts the VCO output to digital data by counting frequency, which linearly increases or decreases as the output voltage of the spectrometer. An executable program named SigREC was written in C language using Microsoft Visual C++ compilers for the data recording. The use of the sound card allows utilization for this signal interfacing purpose of the already existing device that comes with general personal computer, resulting in a low-cost system for computer interfacing that is easily accessible and reliable for PC-based data acquisition. There have been reports of the utilization of the PC sound card for various scientific purposes. For instance, a computer sound card has been used in measuring the speed of a moving object by direct application of the Doppler effect for sound [6], for recording thermal noise of crystal tuning forks by detecting nanomechanical interactions as an alternative to the usual microcantilever-optical levers used in atomic force microscopy (AFM) [7], and for data acquisition in resonant photo-acoustic detection [8].

This chapter describes a deployment of the sound card with VCO in the construction of a telemetric system for applications in chemical analysis. Three wireless communication systems for carrying the data from the VCO circuit to the sound card are compared.

Application of the system for remote monitoring of pH value in an aquarium tank is also described.

2. Experimental Section

2.1. Apparatuses, Software and Instrument

The developed telemetry consists of three major parts: a VCO, a signal transmission system (a transmitter and a receiver) and a computer's sound card, as shown in Figure 1. We used an Intel Pentium 4 computer (1.8 GHz CPU, 128 MB RAM and 40 GB hard drive). The VCO was self-constructed according to the manual of LM331 (Precision Voltage-to-Frequency Converter, National Semiconductor [9]) for transforming dc output signal to audio-wave signal.

A signal generator (GEMS model HY3003, Taiwan) was used for testing the system. The experiment was conducted according to the set up, as shown in Figure 1. Voltage output signals from the signal generator were fed to the VCO before being transmitted through three public transmission systems: two citizen band walkie-talkies (carrier wave 27 MHz), two items of mobile phone (transmitter: Nokia 3530 and receiver: Motorola T190) and a wireless microphone system (TOA, model WA-641C, Japan).

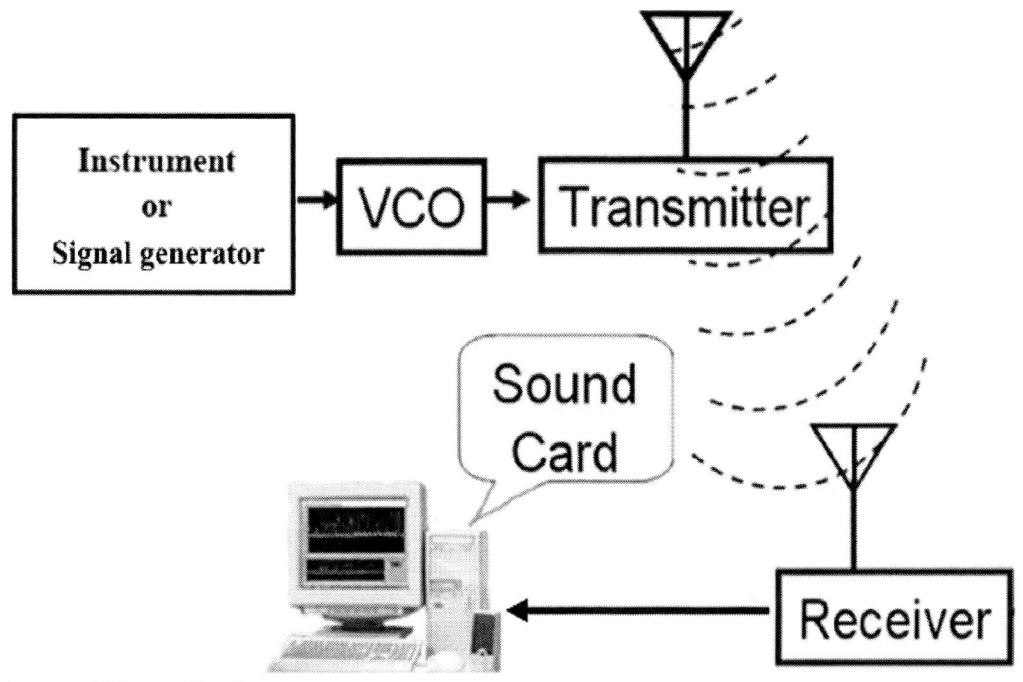

Courtesy Talanta, Elsevier.

Figure 1. Schematic diagram shows the arrangement for signal transmission. The arrows indicate the directions of signal transfer. The dashed lines represent the wireless transmission. VCO: Voltage-Controlled Oscillator.

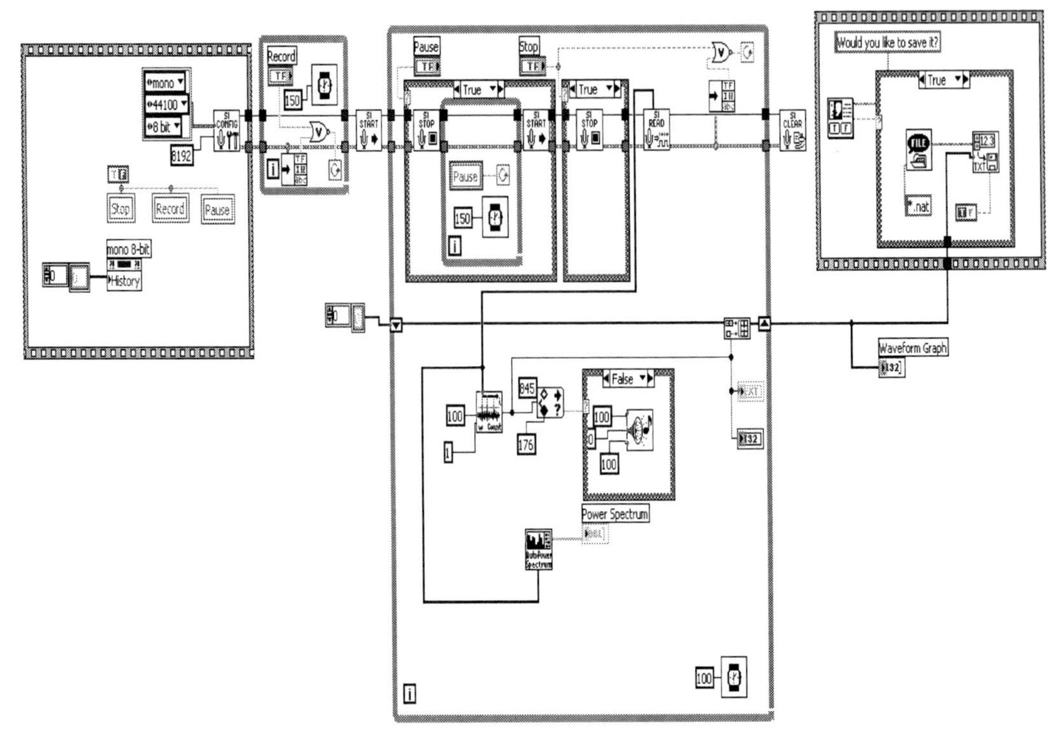

Courtesy Talanta, Elsevier.

Figure 2. Source code of the developed software (SigFLY2005), written by using LabVIEW.

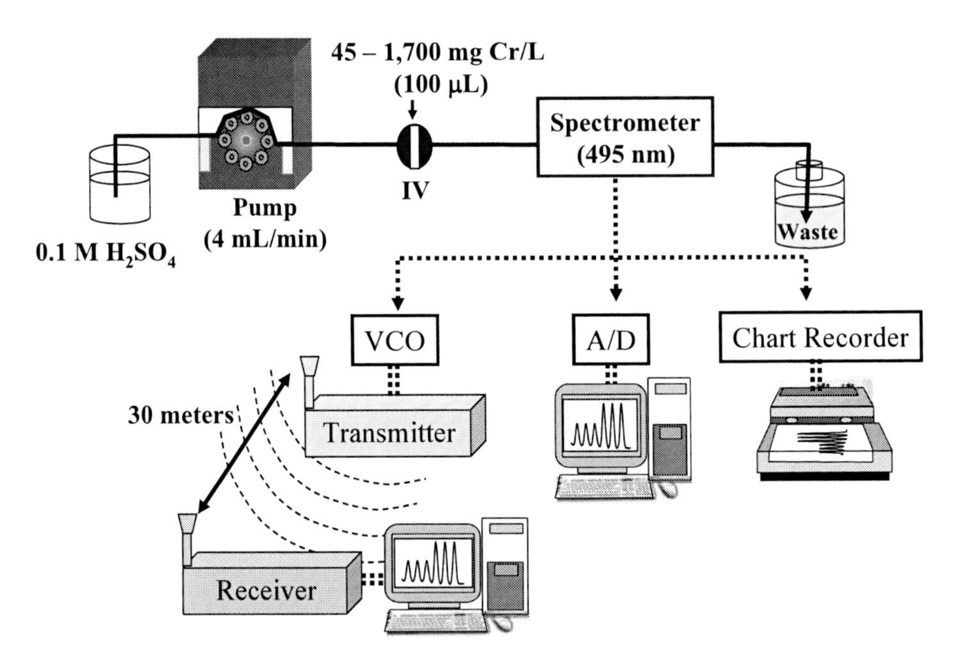

Figure 3. Validation of the telemetry was using a flow injection experimental set-up with three recording systems (chart recorder, conventional A/D and the developed telemetry). A/D: analog-to-digital converter; VCO: voltage-controlled oscillator.

LabVIEW 6.1 was used to develop recording software, called "SigFLY 2005". Figure 2 shows the written source code of this software. The software is able to record audio-type signal via a microphone channel of the sound card at sampling rate 10 Hz. The recorded data is saved in an excel file format that is readily available for further signal processing or analysis by other computer software.

2.2. Validation of the Developed System

To test the developed system, we deployed the system in a flow injection (FI) system as shown in Figure 3. The simple FI system consisted of a peristaltic pump (Ismatec model ISM850, Switzerland) and an injection valve (Rheodyne model 5020, USA). Repetitive injections of potassium dichromate solutions were made. Detection of the absorption signal was performed by using a spectrophotometer (Jenway 6405, UK) with 1 cm-path length flow cell (Phillip, USA). The telemetric system was set to remotely record (in distance of 30 meters) of the signal output from the spectrophotometer. Meanwhile, other two recording systems (strip-chart recorder and A/D card) were typically connected by cables (not wireless) with the output channel of the spectrophotometer to test the accuracy of the telemetric system.

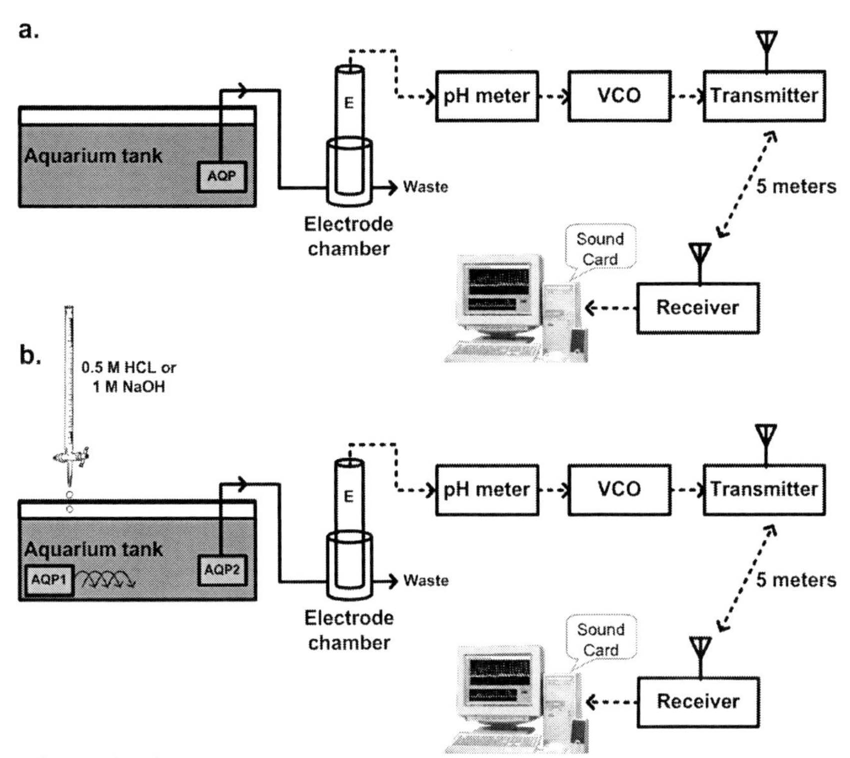

Courtesy Talanta, Elsevier.

Figure 4. Testing and application of the telemetry for pH measurements. (a) Testing for wireless signal transmission from a pH meter. (b) Efficiency testing of two mixing devices (aquarium pump and air-bubbling bars). The solid lines represent the flow of sampling water pumped by an aquarium pump through pH-electrode chamber. The dashed lines represent the transfer of signal. AQP: aquarium pump, E: combined-glass electrode, VCO: voltage-controlled oscillator.

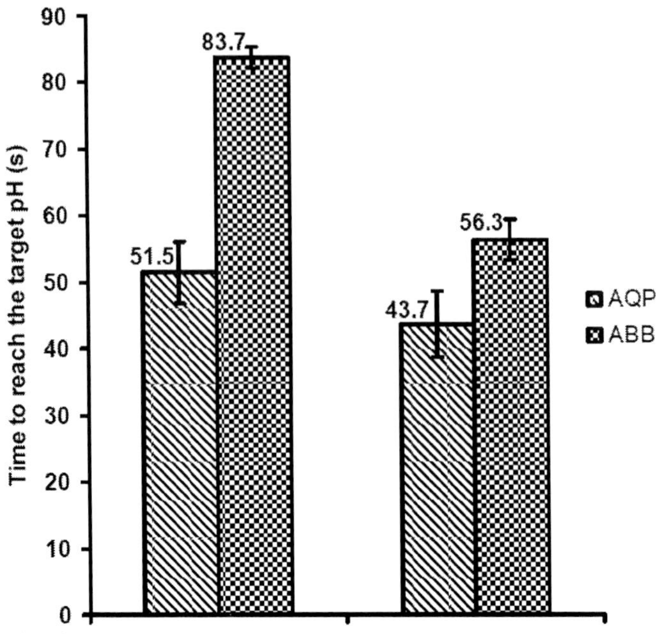

Courtesy Talanta, Elsevier.

Figure 5. Comparison of mixing efficiency between an aquarium pump (AQP) and four air-bubbling bars (ABB). Results were measured in triplicate.

2.3. pH Measurement Testing

The telemetric system was applied for determination of pH as shown in Figure 4. Two experiments were designed to compare the efficiency of two types of mixing apparatus (aquarium pump and air-bubbling bars). An Orion pH meter (model 601 A/digital IONALYZER, USA) was used with a combined-glass electrode (Orion, USA). All the pumps depicted in Figure 4 (designated "AQP") were aquarium water pumps (Lifetech AP 1200, China). We constructed a home-made flow through chamber from Plexiglas. The chamber was fitted with the pH electrode by using a rubber O-ring. Before using, the electrode was calibrated by commercial standard buffers at pH 4, 7 and 10 (Merck, Germany).

2.3.1. Continuous Measurement of pH

This experiment was set up as shown in Figure 4a. A homogeneous liquid sample in the aquarium tank was pumped by the aquarium pump into the pH-electrode chamber. The aquarium pump and the electrode chamber were connected with tubing, so the liquid in the tank can be pumped continuously into the electrode chamber for continuous monitoring of pH value. Measurement of pH was carried out for 24 water samples. Readouts obtained from the telemetric system were compared with the readings on the display of the Orion pH meter.

2.3.2. Comparison of Mixing Efficiency

The telemetric system was tested to monitor pH changes during acid-base adjustment of water inside the tank. The configuration of this experiment was carried out accordingly to the diagram shown in Figure 4b. Adjustment of pH was done by using 0.5 M HCl and 1 M

NaOH. Mixing of water with the acid and base solutions was made by two types of general mixing apparatus that are widely used for fish tanks: aquarium pump and air bubbling bars. We used an aquarium pump (AQP1 in Figure 4b) and 4 bars of air-bubbling bar (ABB) for comparing of their mixing efficiency.

AQP1 was not connected to any of the apparatus; therefore AQP1 sucked and propelled water all the time causing the liquid to mix inside the tank. AQP1 was left on at all times during its efficiency test. Meanwhile, AQP2 was connected to the electrode chamber for pH monitoring (similar to the experiment in Figure 4a). Thus, AQP2 was turned on only when measuring pH.

Figure 4b, adding of 0.5 M HCl (10 mL) to the tank was performed to make the original neutral pH of water become acidic (target pH 6). To make the water become basic (pH 9), addition of 1 M NaOH (3 mL) was carried out. We determined (triplicate experiments) how long AQP and ABB took to get the liquid in the tank to reach the target pH.

3. Results and Discussion

3.1. Comparison of Three Communication Systems

The results in Table 1 show comparison of three well-known communication systems. Apparently, the wireless microphone was the best in terms of sensitivity, linearity and resolution. The reason that the wireless microphone gave the best performance is its transmission provided the clearest voice (lowest noise) when compared to the others. Although the results indicated that the walkie-talkie and the mobile phone did not perform better than the wireless microphone, it does not mean that they are not suitable for making a telemetric system. It only indicates that they are not ideally compatible with the characteristics of the signal provided by the VCO circuit.

However, we were interested in extending the application of using PC sound card. For sake of simplicity, we chose to use the wireless microphone system. With its remote range at 30 meters, the telemetric system is enough for many routine applications and certainly provides a low-cost system to set up scientific experiments.

Table 1. Comparison of three wireless communication systems at a testing distance of five meters

Wireless system	Range (V)	Calibration[a]	Linearity (r^2)	Resolution (step)
1. Walkie-talkie	0-2	$Y = 188(\pm 8)x + 435(\pm 10)$	0.985	400
2. Mobile phone	0-1	$Y = 363(\pm 21)x + 283(\pm 12)$	0.981	300
3. Wireless microphone	0-2	$Y = 753(\pm 4)x + 549(\pm 5)$	0.999	1100

Courtesy Talanta, Elsevier.

[a] y: count/0.1 sec, x: applied dc voltage.

Table 2. Performance test of the telemetric system

Applied voltage (V)	% Accuracy		Precision (n = 5) (%RSD)		Noise level (V)	
	5 m	30 m	5 m	30 m	5 m	30 m
0.2	100.1+0.1	100.6+0.1	0.07	0.02	< 0.003	< 0.003
0.5	100.0+0.1	99.9+0.1	0.08	0.03	< 0.003	< 0.003

Courtesy Talanta, Elsevier.

Table 3. Validation of the telemetric system using a simple flow injection system at a transmission distance of 30 meters

System	Equation[1]	Linearity (r^2)
Jenway (A/D)	$y = 5.46e^{-4}(\pm 2.16e^{-6})x$	0.999
Strip-chart recorder	$y = 5.19e^{-4}(\pm 1.96e^{-6})x$	0.999
Telemetry	$y = 5.37e^{-}4(\pm 2.55e^{-6})x$	0.999

Courtesy Talanta, Elsevier.
x = concentration of Cr (mg Cr/L); y = absorbance at 495 nm.
[1]Calibration curve: 20 points, 5 injections/point.

3.2. Validation of the Telemetric System

In this experiment, the telemetric system was fed with two constant dc signals (0.2 and 0.5 Volt) at two distances (5 and 30 meters). The results in Table 2 show that the accuracy of the remote recording was approximately 100% with low noise level (less than 3 mV). The %RSDs for both distances were less than 0.1%, which is indicating high precision.

The system was then applied to record output signals from a spectrophotometer, which is set up as detector in a simple flow injection system as shown in Figure 3. This experiment was carried out at a distance of 30 meters from the transmitter to the receiver of the telemetric system. Signal recording was made for injections of twenty different concentrations of potassium dichromate solutions (45–1700 mg Cr/L). Also, two typical types of signal recorder (a strip-chart recorder and an A/D card) were used at the same time for testing the accuracy of the telemetric system. The results in Table 3 show linear calibrations obtained from all the three recording systems. A statistical method (ANOVA test) shows no significant difference of these three recording systems at 95% confidence level: $F_{\text{value}} = 0.03$ and $F_{crit} = 3.16$.

3.3. Applications

In general, most of industrial factories have a water treatment plant that probably has more than one pool in the process of treatment. Adjustment of pH by adding acid or base solutions is routinely carried out. Therefore, each pool always requires control of pH especially the one before discharge to environment. We tested the telemetric system in the measurement of pH by two applications. Firstly, the system was tested for remote pH measurement of 24 samples. The results from the telemetric system were exactly the same as the values that showed on the pH meter's screen. That means the telemetric system worked well in term of pH measurement.

Secondly, the telemetric system was used to study of mixing efficiency of two widely-used air mixing systems for aquarium tank. The results in Figure 5 show an aquarium pump (AQP) provided better than four bars of air-bubbling (ABB) in term of mixing efficiency because the mixing system using an aquarium pump always reached the equilibrium pH faster than using four bars of air-bubbling.

3.4. Other Features of the Telemetric System

With the wireless transmission to the sound card, continuous monitoring of pH (sampling rate of 10 Hz) was also achieved using the screen display of the "SigFLY 2005". The software causes an alarm to sound at the computer's speakers when the pH exceeds certain values, both acidic and basic. This alarm function is particularly useful in the water treatment plant, where pH is the major concern of industrial effluents.

4. Conclusion

This chapter describes a homemade telemetry system, which is a combination of a simple voltage-controlled oscillator (VCO), a general wireless microphone system, and a general PC sound card. The output signal from the spectrophotometer was transformed to audible-wave signal, which was then sent to a wireless microphone system to transmit the signal to the receiver for remote recording up to 30 meters. After that, the audible-wave signal was sent to the PC sound card for signal processing.

Although the dynamic distance is shorter than the commercial system based on the technique of pulse code modulation (PCM) [10], the developed system is still practical for routine monitoring in a small factory or a small water treatment plant. The system can also be used in biomedical and neuroscience laboratories, which mostly do not require long-distance recording. Its low cost of construction is another attractive feature compared to commercial systems.

Acknowledgements

This work was financially supported by the Center for Innovation in Chemistry: Post-graduate Education and Research in Chemistry (PERCH-CIC) and the Thailand Research Fund (TRF). The authors would like to thank the TRF Royal Golden Jubilee Ph.D. Program for support to N.A. and K.S.

REFERENCES

[1] Xu, SH; Talwar, SK; Hawley, ES; Li, L; Chapin, JK. *J. Neurosci. Methods.* 2004, 133, 57–63.

[2] Giuliano, F; Rossler, AS; Clement, P; Droupy, S; Alexandre, L; Bernabe, *J. Eur. Urol.* 2005, 48, 145–152.

[3] Joshi, AK; Kowey, PR; Prystowsky, EN; Benditt, DG; Cannom, DS; Pratt, CM; McNamara, A; Sangrigoli, RM. *Am. J. Cardiol.* 2005, 95, 878–881.

[4] Amornthammarong, M; Jareonsutasinee, K; Nacapricha, D. *Lab. Robotic Autom.* 2000, 12, 138–141.

[5] Nacapricha, D; Amornthammarong, N; Sereenonchai, K; Anujarawat, P; Wilairat, P. *Talanta.* 2007, 71, 605–609.

[6] Bensky, TJ; Frey, SE. *Am. J. Phys.* 2001, 69, 1231–1236.

[7] Mariani, T; Lenci, L; Petracchi, D; Ascoli, C. *Meas. Sci. Technol.* 2002, 13, 28–32.

[8] Santiago, G; Slezak, V; Peuriot, AL. *Appl. Phys. B* 2003, 77, 463–465.

[9] http://www.national.com/mpf/LM/LM331.html

[10] Carden, F; Henry, R; Jedlicka, R. *Telemetry Systems Engineering*, Artech House Inc., Norwood, MA, USA, 2002, 25–119.

In: Telemetry: Research, Technology and Applications
Editors: Diana Barculo and Julia Daniels
ISBN 978-1-60692-509-6

Short Communication

TELEMETRY OF BODY TEMPERATURE FOR LONG-TERM RECORDINGS OF BREATHING

Jacopo P. Mortola[1]
McGill University, Dept. Physiology, 3655 Sir William Osler,
Montreal, (PQ) H3G 1Y6 Canada

ABSTRACT

The barometric methodology is a practical and frequently adopted technique for the measurements of the breathing pattern and pulmonary ventilation in behaving animals. However, one of the problems in its application to long-term studies has been the need of monitoring the animal's body temperature and movements; the former is essential for the computation of tidal volume, the latter for the interpretation of the results. Hence, most commonly the technique is used for intermittent measurements of very short (a few minutes) duration. The availability of commercially available externally powered temperature transmitters, chronically implanted in the abdomen of experimental animals has solved this problem. Their small size is suitable to common laboratory species, like mice and rats. Hence, it is now feasible to obtain uninterrupted measurements of the breathing pattern and pulmonary ventilation lasting hours and days.

1. INTRODUCTION

The "barometric technique", was originally proposed for non-invasive measurements of tidal volume (V_T) in infants (Drorbaugh and Fenn, 1955). From V_T and breathing frequency (f) pulmonary ventilation ($\dot{V}_E$) was also computed ($\dot{V}_E = V_T \cdot f$). Pressure oscillations synchronous with breathing can be recorded in a small size chamber in which a subject or an animal is placed. These oscillations are generated by the changes in temperature and humidity as air is passing back and forth between the chamber and the subject's airways. The methodology has been since adapted to many animal models. Indeed, more than half a

[1] jacopo.mortola@mcgill.ca.

century after its original presentation, the technique remains the most practical approach for the computation of $\dot{V}E$ in conscious and behaving animals. Nevertheless, its use is commonly restricted to intermittent measurements of very short (a few minutes) duration, because the chamber needs to be opened intermittently to avoid rebreathing the expired gases. In addition, accurate computations of tidal volume (VT) require the measurements of body temperature (Tb). These, usually, are performed just before or after the $\dot{V}E$ recording, for practical reasons and to avoid interfering with the animal's normal behaviour. Now, some modifications of the technique have solved both problems, therefore eliminating the needs of periodic openings and permitting long-term studies. This communication briefly recalls the principle of the barometric technique; then, it offers some example of recordings in which the telemetric measurements of Tb have been of valuable assistance.

2. The Barometric Technique

Drorbaugh and Fenn (1955) proposed to measure VT simply by recording the pressure oscillations in a small size totally sealed chamber in which the subject was placed. In fact, during inspiration, the bolus of air inhaled (VT) is warmed and humidified from the ambient to the pulmonary values, raising the chamber pressure; the opposite occurs in expiration.

From the measurements of ambient and body temperature and humidity, and knowing the compliance of the chamber (K), it is possible to convert the box pressure oscillation (P) into VT as

$$VT = P \cdot K \cdot [Tb \cdot (Pb - PcH_2O)] / \{ [Tb \cdot (Pb - PcH_2O)] - [Tc \cdot (Pb - PbH_2O)] \},$$

Pb representing barometric pressure, Tc the temperature in the chamber, and PH_2O the water vapour pressure in the chamber (PcH_2O) or in the body (PbH_2O). K is a fixed characteristic of the chamber, measured as the increase in P for a given change in chamber volume, i.e. $K=\delta V/\delta P$. The breathing-related pressure oscillation P is measured by a sensitive pressure transducer, while standard temperature and humidity sensors measure Tc and PcH_2O. The only remaining variable to measure is Tb, and the accuracy of its measurement has an impact also on the accuracy of PbH_2O, because this latter is the saturation PH_2O at Tb. Full details of the implications introduced by small errors in any of these measurements have been discussed elsewhere (Mortola and Frappell, 1998). Here suffices to emphasise that the consequences of errors in the computation of Tb are larger the smaller the Tb-Tc differences. A small Tb-Tc difference occurs often when the technique is applied to newborn or young animals that need high ambient temperatures to limit their heat loss. An error in Tb by 1°C is not uncommon, especially when rectal temperature is taken as representative of Tb. When Tc is at 30-31°C, a 1°C error in Tb introduces a 10-20% error in the computation of VT . Despite a variety of precautions (Szdzuy and Mortola, 2007), when Tc is high the errors in the computation of VT can be substantial.

3. Limitations to Long-term Studies

In its original conception, one obvious limitation was the needs to open the chamber at short time intervals (a few minutes) to avoid rebreathing the expired gases. As mentioned in the preceding paragraph, another important limitation was introduced by the measurement of Tb itself. In fact, for practical reasons and to avoid any interference with the animal's behaviour, measurements by Tb with rectal probes or thermocouples are performed at times separate from the actual $\dot{V}E$ measurements. This may introduce an error especially when Tb and $\dot{V}E$ measurements are conducted at different times of the day or in conditions of altered inspired gases. In fact, the circadian pattern has a major effect on Tb, and so does the inspired gas, especially hypoxia (Frappell et al., 1992; Mortola and Seifert, 2000).

4. Solutions for Long-term Studies

The needs of avoiding rebreathing the expired gas in a sealed chamber have been fulfilled by using a chamber in which the combined action of a positive pressure pump on the entrance and a negative pressure pump at the outlet generated a large flow through very small restrictions. A computer controlled the action of the pumps so that the pressure within the chamber was maintained at the atmospheric value. In this way the chamber effectively behaves as a closed system, despite having a high enough flow for long-term recording in freely moving, undisturbed small animals (Seifert et al., 2000).

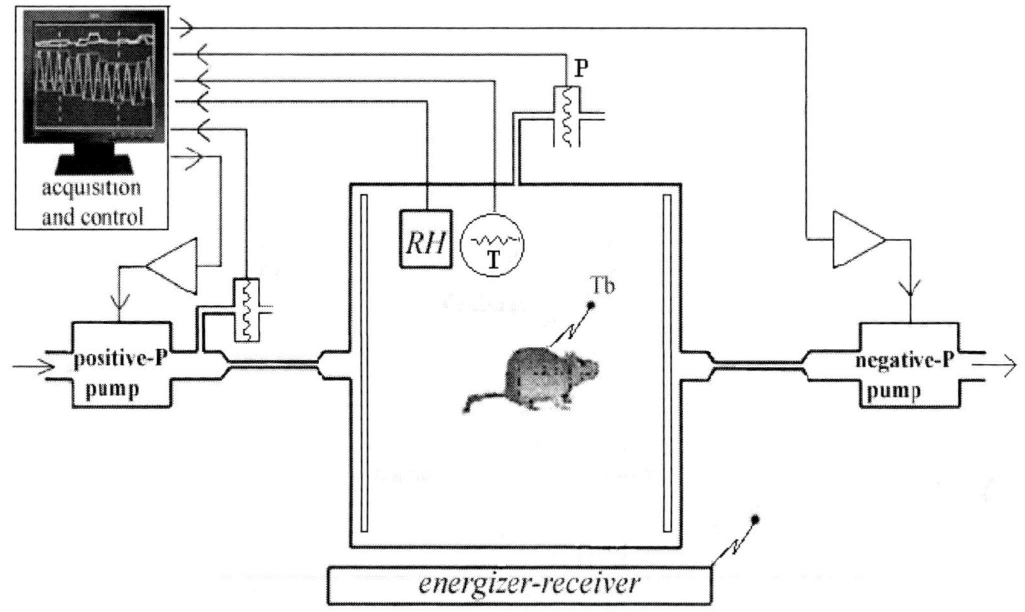

Figure 1. Schema of the setup for recording pulmonary ventilation by the barometric technique. Body temperature (Tb) and activity are recorded by telemetry. P, pressure recording. RH and T are, respectively, relative humidity and temperature in the animal's chamber. For additional details see Seifert et al., 2000, from which the figure has been adapted.

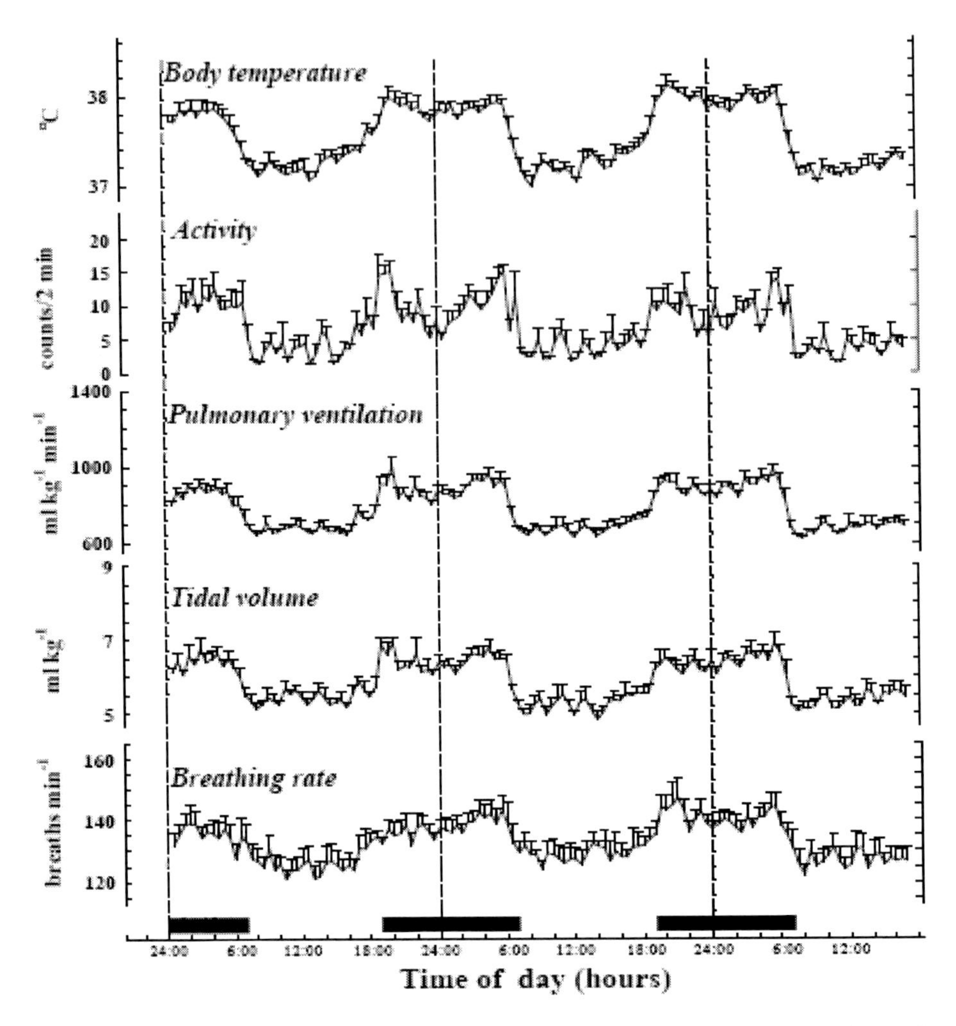

Figure 2. Recordings of body temperature and activity (both by telemetry) and corresponding values of the breathing parameters in rats during continuous measurements for a few days. Vertical dashed lines indicate midnight. Horizontal bars indicate the periods of darkness. (From the data of Seifert and Mortola, 2002b).

Continuous recording of Tb can be obtained by telemetry, with a temperature-sensitive transmitter chronically implanted in the abdomen of the experimental animal. The telemetric measurements of Tb also permit computation of the animal's movements, which could be important to eliminate artefacts and to be aware of the animal's conditions during long-term studies. Nowadays, externally powered transmitters are commercially available at modest prices. Their small sizes are suitable to common laboratory species, like mice and rats. Hence, it is now feasible to obtain uninterrupted measurements of the breathing pattern and $\dot{V}E$ for hours and days.

Figure 1 is a schematic representation of the setup. A temperature transmitter placed in the animal's abdomen a few days before the $\dot{V}E$ measurement monitors Tb breath-by-breath. The transmitter is powered by an external energiser-receiver platform. Simultaneous measurements of the animal's activity are recorded in the form of the total score of counts registered by the radiating coils of the platform over a period of time.

5. Examples and Conclusions

Some examples of such recordings are presented in Figure 2 and Figure 3. These examples were chosen to indicate situations characterised by important changes in Tb. The circadian oscillations have amplitude of about 1°C, and they are paralleled by changes in activity and $\dot{V}E$ (Figure 2). Hypoxia causes hypometabolism, with a rapid and major decrease in Tb, minor effects on activity and a major increase in $\dot{V}E$ (Figure 3).

In conclusion, it is now feasible to perform long-term recordings of breathing pattern and $\dot{V}E$ in conscious and behaving animals, lasting hours or days. The telemetric measurements of Tb often permit simultaneous recording of the activity pattern. This latter can be used also as an indirect index of the state of arousal, which is a useful information for the evaluation of the $\dot{V}E$ data.

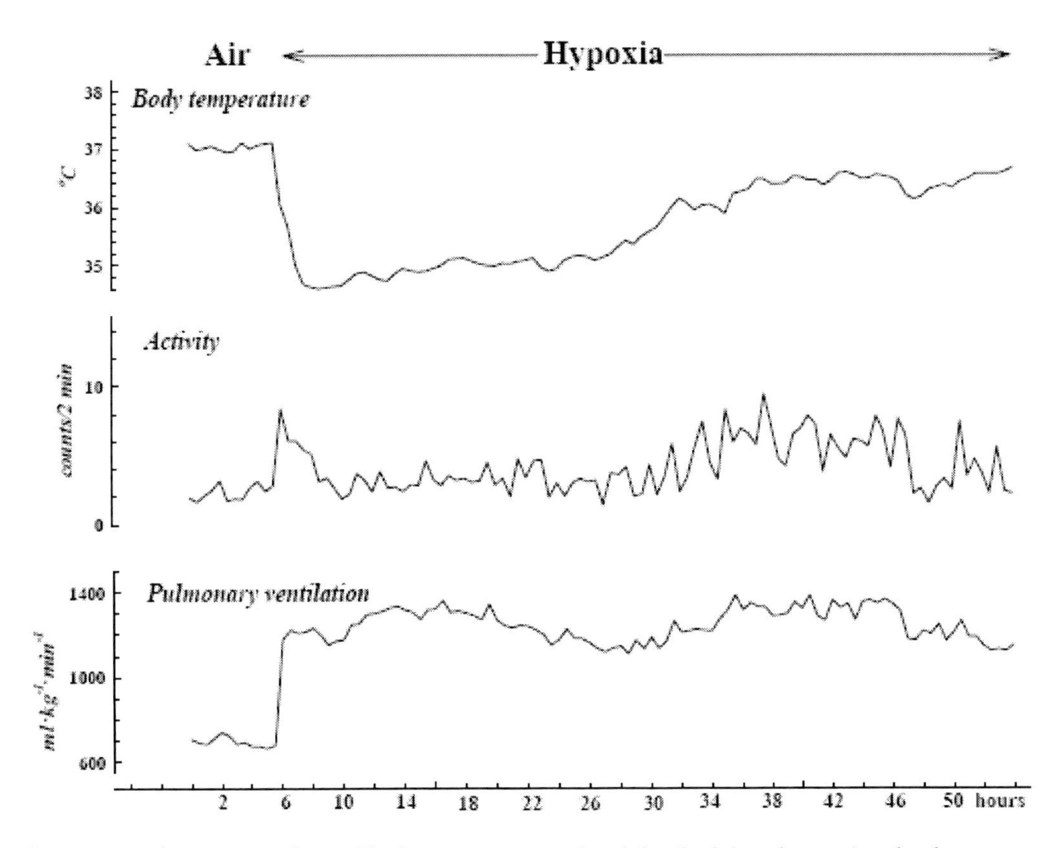

Figure 3. Continuous recordings of body temperature and activity (both by telemetry) and pulmonary ventilation in rats in air and for several hours in hypoxia. Note the drop in body temperature at the onset of hypoxia, and its gradual recovery in the following hours. (From the data of Seifert and Mortola, 2002a).

References

Drorbaugh, J.E., Fenn, W.O. 1955. A barometric method for measuring ventilation in newborn infants. *Pediatrics* 16, 81-87.

Frappell, P., Lanthier, C., Baudinette, R.V., Mortola, J.P. 1992. Metabolism and ventilation in acute hypoxia: a comparative analysis in small mammalian species. *Am. J. Physiol.* 262, R1040-R1046.

Mortola, J.P., Frappell, P.B. 1998. On the barometric method for measurements of ventilation, and its use in small animals. *Can. J. Physiol. Pharmacol.* 76, 937-944.

Mortola, J.P., Seifert, E.L. 2000. Hypoxic depression of circadian rhythms in adult rats. *J. Appl. Physiol.* 88, 365-368.

Seifert, E.L., Knowles, J., Mortola, J.P. 2000. Continuous circadian measurements of ventilation in behaving adult rats. *Respir. Physiol.* 120, 179-183.

Seifert, E.L., Mortola, J.P. 2002a. Circadian pattern of ventilation during prolonged hypoxia in conscious rats. *Respir. Physiol. Neurobiol.* 133, 23-34.

Seifert, E.L., Mortola, J.P. 2002b. Circadian pattern of ventilation during acute and chronic hypercapnia in conscious adult rats. *Am. J. Physiol.* 282, R244-R251.

Szdzuy, K., Mortola, J.P. 2007. Monitoring breathing in avian embryos and hatchlings by the barometric technique. *Respir. Physiol. Neurobiol.* 159, 241-244.

INDEX

A

B

C

D

E

F

G

H

I

J

K

L

M

N

O

P

Q

R

U

V

W